The Artech House Communication and Electronic Defense Library

DIGITAL
SIGNAL
PROCESSING

DIGITAL
SIGNAL
PROCESSING

Murat Kunt

ARTECH HOUSE

Introduction

Generalities

By definition, a signal is the physical representation of information. The signals involved in exchanges of information are complex, and can be altered by undesired distortion and noise. A digital signal is a sequence of numbers. Digital signal processing consists of processing such sequences to extract the useful information.

The field of digital signal processing has important applications in many scientific and technical areas such as telecommunication, acoustics, geophysics, astrophysics, and medicine, to name a few.

With continuing technological developments, the applications and influence of this type of processing are increasing more and more, with no obvious or predictable signs of saturation.

Originally, the different processing operations performed on signals were essentially based on analog techniques and devices. Although continuous progress is being made in analog signal processing, digital signal processing has undergone a dramatic evolution over the past two decades. In the beginning, this evolution was in parallel with the development of computers, which have a very large processing capacity and an efficiency unattainable with other tools. The potential offered by the computer has made possible the use of ever more complex methods, generally impossible to implement by analog techniques. This has produced a chain reaction: with the increasing power of the computer, processing methods can become so complex that they require even more powerful computers.

The use of the computer as a signal processing system offers an additional advantage. It permits a detailed study, by simulation, of analog processing systems before their implementation. Thus, it is possible to verify the properties of these systems quickly and to optimize their parameters, thus avoiding false economical engagements. If computers have provided great flexibility for different processing operations, they have kept a drawback which has slowly become obvious: most processing operations, especially complex ones, cannot be performed instantaneously, that is to say, in real time. It is thus a compromise, as yet unsolved, between processing flexibility and speed. Increasing the first implies decreasing the second: *c'est la vie.*

Organization of the book

This book is composed of two parts. The first four chapters are devoted to the presentation of the different basic methods of digital signal processing. The last four chapters are specialized and present applications.

Chapter 1 is an introduction to digital signals and systems. The principal operations for analysis and synthesis of digital signals are presented. Linear digital systems are defined and their fundamental properties studied.

Chapter 2 is entirely devoted to a comprehensive study of a powerful tool in digital processing: the z-transform. Its use and properties are studied in detail.

The properties and basic applications of the discrete Fourier transform are presented in chapter 3.

Chapter 4 gives the general theory of the fast transformations. The fast Fourier transform and other particular transforms are studied. The principal practical applications of these transformations are also given.

Chapter 5 is an introduction to digital filtering and filters. The most commonly used methods and their implementations are presented.

Digital spectrum analysis is studied in detail in chapter 6. The notions developed and the applicable rules are well illustrated by examples.

A general theory concerning a relatively rich set of nonlinear processing operations is presented in chapter 7. The various applications of this theory are examined and illustrated.

Finally, chapter 8 is devoted to two-dimensional digital signals and systems, in other words, to digital image processing. The concepts presented include the most recent research results.

Chapters 1 to 5 build a sequence that ought to be read in order. However, the three last three chapters, relatively independent of each other, can be read alone.

Conventions

This book is a translation from French of volume XX of the *Traité d'Électricité*. Reference to other volumes of the series is given by number (vol. VI) and, where applicable, by chapter and section. The volumes in the series are listed in the Select Bibliography at the end of this book. References to other books and articles are given by a number enclosed within brackets (for example, [1]) and listed in the Bibliography.

Reference to other parts of this book are given by chapter (chap. 1), section (sect. 1.1), and subsection (sect. 1.1.1).

Key terms are given in *italics* when they first appear in the text and particularly important passages are given in ***bold italics***.

Equations are numbered consecutively by chapter and given as (1.1). Figures and tables are numbered consecutively by chapter in a common sequence, whereby Table 4.10 follows Fig. 4.9, for example.

Publisher's Note

This book was originally published in French as volume XX of the *Traité d'Électricité* by the Presses Polytechnique Romandes, Lausanne, Switzerland, in collaboration with the École Polytechnique Fédérale de Lausanne, under the series editorship of Professor Jacques Neirynck. A complete list of the series, including English translations by Artech House, is given in the Select Bibliography.

Contents

Chapter 1

Digital Signals and Systems

1.1 FUNDAMENTALS

1.1.1 Definitions

A *signal* can be defined as the physical carrier of information. For example, audio signals are variations of the air pressure carrying a message to our ears and visual signals are light waves carrying information to our eyes. From a mathematical point of view, signals are represented by a function of one or more variables. The majority are functions of only one variable, generally *time*. The information carried by a signal is then a variation in time. However, time is not the only variable on which a signal may depend. Variations of the pressure with respect to altitude or variations of the temperature with respect to location are also examples of signals. As previously stated, there are also signals which are functions of many variables. For example, an *image* is characterized by a grey-level value depending on two variables, which represent the coordinate system in the image plane.

1.1.2 Classification of signals

The independent variable in the mathematical representation of a signal may be either a continuous or discrete variable. In the first case, the corresponding signal is called an *analog signal,* whereas in the second case it is called a *discrete signal* or a *sampled signal*. In addition, the amplitude of a signal may also be continuous or discrete. An analog signal whose amplitude is discrete is called a *quantized signal*. A discrete signal whose amplitude is also discrete is called a *digital signal*.

1.1.3 Classification of processing systems

Generally, signals must be processed either to extract some information from them or to make them carry some information. These processes are carried out with what are known as signal processing systems. Such a system acts on an input signal and produces at its output a signal which is more appropriate for the required application.

The processing systems can be classified in the same way as the signals. *Analog systems* are systems that operate on analog signals and produce analog signals. *Sampled systems* operate on sampled signals and produce sampled signals. *Digital systems* are systems that operate on digital signals and produce digital signals.

1.1.4 Adopted terminology

Although this book is entitled *Digital Signal Processing,* the study of digital signals appears only in the last chapter. The methods and the tools presented in this book can also be applied to the class of sampled signals and systems. Nowadays, the importance of these is growing with the technological progress in the area of charge-coupled devices.

From a pedagogical point of view, it is wise to begin with the study of sampled signals and systems in order to better emphasize later the particularities of amplitude quantization. Because of space limitations, the latter have been studied only in very simple cases.

1.1.5 Organization of the chapter

In this chapter, we present the fundamental concepts of digital signals and systems. The definition of elementary signals and the basic operations are introduced. Then, we study the frequency representation of signals using the Fourier transform and the measurement of the similarity between two signals by the correlation function. Linear systems, which are a very important class of digital processing systems, are examined in detail. Their representation by a difference equation and the general solution of this equation are presented together with the subclasses of causal and linear systems.

The problems of sampling and reconstruction of analog signals are also studied because a large majority of the digital signals are obtained by sampling analog signals.

1.2 DIGITAL SIGNALS

1.2.1 Definitions

By definition, a *discrete signal* (or *discrete time signal*) is a sequence of real or complex numbers. If their values are real, the signal is called a real signal, and if the values are complex, the signal is called a *complex signal*. A *digital signal* is a discrete signal whose amplitude is quantized.

1.2.2 Notations

We shall use two types of notation to represent a discrete signal:

$$x(k) \tag{1.1}$$

$$x(k \, \Delta t) \tag{1.2}$$

where the independent variable k is an integer.

Equation (1.1) can be used to represent a signal which is not necessarily a time-varying function. Equation (1.2) implies a periodic repetition of the values by an independent variable, which is usually time. The term Δt is the repetition period of the values and is called the *sampling period*. Equation (1.1) can be obtained from (1.2) by setting $\Delta t = 1$. During the study of digital signals, a large number of operations will be carried out on the index k. Consequently, to simplify the writing, the value $\Delta t = 1$ will frequently be used. In order to return to the general case, one must simply multiply the general form of the index by Δt.

The amplitude of a particular sample will be denoted by an index for an independent variable such as k_0 or k_i. Notations (1.1) and (1.2) are more convenient for formal manipulation than

$$x(k) \text{ with } k = K_1, \ldots K_2 \tag{1.3}$$

or

$$\{x(k)\} \tag{1.4}$$

The graphic representation of a discrete signal is illustrated in Fig. 1.1.

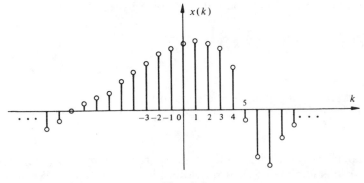

Fig. 1.1

It should be noted that the signal $x(k)$ is defined for integer values of k in this representation. The signal must not be considered to be zero for noninteger values of k: it is not defined for these values.

The domain which contains all the possible values of the amplitude of a signal is called the *dynamic range* of this signal.

1.2.3 Definitions of some elementary signals

Certain signals appear very often in digital signal processing problems. In addition, they are examples of digital signals which have a simple form.

The *unit sample* or the *unit impulse* is defined by

$$d(k) = \begin{cases} 1 & \text{for } k = 0 \\ 0 & \text{for } k \neq 0 \end{cases} \tag{1.5}$$

In the next section, this signal will play a very important role in the study of digital systems. It is shown in Fig. 1.2.

The *unit step* is defined by (Fig. 1.2):

$$\epsilon(k) = \begin{cases} 1 & \text{for } k \geqslant 0 \\ 0 & \text{for } k < 0 \end{cases} \tag{1.6}$$

The expression for a *rectangular signal* is given by

$$\text{rect}_K(k) = \begin{cases} 1 & \text{for } 0 \leqslant k \leqslant K-1 \\ 0 & \text{otherwise} \end{cases} \tag{1.7}$$

For reasons of simplicity, which will become evident later, the rectangular digital signal is defined by starting from the origin in contrast to its analog version.

The *sinusoidal signal* of period N is given by

$$x(k) = A \sin\left[\frac{2\pi}{N}(k + k_0)\right] \tag{1.8}$$

It is shown in Fig. 1.2.

The *exponential signal* has the general form:

$$y(k) = a^k \tag{1.9}$$

This signal increases if the value of the parameter a is greater than 1 and decreases if a is less than 1. It is shown in Fig. 1.2 for $a = 0.75$.

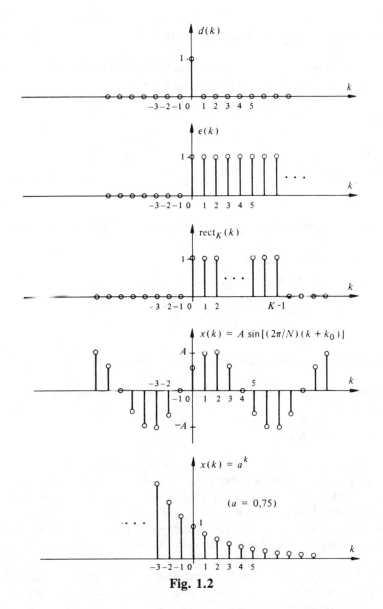

Fig. 1.2

It is also possible to have complex signals. For example, a *complex exponential signal* has the following expression:

$$x(k) = \exp[(\sigma + j\,2\pi f)k] \tag{1.10}$$

where $j = \sqrt{-1}$

1.2.4 Relationships between elementary signals

Some of these signals are related to each other in simple ways. For example, the unit step is related to the unit impulse:

$$\epsilon(k) = \sum_{l=-\infty}^{k} d(l) = \sum_{l=0}^{k} d(l) \tag{1.11}$$

with $k = 0, 1, 2, \ldots$

The inverse equation is simply

$$d(k) = \epsilon(k) - \epsilon(k-1) \tag{1.12}$$

In a similar manner, the rectangular signal can be expressed with

$$\text{rect}_K(k) = \epsilon(k) - \epsilon(k-K) \tag{1.13}$$

Using Euler's equations, signal (1.10) can be written

$$x(k) = e^{\sigma k} (\cos 2\pi f k + j \sin 2\pi f k) \tag{1.14}$$

1.2.5 Definitions: Important classes of digital signals

A signal is called *periodic* of period K if the following equation holds for all values of k:

$$x(k) = x(k+K) \tag{1.15}$$

In general, a periodic signal will be written with the index p as $x_p(k)$. A typical example of a periodic signal is the sinusoidal signal (1.8). The sum of two periodic signals with different periods is not necessarily a periodic signal. Only if the ratio of the periods is a rational number will this be the case. If this condition is not satisfied, the periods are called *incommensurable*. The sum of two or more periodic signals that have incommensurable periods is called a *quasiperiodic signal*.

If a signal is defined for a finite number K of samples, it is called a *finite length signal*. The number K then represents the *length* (or the *duration*) of such a signal, the expression for which is

$$x(k) = \begin{cases} x(k) & \text{for } k_0 \leqslant k \leqslant k_0 + K - 1 \\ 0 & \text{otherwise} \end{cases} \tag{1.16}$$

Deterministic signals are signals that have an evolution with respect to the independent variable which can be exactly predicted by an appropriate mathematical representation. *Random signals,* on the other hand, are signals that have a behavior which cannot be predicted. They are characterized by their statistical and spectral properties. Chapter 6 is devoted to the detailed study of their spectral description.

As with analog signals, it is possible to define the *average* (or *mean*) *power* of a digital signal:

$$P_x = \lim_{K \to \infty} \frac{1}{K} \sum_{k=-K/2}^{K/2} |x(k)|^2 \tag{1.17}$$

and the energy of a digital signal:

$$W_x = \sum_{k=-\infty}^{+\infty} |x(k)|^2 \tag{1.18}$$

where the absolute value delimiters represent the modulus.

1.2.6 Generation of digital signals

Digital signals can be obtained in several ways. One is to generate a series of numbers and to identify each number with a sample of a signal, with respect to a specific order. For example, it is possible to generate a random digital signal by sorting values taken from a table of random numbers.

Another type of digital signal can be obtained by using the numerical values taken at regular time intervals in some physical measurement, or as a function of another discrete variable. The data are collected by automatic or remote control systems when such a method is used. For example, in oceanography, digital measurements of the daily mean value of the impurity in the water are commonly taken.

Another possibility is to use a recurrence relation. For example, using the equation:

$$x(k) = a x(k-1) \tag{1.19}$$

and the initial condition $x(0) = 1$, it is possible to generate the following exponential signal:

$$x(k) = \begin{cases} a^k & \text{for } k \geqslant 0 \\ 0 & \text{for } k < 0 \end{cases} \tag{1.20}$$

In the most frequent case, samples are taken from a continuous function $x_a(t)$. Generally, these samples are taken periodically with a period Δt. We then have

$$x(k \, \Delta t) = x_a(t)\big|_{t = k \, \Delta t} \tag{1.21}$$

It is also possible, at least theoretically, to reconstruct the entire analog signal $x_a(t)$ from the samples of the digital signal $x(k\Delta t)$ if the period Δt satisfies some specific conditions. This problem, known as the *sampling theorem* (F. de Coulon, chap. 8, vol. VI [see Select Bibliography]) is studied in section 1.7.2. This theorem is of great importance insofar as not only digital signals, but analog signals as well, can be processed with digital processing methods and systems. The great flexibility of these methods and systems allows for ever more complex processing, which is usually very difficult, if not impossible, to implement with analog systems.

1.2.7 Example

This example consists of generating a sinusoidal (sine wave) signal. The recurrence equation that can be used is deduced from the well known trigonometric identities:

$$\sin (a + b) = \sin a \cos b + \sin b \cos a$$
$$\cos (a + b) = \cos a \cos b - \sin a \sin b$$

By setting $a = kb$, these equations become

$$\sin (kb + b) = \sin kb \cos b + \cos kb \sin b$$
$$\cos (kb + b) = \cos kb \cos b - \sin kb \sin b$$

It is possible to associate the terms $\sin kb$ and $\cos kb$ to two samples of two signals $x(k)$ and $y(k)$. Thus, the following equation is obtained:

$$x(k + 1) = x(k) \cos b + y(k) \sin b$$
$$y(k + 1) = y(k) \cos b - x(k) \sin b$$

$$(1.22)$$

With the initial conditions $x(0) = 0$ and $y(0) = 1$ and b as parameter, it is possible to generate two signals with

$$x(k) = \sin kb$$
$$y(k) = \cos kb$$

Equations (1.22) are called *simultaneous difference equations*. In these equations, the terms $\sin b$ and $\cos b$ are simply multiplicative coefficients. The general form of the difference equations will be considered in section 1.5.

1.2.8 Definition of pseudorandom signals: Example

In this example, a *pseudorandom signal* is generated. At first glance, it may appear unrealistic, considering the definition given for a random signal, to try to generate a signal of unpredictable behavior with a perfectly deterministic recurrence relation. That is why the prefix "pseudo" has been used in order to indicate that it is an imitation. Pseudorandom signals are periodic signals, but in a given period, their statistical properties and frequency descriptions are the same as those of random signals. The statistical description of random variables is summarized in *Signal Theory and Processing* by F. de Coulon (chap. 5, vol. VI). The spectral description of these signals will be considered in chapter 6 of this book.

Pseudorandom signals are very important in signal processing, particularly in the identification of linear systems and in the simulation of communication systems.

The recurrence relation has the following form:

$$x(k+1) = [A\,x(k)] \bmod P \tag{1.23}$$

where A and P are particular integers. P must be a prime number and A must be a positive *primitive root* [5] of P. The notation *mod P* means modulo P. It is said that $a = b$ modulo N and written $a = b \bmod N$ if $b - a = iN$, where i is the largest integer satisfying this equation. In other terms, a is the remainder of the division of b by N.

The signal generated by (1.23) is periodic with period $P - 1$. Moreover, on a given period each integer $x(k_i)$ between 1 and $P - 1$ appears only once at a practically unpredictable instant, i.e. almost randomly.

If each generated number is divided by P, a pseudorandom signal is obtained with a uniform probability distribution over the interval $[0, 1]$.

A portion of the signal generated with $P = 2^{35} - 31$, $A = 5^5$, and $x(0) = 12345$, is shown in Fig. 1.3. The analysis of 30,000 samples of this signal gives, for the probability density function (chap. 15, vol. VI) $P(x)$ and its integral $F(x)$, the results depicted in the same figure.

1.2.9 Example

Another recurrence relation that is also commonly used allows direct generation of a pseudorandom signal with a uniform distribution over the interval $[0, 1]$ without any preliminary divisions.

This equation has the following form:

$$x(k+1) = [x(k) + x(k-k_0)] \bmod 1 \tag{1.24}$$

where k_0 is a positive integer, which is relatively large (about 50 or more). It is clear that this equation needs k_0 initial conditions. The first k_0 samples of the signal must be randomly chosen within the interval $[0, 1]$ and used as initial conditions. If the complexity of the arithmetical operations is considered, this equation is more convenient than (1.23).

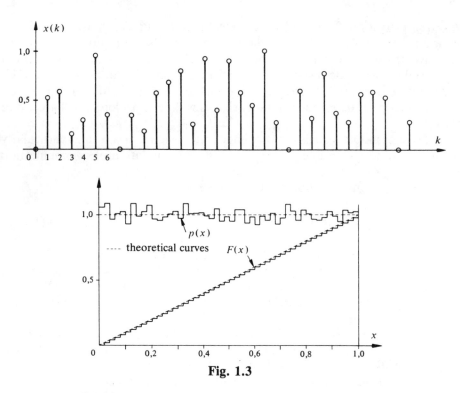

Fig. 1.3

1.2.10 Example: Signals with other distributions

From a pseudorandom signal with a uniform distribution, it is possible to generate, again with the help of a recurrence relation, pseudorandom signals that have other distributions. For example, the nonlinear recursion relation:

$$y(k) = \sqrt{2\sigma^2 \ln \frac{1}{x(k)}} \tag{1.25}$$

may be used to generate a pseudorandom signal $y(k)$, which has a Rayleigh distribution (chap. 15, vol. VI) using a signal $x(k)$, which has a uniform distribution. In fact, the equivalence condition in probability gives

$$p(y)\,dy = p(x)\,dx \tag{1.26}$$

thus,

$$p(y) = p(x)\left|\frac{dx}{dy}\right|_{x=f(y)} \tag{1.27}$$

Computing the term $|(dy)|/|(dx)|$ from the inverse equation of (1.25), we obtain

$$p(y) = \frac{y}{\sigma^2} \exp\left(-\frac{y^2}{2\sigma^2}\right) \qquad \text{avec } y \geqslant 0 \qquad (1.28)$$

One can show, using (1.26), that from the generated signal $y(k)$ and from the uniformly distributed signal $x(k)$, a pseudorandom signal that has a Gaussian distribution can be generated with the following equations:

$$z(k) = y(k) \cos[2\pi x(k+1)]$$
$$z(k+1) = y(k) \sin[2\pi x(k+1)] \qquad (1.29)$$

The generated signal has a Gaussian distribution (chap. 15, vol. VI):

$$p(z) = \frac{1}{\sigma\sqrt{2\pi}} \exp\left(-\frac{z^2}{2\sigma^2}\right) \qquad (1.30)$$

Another method for generating a signal that has a Gaussian distribution will be presented in section 4.7.17.

1.2.11 Elementary operations on signals

In processing systems, signals are combined in several ways. The basic rules are listed below.

- *Sum:* the sum of two signals is obtained by adding two by two the samples for a given value of the independent variable.
- *Product:* the product of two signals is obtained by multiplying two by two the samples for a given value of the independent variable.
- *Product by a constant:* the product of a signal by a constant is obtained by multiplying all the samples of the signal with this constant.
- *Shift:* the signal $y(k)$ is called the *shifted version* or the *delayed version* of another signal $x(k)$, if for any value of k:

$$y(k) = x(k - k_0) \qquad (1.31)$$

where k_0 is a positive or negative integer.

Using the first two rules, a digital signal can be written as the linear combination of shifted unit impulses:

$$x(k) = \sum_{l=-\infty}^{+\infty} x(l) d(k-l) \qquad (1.32)$$

1.2.12 Remark

If some processing of analog signals is simulated with digital signals, the rules for the sum and the product are valid only for signals which are sampled with the same period.

1.3 THE FOURIER TRANSFORM

1.3.1 Introduction

The idea of expressing a complicated function as a linear combination of simple elementary functions is well known (chap. 3, vol. VI). For example, in the interval $[u_1, u_2]$, a function $\theta(u)$ can be written as

$$\theta(u) = \sum_{i=0}^{\infty} \alpha_i \psi_i(u) \tag{1.33}$$

where the functions $\psi_i(u)$ make up a set of simple elementary functions. If these functions are orthogonal, then the coefficients α_i are independent of each other. In this case, the expansion is called a *series expansion of orthogonal functions*. Mathematically, digital as well as analog signals are described by functions, so the series expansion is a valid method for describing both types of signals. According to the definition of a digital signal, the series whose elements are the coefficients α_i can be regarded as a digital signal which is describing the analog signal $\theta(u)$. The Fourier series is certainly the most commonly used of all the orthogonal series expansions in signal processing (R. Boite and J. Neirynck, sect. 7.4, vol. IV; chap. 3 and 4, vol. VI).

It is well known that the Fourier series can be generalized in order to represent functions on an infinite interval. This generalization leads to the integral Fourier transform, the definition of which (sect. 7.3.9, vol. IV) is recalled below:

$$X_a(f) = \int_{-\infty}^{+\infty} x_a(t) \exp(-j2\pi ft)\, dt \tag{1.34}$$

where analog signals are noted with the index a. The inverse transform is

$$x_a(t) = \int_{-\infty}^{+\infty} X_a(f) \exp(j2\pi ft)\, df \tag{1.35}$$

In general, the Fourier transform $X_a(f)$ is a complex function.

1.3.2 Definition

The *Fourier transform of a digital signal* is defined by

$$X(f) = \sum_{k=-\infty}^{+\infty} x(k) \exp(-j2\pi fk) \tag{1.36}$$

In the rest of this book, upper-case letters will be used to denote the transform of a signal. According to this definition, it is clear that the function $X(f)$ is in general a complex function of the continuous and real variable f.

1.3.3 Existence of Fourier transforms

The Fourier transform $X(f)$ exists if the right-hand side of (1.36) is finite, that is, if the series converges. Since the modulus of the factor $\exp(-j2\pi fk)$ is always equal to one, the sufficient condition for the existence of $X(f)$ is

$$\sum_{k=-\infty}^{+\infty} |x(k)| < \infty \tag{1.37}$$

If this condition is satisfied, the series *converges absolutely* towards a continuous function of f. The class of signals that satisfies condition (1.37) is also the class of signals having finite energy. This property is a direct consequence of the relationship:

$$\sum_{k=-\infty}^{+\infty} |x(k)|^2 \leqslant \left[\sum_{k=-\infty}^{+\infty} |x(k)| \right]^2 \tag{1.38}$$

The left-hand side of this inequality is the energy of the signal $x(k)$ (see definition (1.18)). Thus, if condition (1.37) is satisfied, condition (1.38) is also satisfied. The signals satisfying (1.37) are called *absolutely integrable signals*. However, a signal with finite energy is not necessarily absolutely integrable. A typical example of such a signal is the following:

$$x(k) = (1/k)\,\epsilon\,(k+1) \tag{1.39}$$

Condition (1.37) is a sufficient condition for the existence of the transform $X(f)$. That is why this condition is difficult to satisfy. The necessary condition is subject to many different definitions and interpretations. We shall state, without proof, that the transform $X(f)$ defined by (1.36) exists for all signals having finite energy. In this context, these signals are sometimes called *square summable signals*. For real-life signals, this condition is always satisfied.

1.3.4 Property

The periodicity of $X(f)$ is a property which must be noted. The Fourier transform of a digital signal defined by (1.36) is a periodic function of f with period unity. This property is a direct consequence of the definition:

$$X(f+1) = \sum_{k=-\infty}^{+\infty} x(k) \exp[-j2\pi(f+1)k]$$

$$= \sum_{k=-\infty}^{+\infty} x(k) \exp(-j2\pi fk) \exp(-j2\pi k)$$

$$= \sum_{k=-\infty}^{+\infty} x(k) \exp(-j2\pi fk) = X(f) \qquad (1.40)$$

This is in contrast with (1.34), where the Fourier transform $X_a(f)$ of an analog signal is not periodic. Since $X(f)$ is a periodic function, it can be described on any interval of unit length. Generally, the interval $[-\frac{1}{2}, +\frac{1}{2}]$, called the *principal interval* or *principal period,* is used.

1.3.5 The inverse Fourier transform

Like any periodic function, the function $X(f)$ can be expanded as a Fourier series in the interval $[-\frac{1}{2}, +\frac{1}{2}]$. In fact, this expansion is given by definition (1.36). This is equivalent to considering the coefficients of the classical Fourier series expansion as a digital signal. The samples of the signal $x(k)$ can be obtained from the transform $X(f)$ with the well known equation (eq. (7.44), vol. IV):

$$x(k) = \frac{1}{F} \int_{-F/2}^{F/2} X(f) \exp(j2\pi fk) df \qquad (1.41)$$

where the period $F = 1$. Thus, we have

$$x(k) = \int_{-1/2}^{1/2} X(f) \exp(j2\pi fk) df \qquad (1.42)$$

This is the inverse Fourier transform for finite-energy digital signals.

1.3.6 Frequency spectrum: Definitions

Generally, $X(f)$ is a complex function. It can thus be written with its real and imaginary parts:

$$X(f) = \text{Re}[X(f)] + j\,\text{Im}[X(f)] \qquad (1.43)$$

In the case of a real signal $x(k)$, the real and imaginary parts of $X(f)$ are respectively given by

$$\text{Re}\,[X(f)] \;=\; \sum_{k=-\infty}^{+\infty} x(k)\cos 2\pi f k \tag{1.44}$$

and

$$\text{Im}\,[X(f)] \;=\; -\sum_{k=-\infty}^{+\infty} x(k)\sin 2\pi f k \tag{1.45}$$

It is also possible to write this equation with the amplitude and the phase:

$$X(f) \;=\; |X(f)|\,e^{j\,\arg[X(f)]} \tag{1.46}$$

The factor $|X(f)|$, called *amplitude spectrum*, expresses the frequency distribution of the magnitudes of $x(k)$. The factor $\theta_x(f) = \arg[X(f)]$, called the *phase spectrum*, contains the frequency distribution of the phase of $x(k)$. Finally, the factor $|X(f)|^2$, called the *energy spectrum*, represents the frequency distribution of the energy of $x(k)$. It is denoted by $\Phi_x(f)$.

The *bandwidth* of the signal is the interval of the positive frequencies that contains the nonzero values of the amplitude spectrum.

1.3.7 Example

As an example, let us compute the Fourier transform of the following signal $x(k)$ (Fig. 1.4):

$$x(k) \;=\; \text{rect}_N\!\left(k + \frac{N}{2}\right) \;=\; \begin{cases} 1 & \text{for } -N/2 \leqslant k \leqslant (N/2)-1 \\ 0 & \text{otherwise} \end{cases}$$

Thus, we have

$$X(f) \;=\; \sum_{k=-N/2}^{N/2-1} e^{-j2\pi f k} \;=\; \sum_{l=0}^{N-1} e^{-j2\pi f(l-N/2)}$$

$$=\; e^{j\pi f N}\sum_{l=0}^{N-1} e^{-j2\pi f l} \;=\; e^{j\pi f N}\,\frac{1 - e^{-j2\pi f N}}{1 - e^{-j2\pi f}}$$

It is also possible to write this as

$$X(f) \;=\; e^{j\pi f}\frac{(e^{j\pi f N} - e^{-j\pi f N})}{(e^{j\pi f} - e^{-j\pi f})} \;=\; e^{j\pi f}\,\frac{\sin(\pi f N)}{\sin(\pi f)}$$

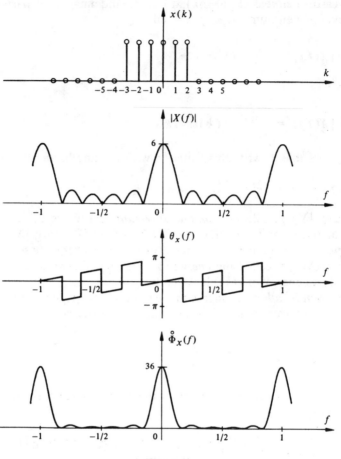

Fig. 1.4

The amplitude spectrum is given by

$$|X(f)| = \left| \frac{\sin(\pi f N)}{\sin(\pi f)} \right|$$

This is shown in Fig. 1.4 for $N = 6$.

The phase spectrum is given by

$$\theta_x(f) = \pi f + \left(1 - \frac{\sin(\pi f N)/\sin(\pi f)}{|\sin(\pi f N)/\sin(\pi f)|} \right) \frac{\pi}{2}$$

This is shown in the interval $[-\pi, \pi]$ in Fig. 1.4 for $N = 6$.
The energy spectrum is easily derived (Fig. 1.4):

$$\overset{\circ}{\Phi}_x(f) = |X(f)|^2 = \left[\frac{\sin(\pi f N)}{\sin(\pi f)} \right]^2$$

1.3.8 Property

Examining (1.43) to (1.48), we can notice that, for a real signal $x(k)$:

$$\text{Re}\,[X(-f)] = \text{Re}\,[X(f)] \tag{1.47}$$

$$\text{Im}\,[X(-f)] = -\,\text{Im}\,[X(f)] \tag{1.48}$$

Thus, the real part of the Fourier transform of a real signal is an even function and the imaginary part is an odd function. By combining these results with (1.46), it can be shown that the magnitude spectrum of a real signal is an even function and the phase spectrum is an odd function (Fig. 1.4).

1.3.9 Shift theorem

Let $y(k)$ be the shifted version of a signal $x(k)$:

$$y(k) = x(k - k_0) \tag{1.49}$$

The Fourier transform of this signal is

$$Y(f) = \sum_{k=-\infty}^{+\infty} y(k)\exp(-j2\pi fk) = \sum_{k=-\infty}^{+\infty} x(k - k_0)\exp(-j2\pi fk) \tag{1.50}$$

Let us use the new variable $l = k - k_0$:

$$Y(f) = \sum_{l=-\infty}^{+\infty} x(l)\exp[-j2\pi f(l + k_0)]$$

$$= \exp(-j2\pi fk_0)\left[\sum_{l=-\infty}^{+\infty} x(l)\exp(-j2\pi fl) \right] \tag{1.51}$$

However, the expression in brackets is, by definition, the Fourier transform $X(f)$ of $x(k)$. Thus,

$$Y(f) = \exp(-j2\pi fk_0)X(f) \tag{1.52}$$

Consequently, to obtain the Fourier transform of a signal shifted by k_0, the Fourier transform of the initial signal must be multiplied by $\exp(-j2\pi f k_0)$. Equation (1.52) is called the *shift theorem*. The amplitude and energy spectra are not altered by such a shift, since the modulus of the factor $\exp(-j2\pi f k_0)$ is one. The changes due to a shift are confined to the phase spectrum.

1.3.10 Transform of a product

Let $z(k)$ be the product of two finite-energy signals, $x(k)$ and $y(k)$:

$$z(k) = x(k)y(k) \tag{1.53}$$

The signals can be replaced by the inverse transforms of their Fourier transforms:

$$z(k) = \int_{-1/2}^{1/2} X(g)e^{j2\pi gk}\,dg \int_{-1/2}^{1/2} Y(g')e^{j2\pi g'k}\,dg'$$

$$= \int_{-1/2}^{1/2}\int_{-1/2}^{1/2} X(g)Y(g')e^{j2\pi k(g+g')}\,dg\,dg' \tag{1.54}$$

With the change of variable $f = g + g'$, we obtain

$$z(k) = \int_{-1/2}^{1/2}\left[\int_{g_0-1/2}^{g_0+1/2} X(g)Y(f-g)\,dg\right]\exp(j2\pi fk)\,df \tag{1.55}$$

where g_0 is a given parameter.

Comparing this equation with the inverse transform (1.42), we obtain the Fourier transform of $z(k)$:

$$Z(f) = \int_{g_0-1/2}^{g_0+1/2} X(g)Y(f-g)\,dg \tag{1.56}$$

This equation is called the *continuous and periodic convolution*.

Typically, the product (1.53) is used when studying very long signals. These signals are often multiplied by a finite-length signal called the *observation window* or simply *window*. A typical example of such a window is the rectangular signal $x(k) = \text{rect}_N(k)$.

Equation (1.56) shows change in the transform of the observed signal. Other examples of windows will be given later.

1.3.11 Comments

In the remainder of this chapter, other properties of the Fourier transform will be established when other ideas and processing operations are introduced.

The Fourier transform as defined here is of great theoretical importance.

It is not directly suitable for practical applications, however. On one hand, an infinite number of samples is required, which cannot be obtained in practice, while on the other hand, the continuous frequency variable f is not compatible with the discrete nature of digital signal processing systems. Considering these remarks, we will define, in chapter 3, the *discrete Fourier transform* (DFT), which, although based on (1.36), is more appropriate for discrete-time and digital signal processing. Its properties will be studied in detail. Then, in chapter 4, other orthogonal transformations will be introduced, particularly those which can be computed with a fast algorithm, including the DFT. The Fourier series expansion of periodic digital signals will also be studied in chapter 3.

1.4 CORRELATION OF SIGNALS

1.4.1 Definitions

In signal processing, it is often necessary to compare two signals. This can be done in several ways. One commonly used possibility is to shift one of the signals with respect to the other and to measure the similarity as a function of the shift. Mathematically, this operation is represented by

$$\overset{\circ}{\varphi}_{xy}(k) = \sum_{l=-\infty}^{+\infty} x(l)\,y(l+k) = \sum_{l=-\infty}^{+\infty} x(l-k)\,y(l) = \overset{\circ}{\varphi}_{yx}(-k) \quad (1.57)$$

where at least one of the two signals, assumed to be real, has finite energy. The signal $\overset{\circ}{\varphi}_{xy}(k)$ is called the *crosscorrelation function* of $x(k)$ and $y(k)$. If these two signals are identical, the signal $\overset{\circ}{\varphi}_{xy}(k)$ is called the *autocorrelation function* of $x(k)$. Since $\overset{\circ}{\varphi}_{xy}(k)$ is a measurement of the similarity between $x(l)$ and $y(l)$, it will reach its maximum for a particular value of k corresponding to the greatest similarity.

The different steps and operations in the computation of a correlation function (1.57) are

- the signal $y(l)$ is shifted by a given amount k;
- the product $x(l)\,y(l+k)$ is computed sample by sample for all values of l;
- the obtained values are added to arrive at $\overset{\circ}{\varphi}_{xy}(k)$.

These steps are repeated as many times as necessary.

1.4.2 Example

Let us examine, as an example, the crosscorrelation function of two signals $x(k)$ and $y(k)$:

$$x(k) = \text{rect}_N(k)$$
$$y(k) = a^k\,\epsilon(k) \quad \text{where } |a| < 1.$$

These two signals are shown in Fig. 1.5. First, it must be pointed out that a shift of $y(l)$ to the left corresponds to positive values of k in (1.57), and that a shift to the right corresponds to negative values of k. Among all the possible values of k, three cases can be defined.

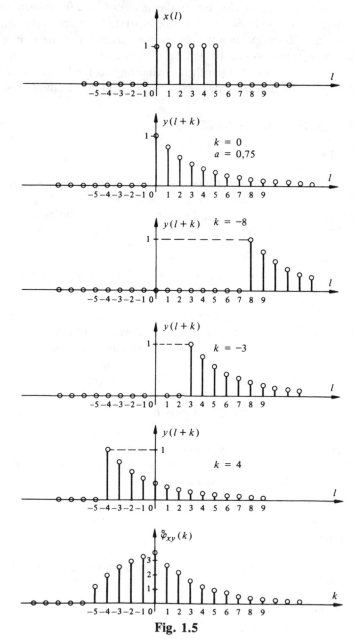

Fig. 1.5

First, for $-\infty < k \leqslant -N$, there is no nonzero product $x(l)y(l + k)$ in $\overset{\circ}{\varphi}_{xy}(k)$. Then,

$$\overset{\circ}{\varphi}_{xy}(k) = 0 \quad \text{for } k \leqslant -N$$

Second, for $-N < k \leqslant 0$, we have

$$\overset{\circ}{\varphi}_{xy}(k) = \sum_{l=-k}^{N-1} a^{l+k} = \sum_{m=0}^{N-1+k} a^m = \frac{1 - a^{N+k}}{1 - a}$$

Finally, for $k > 0$:

$$\overset{\circ}{\varphi}_{xy}(k) = \sum_{l=0}^{N-1} a^{l+k} = a^k \sum_{l=0}^{N-1} a^l = a^k \frac{1 - a^N}{1 - a}$$

The entire function $\overset{\circ}{\varphi}_{xy}(k)$ is shown in Fig. 1.5.

1.4.3 Property

A very useful equation can be derived by computing the Fourier transform of the crosscorrelation function. Starting from definition (1.36), we have (sect. 1.8.8):

$$\overset{\circ}{\Phi}_{xy}(f) = \sum_{k=-\infty}^{+\infty} \overset{\circ}{\varphi}_{xy}(k) e^{-j2\pi fk} = X^*(f)Y(f) \tag{1.58}$$

Accordingly, the Fourier transform of the crosscorrelation function is a simple product of the Fourier transform of one of these signals with the complex conjugate of the Fourier transform of the other (chap. 4, vol. VI).
For the autocorrelation function, we have

$$\overset{\circ}{\Phi}_x(f) = X^*(f) X(f) = |X(f)|^2 \tag{1.59}$$

which is the energy spectrum of $x(k)$.
Thus, the Fourier transform of the autocorrelation function is nothing other than the energy spectrum. The functions $\overset{\circ}{\varphi}_{xy}(k)$ or $\overset{\circ}{\varphi}_x(k)$ can be computed indirectly by the inverse Fourier transform of (1.58) and (1.59).

1.4.4 Parseval's theorem

The value of the autocorrelation function at the origin is

$$\overset{\circ}{\varphi}_x(0) = \sum_{l=-\infty}^{+\infty} x^2(l) = \int_{-1/2}^{1/2} \overset{\circ}{\Phi}_x(f) \, df \tag{1.60}$$

This equation is known as Parseval's theorem. From definition (1.57), it can be easily verified that $\mathring{\phi}_{xy}(k)$ is an even function.

1.4.5 Comments

In practice, formula (1.57) is valid only for finite-length signals. For example, if M and N are the lengths of $x(k)$ and $y(k)$, respectively, then the length of the crosscorrelation function is $N + M - 1$. If one of the signals is not of finite length, the previously defined observation window technique can be used. However, in this case, only an approximation of the function $\mathring{\phi}_{xy}(k)$ is obtained. The quality of this approximation depends on the chosen window. This problem will be discussed in chapter 3.

Definition (1.57) cannot be used for signals having infinite energy because the series does not converge.

For two periodic signals of period N, the crosscorrelation function is defined by

$$\varphi_{xy}(k) = \frac{1}{N} \sum_{l=l_0}^{l_0+N-1} x(l)y(l+k) \tag{1.61}$$

It can easily be verified that this is a periodic function of period N. An estimation of the correlation function for nonperiodic signals having finite power is presented in chapter 6.

1.5 DIGITAL SYSTEMS

1.5.1 Definition

A *digital processing system* acts on a digital input signal and produces a digital output signal. In other words, it establishes a cause-and-effect relationship. The input signal is called the *excitation signal* or simply *excitation*. The output signal is called the *response* of the system to the considered excitation.

In order to analyze a given system or to synthesize a system which must perform a given processing operation, it is necessary to represent it with a mathematical model. In its most general form, such a model is a functional operator (or transformation) S, which acts on an input signal $x(k)$ and transforms it into an output signal $y(k)$ (sect. 2.1.5, vol. IV). This operation is formally represented as

$$y(k) = S[x(k)] \tag{1.62}$$

Such an operator is shown in Fig. 1.6. An operator is simply an equation or a set of equations defining the mathematical relationship between $x(k)$ and $y(k)$.

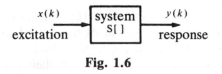

$x(k)$ excitation → system S[ ] → $y(k)$ response

Fig. 1.6

1.5.2 Example

The operator or equation may be the discrete Fourier transform, DFT (see chapter 3). In this particular case, (1.62) can be written with the following form (sect. 7.3.1, vol. IV):

$$X(n) = F[x(k)] \tag{1.63}$$

where F is the discrete Fourier transform operator. The associated system is called *Fourier transformer*.

1.5.3 Comments

Different classes of systems are defined according to the constraints that may be put on the operator S. If relationship (1.62) is a one-to-one mapping, the same excitation always produces the same response. This class includes a great number of the systems that we will study in this book.

If the mapping represented by (1.62) is one-to-many, the system may have the same response to many different excitations. A simple example is a system that computes the mean value of a signal by using the following relationship:

$$m_x = \frac{1}{K} \sum_{k=k_0}^{k_0+K-1} x(k) \tag{1.64}$$

It is obvious that different signals can lead to the same mean value.

Systems corresponding to a many-to-one function (1.63) will not be studied in this book.

1.5.4 Causal systems

An important class of systems is that of *causal systems*. These systems are characterized by the fact that their response never precedes their excitation. That is, if

$$x(k) = 0 \quad \text{for } k < k_0 \tag{1.65}$$

then

$$y(k) = 0 \quad \text{for } k < k_0 \tag{1.66}$$

where $y(k)$ is the response to the excitation $x(k)$. For causal systems, the operator S must not depend on the future values of the excitation. Causal systems are the only realizable systems. However, from a theoretical point of view, it is useful to study some noncausal systems because their particular characteristics make the study of real systems easier.

1.5.5 Linear systems

Another important class of systems is that of *linear systems*. For this class, the constraint on the operator S is called the *principle of superposition*. The operator associated with a linear system must satisfy the following condition:

$$S[ax_1(k) + bx_2(k)] = a\,S[x_1(k)] + b\,S[x_2(k)] \tag{1.67}$$

where a and b are two constants and $y_1(k)$ and $y_2(k)$ are the responses to excitations $x_1(k)$ and $x_2(k)$, respectively. Equation (1.67) indicates that a linear system processes a sum of excitations, as if they were processed separately and summed afterwards. Linear systems are important because they are relatively easy to analyze and to characterize. They lead to elegant and powerful mathematical representations. For this reason, they are also used in the study of complex systems, where the linearity is realized only approximately.

1.5.6 Difference equations

Mathematically, the excitation and the response of a large subset of linear systems satisfy an Nth-order difference equation of the following type:

$$a_0(k)\,y(k) + a_1(k)\,y(k-1) + \ldots + a_N(k)\,y(k-N)$$
$$= b_0(k)\,x(k) + \ldots + b_M(k)\,x(k-M)$$

or

$$\sum_{n=0}^{N} a_n(k)\,y(k-n) = \sum_{m=0}^{M} b_m(k)\,x(k-m) \tag{1.68}$$

where N and M are positive integers.

Such an equation is called a *linear difference equation*. In this equation, the set of coefficients $a_n(k)$ and $b_m(k)$ defines the behavior of the system for a given value of k. Equation (1.68) is the discrete version of the linear differential equations that characterize continuous or analog linear systems. These equations are represented by a relation of the following form:

$$\alpha_0(t) y(t) + \alpha_1(t) \frac{dy(t)}{dt} + \ldots + \alpha_N(t) \frac{d^N y(t)}{dt^N}$$

$$= \beta_0(t) x(t) + \ldots + \beta_M(t) \frac{d^M x(t)}{dt^M} \tag{1.69}$$

A linear difference equation can be derived from a linear differential equation if the derivatives are replaced by their approximations. For example, for the first derivative:

$$\frac{dy(t)}{dt} \cong \frac{y(t) - y(t - \Delta t)}{\Delta t} \tag{1.70}$$

This does not mean, however, that (1.68) is an approximation. It is exact for discrete linear systems and becomes approximate only in the study of linear analog systems that utilize digital methods.

In section 1.6 and chapter 2, methods for finding the general form of a solution to a linear difference equation with constant coefficients $a_n(k)$ and $b_m(k)$ will be studied. For now, however, another method will be used to find a closed form response of a linear system.

1.5.7 Impulse response of a linear system

Linear systems are completely characterized by their response to a unit impulse. Indeed, it was previously shown that a digital signal can be represented by a linear combination of shifted unit impulses (1.32). With the superposition principle (1.67), the following result can be derived:

$$y(k) = S \left[\sum_{l=-\infty}^{+\infty} x(l) d(k-l) \right] = \sum_{l=-\infty}^{+\infty} x(l) S[d(k-l)] \tag{1.71}$$

If the response of the system to the excitation $d(k-l)$ is $g(k, l)$, then

$$y(k) = \sum_{l=-\infty}^{+\infty} x(l) g(k,l) \tag{1.72}$$

This equation indicates that it is possible to express the response of a linear system to any excitation using the response $g(k, l)$ to the excitation $d(k - l)$. This response is called the *impulse response* of a linear system and is a function of l and k.

1.6 LINEAR SHIFT INVARIANT SYSTEMS

1.6.1 Definition

Let $y(k)$ be the response to the excitation $x(k)$. The linear system is called *shift invariant* if the response to $x(k - k_0)$ is $y(k - k_0)$, where k_0 may be any finite integer.

The invariance is related to the shift. If the independent variable k represents time, the invariance is called *time invariance*. In this section, the principal properties of linear invariant systems are studied.

1.6.2 Linear difference equation with constant coefficients

In a linear shift invariant system, the excitation signal $x(k)$ is processed independently of the origin of its independent variable k. Consequently, in (1.68) the coefficients $a_n(k)$ and $b_m(k)$ are constant and independent of k. Then, the following equation is obtained:

$$\sum_{n=0}^{N} a_n y(k - n) = \sum_{m=0}^{M} b_m x(k - m) \tag{1.73}$$

This equation is an Nth-order equation and is called a *linear difference equation with constant coefficients*. The set of coefficients a_n and b_m thus represents a linear shift invariant system. Equation (1.73) is the discrete version of the linear differential equation with constant coefficients. Again, as in the general case, this equation must not be considered as an approximation of equation (1.69). It is exact and becomes approximate if and only if it is used in the study or simulation of the behavior of continuous linear shift invariant systems.

Equation (1.73) can be solved for the kth-sample of the output signal. The following solution is obtained:

$$y(k) = \sum_{m=0}^{M} \frac{b_m}{a_0} x(k - m) - \sum_{n=1}^{N} \frac{a_n}{a_0} y(k - n) \tag{1.74}$$

In contrast with differential equations, it may be noted (1.74) that a linear difference equation with constant coefficients can be solved using elementary arithmetical operations, such as multiplication, addition, and subtraction. In order to start the recursive computation, the first M values of the excitation and the first N values of the response must be known.

1.6.3 Example

Let us consider the first-order difference equation:

$$y(k) = x(k) - ay(k-1) \tag{1.75}$$

with the initial condition $y(k) = 0$ for $k < 0$ and the excitation $x(k) = \epsilon(k)$. Recursively, the following statements are obtained:

$$y(0) = 1 - ay(-1) = 1$$
$$y(1) = 1 - ay(0) = 1 - a$$
$$y(2) = 1 - ay(1) = 1 - a(1-a) = 1 - a + a^2$$
$$y(3) = 1 - ay(2) = 1 - a + a^2 - a^3$$
$$\vdots$$

$$y(k) = 1 - ay(k-1) = 1 - a + a^2 - a^3 + \ldots + (-1)^k a^k$$
$$= \frac{1 - (-a)^{k+1}}{1 + a}$$

Consequently, the response is given by

$$y(k) = \frac{1 - (-a)^{k+1}}{1+a} \epsilon(k)$$

It is clear that with the same excitation, but another initial condition, a different solution to the same equation (1.75) is computed. For example, let us consider the initial condition $y(k) = 0$ for $k < 0$. Equation (1.75) then becomes

$$y(k-1) = a^{-1}[x(k) - y(k)]$$

The following statements are then obtained:

$$y(0) = a^{-1}[x(1) - y(1)] = a^{-1}$$
$$y(-1) = a^{-1}[x(0) - y(0)] = a^{-1} - a^{-2}$$
$$y(-2) = a^{-1}[0 - y(-1)] = -a^{-2} + a^{-3}$$
$$\vdots$$

$$y(k) = (-1)^{k-1}(a^k - a^{k-1})$$

Consequently, the response is

$$y(k) = (-1)^{k-1}(a^k - a^{k-1})\epsilon(-k)$$

This response is totally different from the previous one.

1.6.4 General solution of a difference equation

Generally, a difference equation of type (1.73), as well as linear differential equations, has a family of solutions. A particular solution is obtained only with specific initial conditions. As will be seen in chapter 5, the recursive computation induced by a difference equation is used directly for the practical application of some processing systems.

Another method to obtain the solution of the difference equation (1.73) initially consists of searching for the general solution of the homogeneous equation:

$$\sum_{n=0}^{N} a_n y(k-n) = 0 \tag{1.76}$$

Hence, a particular solution of the equation with the right-hand side of (1.73) is sought. The general solution is then the sum of the general solution of the homogeneous equation and a particular solution of the nonhomogeneous equation. The arbitrary coefficients in the final solution are determined by the initial conditions. With this method, which is of theoretical interest, an analytical solution is directly obtained.

In the next section, another form for the response of a linear shift invariant system is presented, which can be considered as a solution of (1.73) when the response of the system has reached its stationary state. This state is attained when the initial condition has no effect on the response of the system.

1.6.5 Convolution product

The property of invariance immediately appears in the impulse response of a linear system. If $g(k)$ is the response to $d(k)$, then the response to $d(k - k_0)$ is $g(k - k_0)$. In this case, (1.72) becomes (chap. 4, vol. VI):

$$y(k) = \sum_{l=-\infty}^{+\infty} x(l)g(k-l) \tag{1.77}$$

The impulse response $g(k)$ completely characterizes a linear shift invariant system. It must not be forgotten that, even if $g(k)$ characterizes a system, it is at the same time its output signal when the system is excited by $d(k)$. Consequently, it can be considered a signal.

Equation (1.77) is called a *convolution*. The convolution plays a very important role in digital signal processing. Usually, a convolution is written in the following form:

$$y(k) = x(k) * g(k) \tag{1.78}$$

1.6.6 Properties

With a change of variables, it is simple to verify that

$$y(k) = \sum_{l=-\infty}^{+\infty} g(l)\,x(k-l) = g(k) * x(k) \tag{1.79}$$

The convolution is thus a commutative operation. It is easy to see that it is also an associative and distributive operation:

$$x(k) * [y(k) * z(k)] = [x(k) * y(k)] * z(k) \tag{1.80}$$

$$x(k) * [y(k) + z(k)] = x(k) * y(k) + x(k) * z(k) \tag{1.81}$$

Using (1.80), it can be proved that a series of two linear shift invariant systems, with impulse responses $g_1(k)$ and $g_2(k)$, is equivalent to a linear shift invariant system whose impulse response $g(k)$ is given by

$$\begin{aligned} g(k) &= g_1(k) * g_2(k) \\ &= g_2(k) * g_1(k) \end{aligned} \tag{1.82}$$

Using (1.81), it can be shown that two linear shift invariant systems in parallel are equivalent to a linear shift invariant system whose impulse response is the sum of the individual impulse responses.

By comparing (1.57) and (1.77), we see some similarities between a convolution and the correlation function. Equation (1.57) can be written with $l = -l'$, with the following result:

$$\overset{\circ}{\varphi}_{xy}(k) = \sum_{l'=-\infty}^{+\infty} x(-l')\,y(k-l') \tag{1.83}$$

This equation is nothing other than the convolution of $x(-k)$ and $y(k)$:

$$\overset{\circ}{\varphi}_{xy}(k) = x(-k) * y(k) \tag{1.84}$$

Conversely, the convolution of $x(k)$ and $g(k)$ can be considered as the cross-correlation of $x(-k)$ and $g(k)$, or of $g(-k)$ and $x(k)$. It should be noted that the order of the signals in these operations is important because the correlation is not a commutative operation.

1.6.7 Interpretation of the convolution

In signal processing, a great deal of importance is placed on the convolution, in theory as well as in the practical realization of some linear systems. For this reason, the different steps of a convolution are presented here in more detail.

Considering (1.77), the different steps and operations involved in a convolution can be listed as follows:

- the impulse response is reversed about the coordinate axis; $g(l)$ then becomes $g(-l)$;
- it is then shifted by a given amount k;
- the multiplication $x(l) \cdot g(k - l)$ is done sample by sample for all values of l;
- the obtained values are summed to obtain the result $y(k)$.

The last three steps are repeated as many times as necessary.

1.6.8 Example

Let us consider a linear shift invariant system whose impulse response is

$$g(l) = a^l \epsilon(l) \qquad \text{with } |a| < 1$$

and let us compute the response of this system to the excitation:

$$x(l) = \text{rect}_N(l)$$

These two signals are shown in Fig. 1.7. It can be noted from this figure that, for negative values of k, all products of the form $x(l) \cdot g(k - l)$ are zero, and therefore do not contribute to the calculation of $y(k)$. Consequently, $y(k) = 0$ for $k < 0$. Thus, for values of k between 0 and $N - 1$, there is a uniform region. In this region, the response $y(k)$ is

$$y(k) = \sum_{l=0}^{k} a^{k-l} = a^k + a^{k-1} + \ldots + a + 1$$

$$= \frac{1 - a^{k+1}}{1 - a} \qquad \text{for } 0 \leqslant k < N$$

For values of k greater than or equal to N, the nonzero terms of the multiplication $x(l) \cdot g(k - l)$ are in the interval $[0, N - 1]$. Thus,

$$y(k) = \sum_{l=0}^{N-1} a^{k-l} = a^k \frac{1 - a^{-N}}{1 - a^{-1}} \qquad \text{for } k \geqslant N$$

The response $y(k)$ is shown in Fig. 1.7.

The similarities between Figs. 1.7 and 1.5 are not surprising, considering the strong relationship between the correlation function and a convolution.

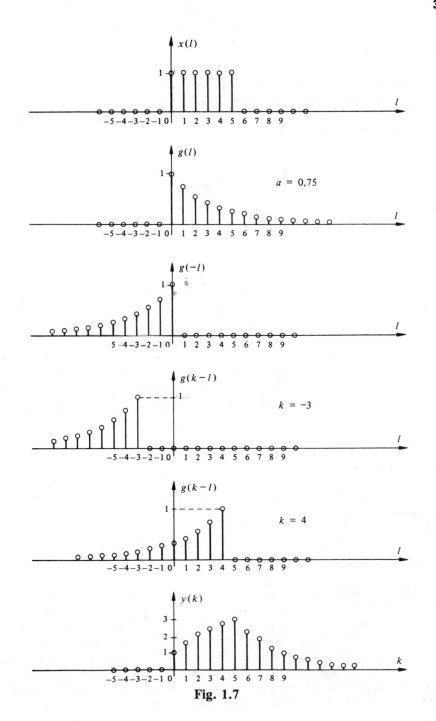

Fig. 1.7

1.6.9 Property

Convolution, as well as correlation, can be computed by means of the Fourier transform using an indirect method.

The Fourier transform for both sides of (1.77) can be calculated with the hypothesis of finite-energy signals:

$$Y(f) = \sum_{k=-\infty}^{+\infty} y(k) e^{-j2\pi fk} = X(f) G(f) \tag{1.85}$$

1.6.10 Remarks

The Fourier transform of a convolution is a simple multiplication of the Fourier transforms. This statement is essential to the study of signal processing methods with linear shift invariant systems. It was shown earlier that the Fourier transform of a signal expresses the spectral distributions of the amplitude and the phase of this signal. Equation (1.85) indicates that the amplitude and phase spectra of the input signal are modified by the system according to the particular form of the function $G(f)$. The modifications to the amplitude spectrum consist of a simple multiplication and the modifications to the phase spectrum are found by simple addition because the Fourier transforms involved in (1.85) are generally complex functions.

1.6.11 Definition

The Fourier transform $G(f)$ of the impulse response $g(k)$ is called *frequency response* or *harmonic response*.

1.6.12 Comments

The attenuation or amplification of the different frequency components of the input signal are determined by the harmonic response of the system. In other words, the system acts as a filter on the frequency distribution of the excitation. For this reason, systems that are usually linear shift invariant are simply called linear filters. When the filtering is performed by a digital linear system, it is called *digital filtering*. The general problem of digital filtering is to work out an impulse response $g(k)$, which, on one hand, has the required frequency response and, on the other hand, can be efficiently implemented.

Generally, the second requirement implies that realizable systems can only approximately approach the ideal frequency response.

The general notion of digital filtering includes a lot of signal processing operations which will be studied in the chapters to follow, especially chapter 5.

1.6.13 Example

In order to illustrate the preceding comments, let us consider a linear shift invariant system the frequency response of which is (Fig. 1.8):

$$G(f) = \begin{cases} 1 & \text{for } -f_c + n < f < f_c + n \\ 0 & \text{for } f_c + n < f < 1 - f_c + n \end{cases}$$

where n is an integer.

This system is called an ideal low-pass filter and its cut-off frequency is f_c. All the components of the input signal, whose frequency is within the interval $f_c < |f| < \frac{1}{2}$ are removed by this filter, while the frequencies for which $|f| <$

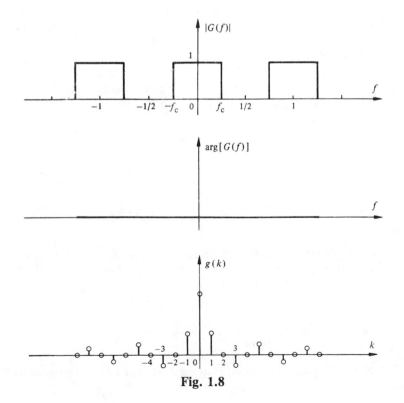

Fig. 1.8

f_c are transferred unchanged to the output. The corresponding impulse response is directly obtained from (1.42):

$$g(k) = \int_{-f_c}^{f_c} e^{j2\pi fk} df$$

$$= \frac{1}{j2\pi k} [\exp(j\,2\pi f_c k) - \exp(-j\,2\pi f_c k)] = \frac{\sin 2\pi f_c k}{\pi k}$$

This response is shown in Fig. 1.8 for $f_c = \frac{1}{4}$. This figure shows that the ideal low-pass filter is not a causal system.

1.6.14 Frequency response of a system defined by a difference equation

With the Fourier transform, it is possible to establish the relationship between a linear difference equation with constant coefficients of type (1.73) and a convolution, which expresses the output signal of a linear shift invariant system defined by the difference equation. For this purpose, we must grant that the initial conditions were applied for $k = -\infty$ and that the response of the system has reached its stationary state.

With this hypothesis, (1.73) is valid for any value of k. Moreover, if the input and output signals are assumed to be of finite energy and if the system is stable (sect. 1.6.19), it is possible to compute the Fourier transform for both sides of (1.73):

$$\sum_{k=-\infty}^{+\infty} \sum_{n=0}^{N} a_n y(k-n)e^{-j2\pi fk} = \sum_{k=-\infty}^{+\infty} \sum_{m=0}^{M} b_m x(k-m)e^{-j2\pi fk}$$

$$\sum_{n=0}^{N} a_n \sum_{k=-\infty}^{+\infty} y(k-n)e^{-j2\pi fk} = \sum_{m=0}^{M} b_m \sum_{k=-\infty}^{+\infty} x(k-m)e^{-j2\pi fk} \quad (1.86)$$

In this equation, the two sums with respect to k are the Fourier transforms of the signals $y(k)$ and $x(k)$, shifted by m and n, respectively. Using the shift theorem (1.52), it can be written:

$$\sum_{n=0}^{N} a_n e^{-j2\pi fn} Y(f) = \sum_{m=0}^{M} b_m e^{-j2\pi fm} X(f) \quad (1.87)$$

If the system is assumed to be of order N (at least N nonzero coefficients), then

$$Y(f) = \frac{\sum_{m=0}^{M} b_m e^{-j2\pi fm}}{\sum_{n=0}^{N} a_n e^{-j2\pi fn}} X(f) \qquad (1.88)$$

Because (1.85) is unique for a given system, comparing (1.85) and (1.88) leads to the following statement:

$$G(f) = \frac{\sum_{m=0}^{M} b_m e^{-j2\pi fm}}{\sum_{n=0}^{N} a_n e^{-j2\pi fn}} \qquad (1.89)$$

This statement (1.89) indicates the way in which it is possible to derive the frequency response for the system from the coefficients of the difference equations.

1.6.15 Property

If the Fourier transforms are replaced in (1.85) by their expressions, the following equation can be written:

$$Y(f) = \sum_{m=-\infty}^{+\infty} g(m) e^{-j2\pi fm} \sum_{l=-\infty}^{+\infty} x(l) e^{-j2\pi fl}$$

$$= \sum_{m=-\infty}^{+\infty} \sum_{l=-\infty}^{+\infty} g(m) x(l) e^{-j2\pi f(m+l)} \qquad (1.90)$$

With the change of variable $m + l = k$, the following statement is obtained:

$$Y(f) = \sum_{k=-\infty}^{+\infty} \left[\sum_{l=-\infty}^{+\infty} x(l) g(k-l) \right] e^{-j2\pi fk} \qquad (1.91)$$

The expression in brackets in the preceding equation is nothing other than the inverse Fourier transform of $Y(f)$. We have, therefore,

$$y(k) = \sum_{l=-\infty}^{+\infty} x(l) g(k-l) \qquad (1.92)$$

This is the convolution (1.77), which was established earlier by another method. The result obtained here proves two points. First it was shown that the solution of a difference equation is a convolution when the system is in the stationary state and, second, that a convolution in the time domain corresponds to a simple multiplication in the spectral domain. This is the converse statement of (1.56).

1.6.16 Remarks

The relationship between a difference equation of type (1.73) and a convolution leads to an attractive method for linking the coefficients of the difference equations to the samples of the impulse response: the frequency response must be computed with equation (1.89) and we must take its inverse Fourier transform. However, the reciprocal method is not always possible. The coefficients a_n and b_m of the difference equations can be derived from the impulse response if and only if the frequency response can be written as a ratio of polynomials of $\exp(-j2\pi f)$. In this unique case, the coefficients a_n and b_m are deduced by identifying them with the coefficients of the two polynomials. This proves that the convolution is the most general representation of a linear shift invariant system. This remark also proves that the subset of linear shift invariant systems characterized by a difference equation contains frequency responses, which are necessarily ratios of two polynomials of exponentials. This is not always the case for a frequency response obtained by the Fourier transform of the impulse response of a system characterized by a convolution. However, in practice, the majority of the frequency responses can be written as a ratio of two polynomials. Consequently, the set of linear shift invariant systems characterized by a difference equation contains a significant number of elements.

If these problems are studied using the Fourier transform, computation with the polynomials in $\exp(-j2\pi f)$ becomes difficult. Therefore, the Fourier transform is not commonly used for the analysis and synthesis of systems. This transformation is more oriented toward the analysis and representation of signals, due to the physical significance of the frequency included in it. The study of systems requires an additional dimension that the Fourier transform does not contain. For these reasons, a more general transformation will be studied in the next chapter, from which a particular case can be considered as the Fourier transform. This new transformation is called the z-transform, and it is oriented toward system analysis and synthesis.

Finally, it must be pointed out that the impulse response, which entirely characterizes a linear shift invariant system, may generally have an infinite number of samples. The corresponding system is then represented by this set of samples. However, the relationship between a convolution and the difference equation shows that the same system with the same impulse response can be characterized by a finite number ($N + M + 2$) of coefficients a_n and b_m.

1.6.17 Causal linear shift invariant systems

A linear shift invariant system may be constrained to be causal. From the previously given definition (1.5.4) of causal systems, it can be deduced that the impulse response of a causal linear shift invariant system is necessarily zero for $k < 0$. Let us consider two excitations $x_1(k)$ and $x_2(k)$ identical for $k < k_0$, and different for $k \geqslant k_0$. The responses of a linear shift invariant system are

$$y_1(k) = \sum_{l=-\infty}^{+\infty} x_1(l)g(k-l)$$

$$y_2(k) = \sum_{l=-\infty}^{+\infty} x_2(l)g(k-l) \tag{1.93}$$

If the system is causal as well, then $y_1(k) = y_2(k)$ for $k < k_0$. These sums can be divided into two parts:

$$y_1(k) = \sum_{l=-\infty}^{k_0-1} x_1(l)g(k-l) + \sum_{l=k_0}^{+\infty} x_2(l)g(k-l)$$

$$y_2(k) = \sum_{l=-\infty}^{k_0-1} x_2(l)g(k-l) + \sum_{l=k_0}^{+\infty} x_2(l)g(k-l) \tag{1.94}$$

Since for $k < k_0$ the responses $y_1(k)$ and $y_2(k)$ must be identical, we have

$$\sum_{l=k_0}^{+\infty} x_1(l)g(k-l) = \sum_{l=k_0}^{+\infty} x_2(l)g(k-l)$$

$$\sum_{l=k_0}^{+\infty} [x_1(l) - x_2(l)]g(k-l) = 0 \qquad \text{for } k < k_0 \tag{1.95}$$

By hypothesis, the difference $x_1(l) - x_2(l)$ is not zero for $l \geqslant k_0$. In order to satisfy this condition, we must have

$$g(k-l) = 0 \qquad \text{for } k < k_0 \quad \text{and} \quad l \geqslant k_0 \tag{1.96}$$

Finally, setting $m = k - l$, the following result is obtained:

$$g(m) = 0 \qquad \text{for } m < 0 \tag{1.97}$$

The converse can also be proved, that is, if the impulse response $g(k)$ of a linear shift invariant system is zero for $k < 0$, then the system is causal. Some authors, however, regard (1.97) as the definition of a causal system.

For causal linear shift invariant systems, the general form of the convolution, which expresses the response of this system to an excitation, can be modified with respect to (1.97). Thus, for a causal system, the following expressions are obtained:

$$y(k) = \sum_{l=-\infty}^{k} x(l) g(k-l) \tag{1.98}$$

or, equivalently,

$$y(k) = \sum_{l=0}^{+\infty} g(l) x(k-l) \tag{1.99}$$

If the impulse response has a finite duration L, this equation becomes

$$y(k) = \sum_{l=0}^{L-1} g(l) x(k-l) \tag{1.100}$$

1.6.18 Definition

The impulse response $g(k)$ of a linear shift invariant system is a signal (output signal for a particular excitation). The notion of *causality* can thus be used for systems as well as for signals. More generally, a signal $x(k)$ is called *causal* if

$$x(k) = 0 \qquad \text{for } k < 0 \tag{1.101}$$

The response of a causal linear shift invariant system to a causal excitation thus has the following form:

$$y(k) = \sum_{l=0}^{k} x(l) g(k-l) = \sum_{l=0}^{k} g(l) x(k-l) \tag{1.102}$$

This equation is widely used in practice because it involves a finite number of samples.

1.6.19 Stable linear shift invariant systems

Another constraint that is important in practice, which can be imposed on linear systems, is *stability*. By definition, a system is called *stable* if its response to a finite dynamic-range signal is also a finite dynamic-range signal.

1.6.20 Theorem

A linear shift invariant system is stable if and only if its impulse response satisfies the following condition:

$$T = \sum_{k=-\infty}^{+\infty} |g(k)| < \infty \qquad (1.103)$$

1.6.21 Proof

If the excitation $x(k)$ has a finite dynamic range, then $x(k) < A$ for any k, where A is a finite positive number. The modulus of the response is then given by

$$|y(k)| - \left| \sum_{l=-\infty}^{+\infty} g(l)x(k-l) \right| < A \sum_{l=-\infty}^{+\infty} |g(l)| \qquad (1.104)$$

Thus, if (1.103) is satisfied, the system is stable. In order to prove that condition (1.103) is necessary, let us consider an excitation signal:

$$x(k) = \begin{cases} +1 & \text{if } g(-k) \geq 0 \\ -1 & \text{if } g(-k) < 0 \end{cases} \qquad (1.105)$$

The dynamic range of this signal is finite. The output signal of the system for $k = 0$ is

$$y(0) = \sum_{l=-\infty}^{+\infty} g(l)x(-l) = \sum_{l=-\infty}^{+\infty} |g(l)| = T \qquad (1.106)$$

Consequently, the dynamic range of the response is not finite if condition (1.103) is not satisfied.

1.6.22 Recursive and nonrecursive implementations

In the previous sections, we have seen the different properties and representations of linear shift invariant systems. A large number of processing operations are performed with these systems. We have seen, in particular, how these systems are characterized by an Nth-order linear difference equation with constant coefficients:

$$\sum_{n=0}^{N} a_n(k)y(k-n) = \sum_{m=0}^{M} b_m(k)x(k-m) \qquad (1.107)$$

Furthermore, the output signal was written as a convolution:

$$y(k) = \sum_{l=-\infty}^{+\infty} g(l)x(k-l) \tag{1.108}$$

Different forms of this equation are also given for causal signals and systems. Some of these convolutions as well as (1.107) are directly applicable to signal processing systems. In this section, we define two classes of systems that are distinguishable by their properties which will be examined later. This division is based on the duration of the impulse response $g(k)$. In general, this duration may be finite or infinite. If the length is finite, the corresponding system is chararacterized by a *zeroth*-order difference equation. In fact, if we set $M = 0$ in (1.107), then

$$y(k) = \sum_{m=0}^{M} \frac{b_m}{a_0} x(k-m) \tag{1.109}$$

If this equation is compared with (1.100), we can derive the impulse response as

$$g(m) = \frac{b_m}{a_0} \quad \text{for} \quad m = 0, ..., M \tag{1.110}$$

A system like that of (1.100) is called *nonrecursive* or *transversal*. In this case, the frequency response of the system is a polynomial in $\exp(-j2\pi f)$.

If the duration of the impulse response is infinite, the order N of the corresponding difference equation, in any case, must be greater than zero. A system that verifies (1.107) with $N > 0$ is called *recursive*. In this case, the frequency response of the system is a ratio of two polynomials in $\exp(-j2\pi f)$.

1.6.23 Remarks

The system is characterized by a finite number of coefficients a_n and b_m with $n = 0, \ldots, N$ and $m = 0, \ldots, M$, even if an infinite number of samples is required to represent the impulse response. The relationship between these coefficients and the impulse response $g(k)$ will be established in the next chapter using the z-transform.

Finally, it must be mentioned that in a system which was not designed recursively the sample $y(k)$ of the output signal does not depend on the previous sample of the same signal. This property is a consequence of the particular form of convolution (1.109). This is not the case for a recursive system (1.107). The difference between these two classes of systems is very important because in practice it is not possible to perform a processing operation with an infinite number of samples. Thus, the difference equation (1.107) enables us to perform a processing operation with a finite number of coefficients, even if the corresponding system has an impulse response of infinite duration. This is why the general form of convolution (1.108) is of little practical interest, but of great importance in theoretical studies.

Table 1.9 Principal properties of the Fourier transform of digital signals.

$x(k)$	$X(f)$ periodic of period 1
$x(k) = \int_{-1/2}^{1/2} X(f) e^{j2\pi fk} df$	$X(f) = \sum_{k=-\infty}^{+\infty} x(k) e^{-j2\pi fk}$
$x(k)$ real	$\mathrm{Re}\,[X(-f)] = \mathrm{Re}\,[X(f)]$ even function
$x(k)$ real	$\mathrm{Im}\,[X(-f)] = -\mathrm{Im}\,[X(f)]$ odd function
$x(k-k_0)$	$X(f)\,e^{-j2\pi fk_0}$
$z(k) = x(k)\,y(k)$	$Z(f) = X(f) * Y(f)$
$\overset{\circ}{\varphi}_{xy}(k) = \sum_{l=-\infty}^{+\infty} x(l)\,y(l+k)$	$\overset{\circ}{\Phi}_{xy}(f) = X^*(f)\,Y(f)$
$x(-k)$ real	$X^*(f)$
$z(k) = \sum_{l=-\infty}^{+\infty} x(l)\,y(k-l)$	$Z(f) = X(f)\,Y(f)$
$z(k) = x(k)\,y(k)$	$Z(f) = \int_{-1/2}^{1/2} X(g)\,Y(f-g)\,dg$

1.7 SAMPLING AND RECONSTRUCTION OF ANALOG SIGNALS

1.7.1 Introduction

As mentioned previously (sect. 1.2.6), the most commonly used method for generating a discrete signal is to sample an analog signal $x_a(t)$. Chapter 8 of *Signal Theory and Processing* by F. de Coulon (vol. VI) is devoted to this subject. The principal results will be reviewed here with neither details nor proofs.

In general, the samples are taken periodically with a period Δt called sampling period (sect. 1.2.2). For all Δt, the signal obtained after sampling will always be a discrete (or sampled) signal. Only if the signal $x_a(t)$ needs to be reconstructed, must we then set a constraint on the choice of Δt. This constraint is a direct consequence of the sampling theorem.

1.7.2 The sampling theorem

An analog signal $x_a(t)$, whose bandwidth is limited to $F(\mathrm{Hz})$, can be reconstructed exactly from its samples $x_a(k\Delta t)$ if and only if these samples were taken with a period $\Delta t \leqslant 1/(2F)$.

1.7.3 Remarks

The sampling theorem is of great importance. It is a bridge between two complementary fields: digital signal processing and analog signal processing. It allows us to use all digital processing methods and systems for analog signal processing. The efficiency and the increasing flexibility of digital processing techniques enable us to use ever more complex operations, the implementation of which is difficult, if not impossible, using analog methods. The same theorem will be used in chapter 3 for the definition of the discrete Fourier transform (DFT) in the frequency domain.

1.7.4 Effects of ideal sampling

An *ideal sampling* is obtained when the analog signal $x_a(t)$ is multiplied by a periodic series of Dirac impulses of period Δt. Thus, we have (Fig. 1.10):

$$x_e(t) = x_a(t) \cdot e(t) \tag{1.111}$$

with

$$e(t) = \sum_{k=-\infty}^{+\infty} \delta(t - k\Delta t) = \delta_{\Delta t}(t)$$

In the frequency domain, the Fourier transform $X_e(f)$ of the sampled signal is given by (chap. 8, vol. VI):

$$X_e(f) = X_a(f) * E(f) = \frac{1}{\Delta t} \sum_{n=-\infty}^{+\infty} X_a\left(f - \frac{n}{\Delta t}\right) \tag{1.112}$$

The function $X_e(f)$ is thus obtained by the periodic repetition of the Fourier transform $X_a(f)$ (Fig. 1.10) of period $1/\Delta t$. The sampling, therefore, introduces a series of secondary spectra, which are proportional to the spectrum of the analog signal $x_a(t)$. The original signal can be obtained if and only if the secondary spectra are removed by appropriate filtering. This is typically performed by an analog low-pass filter. Nevertheless, even with an ideal filter, the spectrum $X_a(f)$ can be reconstructed if and only if it has zero values for frequencies greater than $F = 1/(2\Delta t)$. An overlapping (or aliasing) takes place whenever this condition is violated. The sampling theorem (also known as Shannon's theorem) is based on this condition (sect. 1.7.2 and Fig. 1.10).

In practice, this condition is never satisfied. The resulting unavoidable aliasing (Fig. 1.10) introduces an error into the estimation of the transform $X_a(f)$ from $X_e(f)$. Therefore, the value of Δt is decreased in order to divert the

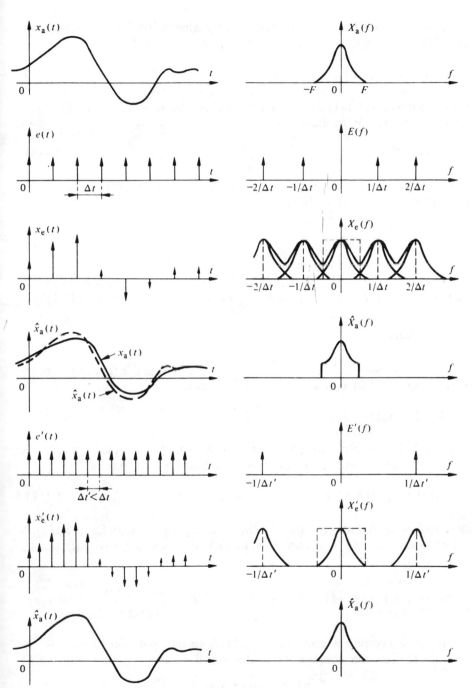

Fig. 1.10

secondary spectra. The error introduced by the aliasing of $X_a(f)$ then becomes negligible. In the digital case, the equation equivalent to (1.111) is

$$x(k \Delta t) = x_a(t)\big|_{t=k\Delta t} \qquad (1.113)$$

This equation and (1.111) are two different expressions of the same operation. Consequently, we can formally write

$$X(f) = X_e(f) \qquad (1.114)$$

where $X(f)$ is the Fourier transform of the digital signal $x(k\Delta t)$. Using (1.36), the following expression is obtained:

$$X(f) = \sum_{k=-\infty}^{+\infty} x(k\Delta t) e^{-j2\pi f k \Delta t} \qquad (1.115)$$

In this way, the sampling period Δt, which is not always unity is taken into account. It can be easily verified that $X(f)$ is periodic of period $1/\Delta t$ (1.115).

1.7.5 Reconstruction

If the sampling theorem is satisfied, the analog signal $x_a(t)$ can be reconstructed with an ideal analog low-pass filter. Thus,

$$X_a(f) = X_e(f) \cdot G_i(f) \qquad (1.116)$$

where $G_i(f)$ is the harmonic response of the ideal analog low-pass filter. In the time domain, the equivalent equation is the following convolution:

$$x_a(t) = x_e(t) * g_i(t) \qquad (1.117)$$

where $g_i(t)$ is the impulse response of the ideal analog low-pass filter. For such a filter, whose cut-off frequency f_c is $1/(2\Delta t)$, the expression of $g_i(t)$ can be derived from (1.35):

$$g_i(t) = \int_{-\infty}^{+\infty} G_i(f) e^{j2\pi ft} df = \Delta t \int_{-f_c}^{f_c} e^{j2\pi ft} df = \frac{\sin(\pi t/\Delta t)}{(\pi t/\Delta t)} \qquad (1.118)$$

If this expression is replaced in (1.117), an interpolation formula is obtained for the reconstruction of the analog signal $x_a(t)$:

$$x_a(t) = \sum_{k=-\infty}^{+\infty} x(k\Delta t) \frac{\sin[(\pi/\Delta t)(t-k\Delta t)]}{(\pi/\Delta t)(t-k\Delta t)} \qquad (1.119)$$

1.7.6 Comments

Equation (1.119) can be considered as a decomposition by a series of functions of the form of (1.33), with

$$\psi_k(t) = \frac{\sin[(\pi/\Delta t)(t - k\Delta t)]}{(\pi/\Delta t)(t - k\Delta t)} \tag{1.120}$$

The coefficients α_k are obtained simply by sampling the analog signal. This is one of the advantages of this decomposition. Another advantage is that the form of the convolution is the same in the analog case as in the digital case. This means that if

$$y(t) = x(t) * g(t) \tag{1.121}$$

then

$$y(k\,\Delta t) = x(k\Delta t) * g(k\,\Delta t) \tag{1.122}$$

as in (1.77).

If the period Δt satisfies the conditions imposed by the sampling theorem, this property allows the digital simulation of analog linear shift invariant systems.

One drawback of this representation (1.119) is its restriction to band-limited signals. Another is that the ideal filter (1.118) is not causal and, therefore, cannot be obtained in practice.

Other interpolators are then used in order to approximate the ideal solution (1.118).

Table 1.11 Comparison between the functional equations of digital and analog signal processing.

Equation	Digital case	Analog case
Signal	$x(k)$	$x(t)$
Fourier Transform	$X(f) = \sum\limits_{k=-\infty}^{+\infty} x(k)\, e^{-j2\pi fk}$	$X(f) = \int\limits_{-\infty}^{+\infty} x(t)\, e^{-j2\pi ft}\, dt$
	$x(k) = \int\limits_{-1/2}^{1/2} X(f)\, e^{j2\pi fk}\, df$	$x(t) = \int\limits_{-\infty}^{+\infty} X(f)\, e^{j2\pi ft}\, df$
Correlation	$\overset{\circ}{\varphi}_{xy}(k) = \sum\limits_{l=-\infty}^{+\infty} x(l)\, y(l+k)$	$\overset{\circ}{\varphi}_{xy}(\tau) = \int\limits_{-\infty}^{+\infty} x(t)\, y(t+\tau)\, dt$
	$\varphi_{xy}(k) = \frac{1}{K} \sum\limits_{l=k_0}^{k_0+k-1} x(l)\, y(l+k)$	$\varphi_{xy}(\tau) = \frac{1}{T} \int\limits_{\theta}^{\theta+T} x(t)\, y(t+\tau)\, dt$
Convolution	$y(k) = \sum\limits_{l=-\infty}^{+\infty} x(l)\, g(k-l)$	$y(t) = \int\limits_{-\infty}^{+\infty} x(u)\, g(t-u)\, du$

1.8 EXERCISES

1.8.1 Let $z(k)$ be the signal generated by the following recursive relations:

$$z(k) = \sqrt{-2 \ln x(k)} \, \cos[2\pi x(k+1)]$$
$$z(k+1) = \sqrt{-2 \ln x(k)} \, \sin[2\pi x(k+1)]$$

where $x(k)$ is a uniformly distributed signal on the interval $[0, 1]$, with

$$p[x(k), x(k+1)] = p[x(k)] \, p[x(k+1)] = 1$$

Find the distribution of the signal $z(k)$.

1.8.2 Find the recursive relation needed to generate a digital saw-toothed signal. How must this equation be modified in order to generate a triangular periodic signal?

1.8.3 Show that a linear digital system is causal if its impulse response is zero for negative values of the independent variable k.

1.8.4 Show that the Fourier transform is a linear transformation.

1.8.5 What is the duration of the following signal:

$$y(k) = \text{rect}_N(k) * \text{rect}_M(k)$$

Give the analytic form of $y(k)$ for $N = M$.

1.8.6 What is the Fourier transform of the following signal:

$$x(k) = a^k \, \epsilon(k) \quad \text{with } a \text{ real}$$

Give the values of a for which the transform exists.

1.8.7 Show that $\overset{\circ}{\varphi}_{xy}(k) = \overset{\circ}{\varphi}_{yx}(-k)$.

1.8.8 Show that the cross-spectral energy can be obtained from

$$\overset{\circ}{\Phi}_{xy}(f) = X^*(f) \, Y(f)$$

1.8.9 Show that in the frequency domain, (1.77) corresponds to

$$Y(f) = X(f) G(f).$$

1.8.10 Let $x(k\Delta t)$ be a signal and $X(f)$ its Fourier transform. What is the Fourier transform of the signal $y(t\Delta t')$ obtained by inserting $K - 1$ zero samples between each sample of the signal $x(k\Delta t)$ with step $\Delta t' = \Delta t/K$, where K is an integer?

What is the relationship between $x(k\Delta t)$ and the signal obtained by filtering $y(k\Delta t')$ with an ideal low-pass filter of cut-off frequency $f_c = 1/2 \, \Delta t$?

1.8.11 Give the frequency response of the linear shift invariant system, the equation for which is

$$y(k) = \frac{1}{M} \sum_{m=0}^{M-1} x(k-m)$$

1.8.12 Give the frequency response of a linear shift invariant system, the equation for which is

$$y(k) - ay(k-1) = x(k)$$

1.8.13 Give the frequency response of a linear shift invariant system, the equation for which is

$$y(k) - ay(k-1) = x(k) - bx(k-1)$$

How must a and b be chosen in order to have a constant response for any frequency?

1.8.14 Let $x(k)$ be a signal and $X(f)$ its Fourier transform. What is the signal, the Fourier transform of which is $X(f - f_0)$, where f_0 is a constant?

1.8.15 Give the frequency response of a linear shift invariant system, the equation for which is

$$y(k) = x(k) - x(k-L)$$

where L is a constant. Give the graphical representation of $|G(f)|$.

Chapter 2

The z-Transform

2.1 INTRODUCTION

2.1.1 Generalities

The Fourier transform is a useful tool, in both theory and practice, for digital and analog signal processing. However, some problems, especially those oriented toward the analysis and synthesis of processing systems (for example, in digital filtering), soon exhaust the possibilities of the Fourier transform. The z-transform provides a solution to the largely theoretical need for a more powerful processing tool. This transformation can be considered as a generalization of the Fourier transform, which, as shown in chapter 1, is only a special case of the z-transform. By its general nature, the z-transform can, for example, represent an infinitely long digital signal by a finite set of numbers. These numbers, which completely characterize the signal, can thus be used to reconstruct the signal in its entirety.

2.1.2 Organization of the chapter

In this chapter, we present the z-transform, the conditions for its existence, and the different methods for the computation of its inverse transform. Next, we study some principal properties of the z-transform, which are currently used in digital signal processing. These properties are similar to those of the Fourier transform, but, again, are more general. The relationships that exist between the z-transform and the Fourier and Laplace transforms are also presented. Knowledge of these relationships enables us to examine the sampling of an analog signal from a more general point of view.

When the properties of a signal and those of its z-transform are linked together, we obtain another representation of the signals. By using this representation, the Fourier transform of a signal can be obtained geometrically.

Finally, the transfer function of a linear shift invariant system is defined as the z-transform of its impulse response. The conditions of stability and causality in the z plane are studied. Two particular cases of a difference equation are considered in detail to illustrate the different notions developed in the chapter and to show the possibilities offered by simple systems.

2.2 THE z-TRANSFORM

2.2.1 Definition

The z-transform $X(z)$ of a signal $x(k)$ is defined by

$$X(z) = \sum_{k=-\infty}^{+\infty} x(k)z^{-k} \qquad (2.1)$$

where z is a complex variable and $X(z)$ a complex function of the variable z. From a practical point of view, the following notation is often used:

$$X(z) = Z[x(k)] \qquad (2.2)$$

2.2.2 Comments

Definition (2.1) is frequently called *the two-sided z-transform* because the summation involves all integers k. The one-sided z-transform is used in the study of causal signals and systems, which is defined by

$$X(z) = \sum_{k=0}^{+\infty} x(k)z^{-k} \qquad (2.3)$$

It is clear that definition (2.1) reduces to (2.3) for a causal signal. In the remainder of this book, the two-sided z-transform will be studied, with definition (2.3) considered as a special case of (2.1).

2.2.3 Existence of the z-transform

As in the case of the Fourier transform, the problem of convergence of a type (2.1) series must be considered. For a given signal, the set of z values for which the series (2.1) converges is called its *region of convergence*. In order to find the region of convergence, the Cauchy criterion for the convergence of a power series can be used:

$$\sum_{k=0}^{\infty} u_k = u_0 + u_1 + u_2 + u_3 + \dots \qquad (2.4)$$

This criterion states that a series of the form (2.4) converges if the following condition is satisfied:

$$\lim_{k \to \infty} |u_k|^{1/k} < 1 \qquad (2.5)$$

In order to apply this criterion, the series (2.1) can be divided into two series, the first summation denoted by $X_1(z)$ and the second by $X_2(z)$:

$$X(z) = \sum_{k=-\infty}^{-1} x(k)z^{-k} + \sum_{k=0}^{+\infty} x(k)z^{-k} \qquad (2.6)$$

Applying the criterion (2.5) to the series $X_2(z)$ leads to

$$\lim_{k \to \infty} |x(k)z^{-k}|^{1/k} < 1$$
$$\lim_{k \to \infty} |x(k)|^{1/k} |z^{-1}| < 1 \qquad (2.7)$$

Let R_{x-} be the limit:

$$\lim_{k \to +\infty} |x(k)|^{1/k} = R_{x-} \qquad (2.8)$$

Then, the series $X_2(z)$ converges for $|z| > R_{x-}$.

With the change of variable $l = -k$, it can be similarly shown that the series $X_1(z)$ converges for $|z| < R_{x*}$, where R_{x*} is the limit:

$$R_{x+} = 1/[\lim_{l \to +\infty} |x(-l)|^{1/l}] \qquad (2.9)$$

Thus, the series (2.1) generally converges in the complex z plane inside a ring given by

$$0 \leqslant R_{x-} < |z| < R_{x+} \leqslant +\infty \qquad (2.10)$$

This is illustrated on figure 2.1. The limits R_{x-} and R_{x*} characterize the signal $x(k)$. It is obvious that if $R_{x-} > R_{x*}$, the series (2.1) does not converge.

2.2.4 Example

Let us consider the following signal:

$$x(k) = \epsilon(k)$$

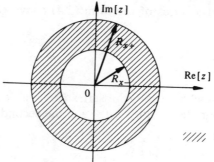

Fig. 2.1

The z-transform is given by

$$X(z) = \sum_{k=0}^{+\infty} 1\, z^{-k} = \frac{1}{1 - z^{-1}} \qquad \text{for } |z| > 1$$

The computation of the sum is simplified in this case to a complex geometrical series of kernel z^{-1}. With (2.8) and (2.9), we can easily find that $R_{x-} = 1$ and $R_{x*} = +\infty$.

2.2.5 Example

Let us consider the following signal:

$$x(k) = a^k \epsilon(k)$$

Its z-transform is given by

$$X(z) = \sum_{k=-\infty}^{+\infty} a^k \epsilon(k) z^{-k} = \sum_{k=0}^{+\infty} a^k z^{-k} = \sum_{k=0}^{+\infty} (az^{-1})^k$$

A geometrical series of kernel (az^{-1}) is obtained in this case. Thus,

$$X(z) = \frac{1}{1 - az^{-1}} \qquad \text{for } |z| > |a|$$

Equations (2.8) and (2.9) give $R_{x-} = |a|$ and $R_{x*} = +\infty$.

2.2.6 Example

As a last example, let us consider the following signal:

$$x(k) = a^k$$

With (2.8) and (2.9), the following result is obtained:

$$R_{x-} = R_{x+} = |a|$$

The inequality (2.10) for $|z|$ is strict (i.e., without equality). Thus, the z-transform of the signal does not exist because, for convergence, the following two inequalities must be simultaneously verified:

$$|a| < |z| < |a|$$

2.2.7 Properties

From the previously presented study of convergence, the following general properties can be derived. In expression (2.6), the series $X_2(z)$ represents the z-transform of a causal signal, or the one-sided z-transform. In general, this transform converges outside a circle of radius R_{x-} in the z plane. The radius R_{x-} is given by (2.8). Similarly, in expression (2.6), the series $X_1(z)$ represents the z-transform of a signal that is zero for $k \geq 0$. For such a signal, which can be called *anticausal*, the z-transform converges inside a circle of radius R_{x*} in the z plane, R_{x*} being given by (2.9).

It is interesting to establish the general rules for the convergence of the z-transform of a finite length signal. For such a signal, we have

$$X(z) = \sum_{k=k_1}^{k_2} x(k) z^{-k} \tag{2.11}$$

where k_1 and k_2 are positive or negative finite integers. If the modulus of each sample is finite within the interval $[k_1, k_2]$, the series (2.11) converges for all the values of z, except possibly $z = 0$ or $z \to +\infty$. For these two values, three cases can be distinguished according to the values of k_1 and k_2.

If k_1 and k_2 are both positive, the series (2.11) does not converge for $z = 0$, even if

$$|x(k)| < \infty \qquad \text{for } k_1 < k < k_2 \tag{2.12}$$

because for $k > 0$, the term z^{-k} diverges for $z = 0$.

If k_1 is negative and k_2 positive, the series (2.11) diverges for both $z = 0$ and $z \rightarrow +\infty$, even if condition (2.12) is satisfied.

Finally, if k_1 and k_2 are both negative, the series (2.11) diverges for z tending toward infinity, even if condition (2.12) is satisfied.

2.2.8 Definition

The series defined by relation (2.1) is known as a *Laurent series* [6]. Such a series defines a function $X(z)$, which is analytic. Consequently, in the region of convergence, the z-transform $X(z)$ and all its derivatives are continuous functions of z.

2.3 THE INVERSE z-TRANSFORM

2.3.1 Integral equation

The inversion of the z-transform can be obtained by applying the Cauchy theorem to the integration along a contour in the complex plane. From the Cauchy theorem, it can be deduced that the integral I defined by

$$I = \frac{1}{2\pi j} \oint_\Gamma z^{l-1} dz \tag{2.13}$$

where Γ is a closed contour which surrounds the origin of the z plane, is given by

$$I = \begin{cases} 1 & \text{for } l = 0 \\ 0 & \text{for } l \neq 0 \end{cases} \tag{2.14}$$

If both members of relation (2.1) are multiplied by $z^{l-1}/2\pi j$, and if the integral is computed along a contour surrounding the origin and contained in the region of convergence, then the following result is obtained:

$$\frac{1}{2\pi j} \oint_\Gamma X(z) z^{l-1} dz = \frac{1}{2\pi j} \oint_\Gamma \sum_{k=-\infty}^{+\infty} x(k) z^{-k+l-1} dz \tag{2.15}$$

Because the integral is computed in the region of convergence, the series converges. For this reason, the order of integration and summation can be inverted:

$$\frac{1}{2\pi j} \oint_\Gamma X(z) z^{l-1} dz = \sum_{k=-\infty}^{+\infty} x(k) \frac{1}{2\pi j} \oint_\Gamma z^{-k+l-1} dz \tag{2.16}$$

Using theorem (2.13), the following result is obtained:

$$x(l) = \frac{1}{2\pi j} \oint_{\Gamma} X(z) z^{l-1} dz \qquad (2.17)$$

This is the expression of the *inverse z-transform*, valid for all values of *l*. The integration contour Γ must be contained in the region of convergence; it must be closed and it must surround the origin of the *z* plane in the positive sense (counterclockwise).

The inverse z-transform can be computed in several ways. The principal methods are given in the following paragraphs.

2.3.2 Direct integration by the residues

The integral (2.17) can be computed with the usual methods for residues. The Cauchy theorem for integration along a contour shows that

$$x(k) = \frac{1}{2\pi j} \oint_{\Gamma} X(z) z^{k-1} dz = \sum \text{ with residues of } X(z) z^{k-1} \text{ in } \Gamma \quad (2.18)$$

The residue corresponding to a pole $z = a$ of order q of the function $[X(z) z^{k-1}]$ is given by

$$\text{Res}_a^q = \lim_{z \to a} \frac{1}{(q-1)!} \frac{d^{q-1}}{dz^{q-1}} [X(z) z^{k-1} (z-a)^q] \qquad (2.19)$$

For simple poles (of order $q = 1$), this expression has the following special form:

$$\text{Res}_a^1 = \lim_{z \to a} [(z-a) X(z) z^{k-1}] \qquad (2.20)$$

Thus, the signal $x(k)$ is the sum of all the residues computed by using (2.18) at the poles of the function $[X(z) z^{k-1}]$ in the contour Γ.

2.3.3 Example

To illustrate the method shown in section 2.3.2, let us consider the example of the z-transform, which was computed in section 2.2.4:

$$X(z) = \frac{1}{1 - z^{-1}} \qquad \text{for} \quad |z| > 1$$

If this expression is substituted into (2.17), then

$$x(k) = \frac{1}{2\pi j} \oint_\Gamma \frac{z^{k-1}}{1-z^{-1}} dz = \frac{1}{2\pi j} \oint_\Gamma \frac{z^k}{z-1} dz$$

where the contour Γ may be taken as a circle of radius greater than one.

For $k \geqslant 0$, this contour has only one simple pole (of order 1) at the point $z = 1$. Using relation (2.20), the residue at this pole can be seen as

$$\text{Res}_1^1 = \lim_{z \to 1} (z-1) \frac{z^k}{(z-1)} = 1 \qquad k \geqslant 0$$

Thus, $x(k) = 1$ for all $k \geqslant 0$.

For $k < 0$, the function $z^k/(z - 1)$ has poles of order k at the point $z = 0$. We can compute $x(k)$ for $k < 0$ recursively. For $k = -1$, a simple pole is obtained at $z = 0$ and another simple pole at $z = 1$. The residue at the pole $z = 0$ is given by

$$\text{Res}_0^1 = \lim_{z \to 0} z \frac{z^{-1}}{z-1} = -1$$

The residue at the pole $z = 1$ is given by

$$\text{Res}_1^1 = \lim_{z \to 1} (z-1) \frac{z^{-1}}{(z-1)} = 1$$

The sum of the residues is thus zero, which means that $x(-1) = 0$. Generally, for a pole of order k at the point $z = 0$:

$$\text{Res}_0^k = \frac{1}{(k-1)!} \lim_{z \to 0} \frac{d^{k-1}}{dz^{k-1}} \left[z^k \frac{z^{-k}}{z-1} \right] = -1$$

and, similarly,

$$\text{Res}_1^1 = \lim_{z \to 1} (z-1) \frac{z^k}{(z-1)} = 1$$

Thus, for any $k < 0$, the sum of the residues at the poles is zero, which means that $x(k) = 0$ for $k < 0$. The function $x(k)$ for all k is then, finally,

$$x(k) = \begin{cases} 1 & \text{for } k \geqslant 0 \\ 0 & \text{for } k < 0 \end{cases} = \epsilon(k)$$

This is the signal whose z-transform was computed in section 2.2.4.

2.3.4 Development as a power series

Because the z-transform $X(z)$ is an analytical function of z in the region of convergence, it can be developed in a Taylor series expansion around the point z^{-1}. Considering that (2.1) is a series of the same type, the samples $x(k)$ of the signal can be obtained by identifying them with the coefficients in the Taylor series expansion.

2.3.5 Example

Let us consider the z-transform given by

$$X(z) = \exp(z^{-1})(1 + z^{-1})$$

The Taylor series expansion gives

$$X(z) = \exp(z^{-1})(1 + z^{-1}) = \sum_{k=0}^{+\infty} \frac{z^{-k}(k+1)}{k!}$$

If this series is compared with definition (2.1), we can derive the corresponding signal:

$$x(k) = \frac{k+1}{k!} \epsilon(k)$$

2.3.6 Expansion by division

A large class of z-transforms that are frequently encountered in practice can be expressed as a quotient of two polynomials in z or z^{-1}:

$$X(z) = \frac{P(z)}{Q(z)} \tag{2.21}$$

The roots of the polynomial $P(z)$ are also the roots of $X(z)$. The roots of $Q(z)$ are the poles of $X(z)$. Furthermore, $X(z)$ can have poles for $z \to \infty$ or for $z = 0$. It is clear that because $X(z)$ does not converge at a pole, the region of convergence of $X(z)$ contains no poles.

If $P(z)$ is an Mth degree polynomial of z and $Q(z)$ is an Nth degree polynomial of z, then relation (2.21) can be written in the following form:

$$X(z) = C \frac{\prod\limits_{m=1}^{M} (z - z_m)}{\prod\limits_{n=1}^{N} (z - p_n)} \tag{2.22}$$

where z_m denotes the M zeros of $P(z)$ and, therefore, of $X(z)$; p_n denotes the N zeros of $Q(z)$ or the poles of $X(z)$; and C is a constant. With this equation, the z-transform $X(z)$ can be simply obtained from its poles and zeros. The method considered here to obtain the signal $x(k)$ whose z-transform is $X(z)$ consists simply of performing the division (2.21).

2.3.7 Example

Let $X(z)$ be the z-transform of a signal given by

$$X(z) = z^{-1}/(1 - 1.414 z^{-1} + z^{-2}) \quad \text{for } |z| > 1 \tag{2.23}$$

The region of convergence is then the outside of the unit circle in the z plane.
The roots of the denominator and $z = 0$ are the three poles of this transform. Thus,

$$p_1 = \frac{1}{1.414} (1 - j) \qquad p_2 = \frac{1}{1.414} (1 + j) \qquad \text{and} \qquad p_3 = 0$$

Figure 2.2 shows the position of the poles, usually represented by crosses, in the z plane and the region of convergence.

To obtain the signal $x(k)$, the division in (2.23) is performed:

$$
\begin{array}{l}
z^{-1} \\
\underline{z^{-1} - 1.414 z^{-2} \quad + z^{-3}} \\
\qquad 1.414 z^{-2} \quad - z^{-3} \\
\qquad \underline{1.414 z^{-2} - 2 z^{-3} + 1.414 z^{-4}} \\
\qquad\qquad z^{-3} - 1.414 z^{-4} \\
\qquad\qquad \underline{z^{-3} - 1.414 z^{-4} + z^{-5}} \\
\qquad\qquad\qquad - z^{-5} \\
\qquad\qquad\qquad \underline{- z^{-5} + 1.414 z^{-6} - z^{-7}} \\
\qquad\qquad\qquad\qquad - 1.414 z^{-6} + z^{-7}
\end{array}
$$

$$\left| \, 1 \quad - 1.414 z^{-1} + z^{-2} \right.$$
$$z^{-1} + 1.414 z^{-2} + z^{-3} + 0 - z^{-5} - 1.414 z^{-6} + \dots$$

The result of the division is a power series in z^{-1}. If this result is compared to definition (2.1) and if the coefficients in the expansion are identified with the corresponding coefficients in the definition, the samples of the signal are obtained. This is shown in figure 2.2.

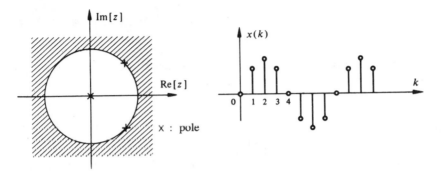

Fig. 2.2

$x(0) = 0$

$x(1) = 1$

$x(2) = 1.414$

$x(3) = 1$

$x(4) = 0$

$x(5) = -1$

.
.
.

The signal is obtained in digital form, without any information about its analytical form or its composition. This is one drawback of this method. However, its speed and its great simplicity weigh heavily in its favor. The method of expansion by division can be applied manually with a pocket calculator or by a computer program.

2.3.8 Property of linearity

Let us consider a signal $x(k)$, which is a linear combination of two signals $x_1(k)$ and $x_2(k)$:

$$x(k) = a\,x_1(k) + b\,x_2(k) \qquad (2.24)$$

The z-transform of this signal is given by

$$X(z) = \sum_{k=-\infty}^{+\infty} [ax_1(k) + bx_2(k)]z^{-k}$$

$$= a\sum_{k=-\infty}^{+\infty} x_1(k)z^{-k} + b\sum_{k=-\infty}^{+\infty} x_2(k)z^{-k} = aX_1(z) + bX_2(z) \quad (2.25)$$

The region of convergence of $X(z)$ is, at least, the union of the regions of convergence of $X_1(z)$ and $X_2(z)$. Therefore,

and
$$R_{x-} = \max [R_{x_1-}, R_{x_2-}]$$

$$R_{x+} = \min [R_{x_1+}, R_{x_2+}] \quad (2.26)$$

The region of convergence of $X(z)$ may be larger if any zeros introduced by the linear combination cancel some poles of $X_1(z)$ or $X_2(z)$.

2.3.9 Development by partial fractions

The idea of decomposing a complicated function into a sum of simple functions is also used in the inversion of a z-transform. This is very useful for rational z-transforms of the type (2.21). They can then be decomposed into partial fractions, and the inverse transform is thus the sum of the inverses of the simple terms using the property of linearity of the z-transform.

A z-transform that is the quotient of two polynomials $P(z)$ and $Q(z)$ of degree M and N, respectively, can be written as follows:

$$X(z) = \frac{P(z)}{Q(z)} = S(z) + \frac{P_0(z)}{Q_0(z)} \quad (2.27)$$

where $S(z)$ is a polynomial of degree $M - N$ and the degree of $P_0(z)$ is less than the degree of $Q_0(z)$. If $M \geqslant N$, the coefficients of the polynomial $S(z)$ can be obtained from the division of $P(z)$ by $Q(z)$. If $M < N$, then $S(z) = 0$.

The quotient of two polynomials, where the degree of the numerator is less than the degree of the denominator, can be decomposed into partial fractions as

$$\frac{P_0(z)}{Q_0(z)} = \sum_{i=1}^{N} \frac{\alpha_i}{z - p_i} \quad (2.28)$$

where p_i denotes the simple zeros of $Q_0(z)$ and, therefore, the simple poles of $X(z)$. To obtain the coefficients α_i, the two members of (2.28) must first be multiplied by $(z - p_i)$ and then z is set equal to p_i. The following expression is obtained:

$$\alpha_i = (z - p_i) \frac{P_0(z)}{Q_0(z)}\bigg|_{z = p_i} \qquad (2.29)$$

If the quotient $P_0(z)/Q_0(z)$ has the form (2.22), the coefficient α_i is given by

$$\alpha_i = \frac{\displaystyle\prod_{m=1}^{M_0} (p_i - z_m)}{\displaystyle\prod_{\substack{n=1 \\ n \neq i}}^{N} (p_i - p_n)} \qquad (2.30)$$

where M_0 is the degree of the polynomial $P_0(z)$. This equation is more appropriate for practical calculations than (2.29). If the poles of $X(z)$ or the zeros of $Q_0(z)$ are multiple or degenerate, the decomposition (2.28) must be modified. For example, if p_n is a qth-order pole, the decomposition has the following form:

$$\frac{P_0(z)}{Q_0(z)} = \sum_{\substack{i=1 \\ i \neq n}}^{N} \frac{\alpha_i}{z - p_i} + \sum_{j=1}^{q} \frac{\beta_j}{(z - p_n)^j} \qquad (2.31)$$

To isolate a coefficient β_j on the right-hand side of this equation and to avoid the poles at the same time, the two sides are first multiplied by $(z - p_n)q$. Then, the derivative with respect to z is taken $(q - j)$ times. Finally, the expression is divided by $(q - j)!$ to balance the multiplicative coefficients coming from the derivative and the final expression is then evaluated at $z = p_n$:

$$\beta_j = \frac{1}{(q - j)!} \frac{d^{q-j}}{dz^{q-j}} \left[(z - p_n)^q \frac{P_0(z)}{Q_0(z)} \right]\bigg|_{z = p_n} \qquad (2.32)$$

2.3.10 Comments

It must be pointed out that in the first sum of relation (2.31), the simple contribution of the multiple pole has been excluded because it is contained in the second summation. The coefficients α_i and β_j are sometimes called *residues* because they are analogous to the Taylor or Laurent series expansions.

After computing the decomposition (2.28) or (2.31), the inverse z-transform of each term can be obtained by using the region of convergence of $X(z)$ and a table for the z-transform of simple functions. Such a table can be found, for example, in references [7] or [8]. Furthermore, if the poles of $X(z)$ are simple, then each term in the decomposition (2.28) corresponds to an exponential signal, as in example 2.2.5. The particular form of each term is determined not only by the coefficients α_i and the poles p_i, but also by the region of convergence. These operations will be simplified by using the fundamental properties of the z-transform, which will be studied in section 2.4.

2.3.11 Example

Let us consider the z-transform given by

$$X(z) = \frac{1}{1 - 3z^{-1} + 2z^{-2}} \qquad \text{for } |z| > 2$$

The degree of the denominator is greater than the degree of the numerator, so the polynomial $S(z)$ in (2.27) is identically zero. The poles of $X(z)$ are given by the roots of the equation:

$$2z^{-2} - 3z^{-1} + 1 = 0$$

Two simple poles $z^{-1} = 1$ and $z^{-1} = 1/2$ are obtained. Thus,

$$X(z) = \frac{1/2}{(z^{-1} - 1)(z^{-1} - 1/2)} = \frac{\alpha_1}{z^{-1} - 1} + \frac{\alpha_2}{z^{-1} - 1/2}$$

Using (2.29), we find $\alpha_1 = 1$ and $\alpha_2 = -1$. Consequently,

$$X(z) = \frac{1}{z^{-1} - 1} - \frac{1}{z^{-1} - 1/2} = \frac{2}{1 - 2z^{-1}} - \frac{1}{1 - z^{-1}}$$

Both of these partial fractions are just a particular case of example 2.2.5. Comparing this term with example 2.2.5 and taking into account the region of convergence $|z| > 2$, the following inverse z-transform is obtained:

$$x(k) = 2 \cdot 2^k \epsilon(k) - \epsilon(k)$$
$$= (2^{k+1} - 1)\epsilon(k)$$

Here, the denominator of $X(z)$ has been considered as a polynomial in z^{-1}. The same result is obtained if $X(z)$ is considered as a quotient of polynomials in z. It is seen, by inspection, that

$$X(z) = \frac{1}{1 - 3z^{-1} + 2z^{-2}} = \frac{z^2}{z^2 - 3z + 2}$$

2.4 PRINCIPAL PROPERTIES OF THE z-TRANSFORM

2.4.1 Introduction

The knowledge of some principal properties of the z-transform makes signal processing problems much easier. For example, we saw previously that the property of linearity of the z-transform can be used to derive a powerful and general method for the inversion of transforms which are the quotient of two polynomials of z or z^{-1}. In this section, other frequently used properties of the z-transform are presented and proved. They are summarized in table 2.6 at the end of the section.

2.4.2 Shift of a signal

Let us consider a signal $x(k)$, whose z-transform is $X(z)$. Let $y(k)$ be the shifted version of the signal $x(k)$:

$$y(k) = x(k - k_0)$$ (2.33)

The z-transform of $y(k)$ is given by

$$Y(z) = \sum_{k=-\infty}^{+\infty} y(k)z^{-k} = \sum_{k=-\infty}^{+\infty} x(k - k_0)z^{-k}$$ (2.34)

With the change of variable $l = k - k_0$, the following expression is obtained:

$$Y(z) = \sum_{l=-\infty}^{+\infty} x(l)z^{-(l+k_0)} = z^{-k_0} \sum_{l=-\infty}^{+\infty} x(l)z^{-l}$$ (2.35)

In this expression, the last sum over l is nothing other than the z-transform of $x(k)$. Thus,

$$Y(z) = z^{-k_0} X(z) \qquad \text{for} \quad R_{x-} < |z| < R_{x+}$$ (2.36)

Thus, the z-transform of a signal shifted by k_0 is the product of the z-transform of the original signal with the term z^{-k_0}. Equations (2.33) and (2.36) comprise the generalization of the shift theorem presented in section 1.3.9.

2.4.3 Comments

From (2.36), we can see that the product of a z-transform with z^{-1} corresponds to a unit shift ($k_0 = 1$). For this reason, z^{-1} is referred to as the *unit shift operator*. Figure 2.3 shows the schematic representation of this operator.

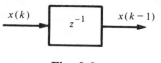

Fig. 2.3

Nevertheless, in such a representation, which is frequently used, we must not assume that z^{-1} acts directly on the input signal. It is used only to represent (2.36) in the particular case $k_0 = 1$.

From (2.36), it can also be pointed out that the region of convergence of $Y(z)$ is the same as the region of convergence of $X(z)$, except for $z = 0$ or $z \to \infty$. For k_0 positive, (2.36) introduces a pole of order k_0 at the origin and a zero for $z \to \infty$. For k_0 negative, the same equation introduces a multiple zero at the origin and poles for $z \to \infty$.

It is clear that (2.33) and (2.36) can be used inversely to show that relation (2.33) can be deduced from a z-transform of the form (2.36).

2.4.4 Change of complex scale

Let us consider a signal $x(k)$ and its z-transform $X(z)$, which converges within $R_{x-} < |z| < R_{x*}$. If, in the complex plane, the change of variable $w = az$, where a may be a complex number, is performed, the region of convergence of $X(w/a)$ is modified as follows:

$$|a| R_{x-} < |w| < |a| R_{x+} \tag{2.37}$$

If, in the z plane, $X(z)$ has a pole for $z = p$, then, in the w plane, $X(w/a)$ will have a pole at $w = ap$. This change of variable allows us to modify the positions of the poles and the zeros by the factor a, which may be a complex number.

If a is a positive real number, then, after the change of variable $w = az$, the positions of the poles and the zeros approach (for $(a < 1)$) or move away from (for $(a > 1)$) the origin along a radial line.

If a has the form $\exp(-jb)$, the same change of variable corresponds to a rotation of b radians. The position of the poles and the zeros is modified along the circles centered at the origin. This is illustrated in figure 2.4.

2.4.5 Interpretation

It is interesting to note the effect of this change of variable on the signal. Using the definition of the inverse z-transform, we have

$$x(k) = \frac{1}{2\pi j} \oint_\Gamma X(z) z^{k-1} dz \tag{2.38}$$

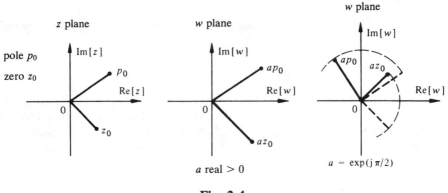

Fig. 2.4

If z is replaced by w/a, then,

$$x(k) = \frac{1}{2\pi j} \oint_\Gamma X\left(\frac{w}{a}\right)\left(\frac{w}{a}\right)^{k-1} \frac{dw}{a} \tag{2.39}$$

thus,

$$a^k x(k) = \frac{1}{2\pi j} \oint_\Gamma X\left(\frac{w}{a}\right) w^{k-1} dw \tag{2.40}$$

or, consequently,

$$X(z/a) = Z[a^k x(k)] \tag{2.41}$$

The change of variable $w = az$ thus corresponds to the product of the signal $x(k)$ and an exponential signal a^k. The inverse can also be proved. If a signal $y(k)$ has the form:

$$y(k) = a^k x(k) \tag{2.42}$$

then its z-transform is given by

$$Y(z) = \sum_{k=-\infty}^{+\infty} a^k x(k) z^{-k} = \sum_{k=-\infty}^{+\infty} x(k)\left(\frac{z}{a}\right)^{-k} = X\left(\frac{z}{a}\right) \tag{2.43}$$

This property allows for the modification of the position at the poles and zeros of the z-transform by multiplying the signal by an exponential signal a^k. The value of the parameter a depends on the particular modification chosen.

2.4.6 Derivative of the z-transform

Using definition (2.1) the derivative of the z-transform can be written

$$\frac{dX(z)}{dz} = \sum_{k=-\infty}^{+\infty} (-k) x(k) z^{-k-1} \tag{2.44}$$

If both sides of this equation are multiplied by $-z$ the following result is obtained

$$-z \frac{dX(z)}{dz} = \sum_{k=-\infty}^{+\infty} [k \, x(k)] z^{-k} \tag{2.45}$$

The right-hand side of this equation is the z-transform of the signal $kx(k)$. Consequently, the derivative of a z-transform multiplied by $-z$ is the z-transform of the product of the original signal and the signal $y(k) = k$. The inverse can also be proved with the properties of the power series.

2.4.7 Correlation of two signals

The crosscorrelation function of two signals $x(k)$ and $y(k)$ was defined (1.57) as

$$\overset{\circ}{\varphi}_{xy}(k) = \sum_{l=-\infty}^{+\infty} x(l) y(l+k) \tag{2.46}$$

The z-transform of this equation leads to a simple product:

$$\overset{\circ}{\Phi}_{xy}(z) = \sum_{k=-\infty}^{+\infty} \sum_{l=-\infty}^{+\infty} x(l) y(l+k) z^{-k} \tag{2.47}$$

Setting $m = l + k$ leads to

$$\overset{\circ}{\Phi}_{xy}(z) = \sum_{l=-\infty}^{+\infty} \sum_{m=-\infty}^{+\infty} x(l) y(m) z^{-m} z^{l}$$

$$= \sum_{l=-\infty}^{+\infty} x(l) z^{l} \cdot \sum_{m=-\infty}^{+\infty} y(m) z^{-m} \tag{2.48}$$

The first sum in this expression is the z-transform of the signal $x(k)$ inverted about the ordinate axis. From the definition of the z-transform, we have

$$X(z) = Z[x(k)] \tag{2.49}$$

If $x_1(k) = x(-k)$, then,

$$X_1(z) = \sum_{k=-\infty}^{+\infty} x(-k) z^{-k} = \sum_{l=-\infty}^{+\infty} x(l) z^{l} = X\left(\frac{1}{z}\right) \tag{2.50}$$

Thus,

$$\overset{\circ}{\Phi}_{xy}(z) = X(1/z) \, Y(z) \tag{2.51}$$

The region of convergence of $\overset{\circ}{\Phi}_{xy}(z)$ consists of at least the intersection of the regions of convergence of $X(1/z)$ and $Y(z)$. It can be larger if the poles of one transform are balanced by the zeros of the other. The region of convergence of $X(1/z)$ can be easily deduced from the region of convergence of $X(z)$. Using equations (2.8), (2.9), and (2.50), it is easy to prove that the region of convergence of $X(1/z)$ is given by

$$\frac{1}{R_{x+}} < |z| < \frac{1}{R_{x-}} \tag{2.52}$$

where $[R_{x-}, R_{x*}]$ is the region of convergence of $X(z)$.

2.4.8 Example

Let us consider the signal $x(k) = \text{rect}_N(k)$ and try to find its autocorrelation function.

The z-transform of $x(k)$ is given by

$$X(z) = \sum_{k=0}^{N-1} z^{-k} = \frac{1 - z^{-N}}{1 - z^{-1}} \qquad \text{for } |z| > 0$$

This can also be written as

$$X\left(\frac{1}{z}\right) = \frac{1 - z^{N}}{1 - z} \qquad \text{for } |z| > 0$$

Thus,

$$\overset{\circ}{\Phi}_x(z) = \frac{z(1 - z^{-N})(z^{N} - 1)}{(z - 1)^{2}} \qquad \text{for } |z| > 0$$

The theorem of residues can be used to compute the inverse transform. To apply this theorem, we must find the poles of the function:

$$\overset{\circ}{\Phi}_x(z) z^{k-1} = \frac{(z^{k} - z^{k-N})(z^{N} - 1)}{(z - 1)^{2}} \tag{2.53}$$

To avoid the calculation of the residue at the pole $z = 1$, a circle centered at the origin and of radius less than one can be chosen as the integration contour. In this case, there are poles at the origin that are dependent on the value of k. Because $\overset{\circ}{\Phi}_x(k)$ is an even function, only positive values of k must be considered. For $k \geqslant N$, there are no poles within the integration contour. Consequently,

$$\overset{\circ}{\Psi}_x(k) = 0 \qquad \text{for } |k| \geqslant N$$

In expression (2.53) there is a pole of order $N - k$ at the origin for $k < N$. The corresponding residue is given by

$$\text{Res}_0^{N-k} = \frac{1}{(N - k - 1)!} \lim_{z \to 0} \frac{d^{N-k-1}}{dz^{N-k-1}} [z^{N-k} \overset{\circ}{\Phi}_x(z) z^{k-1}]$$

With a recursive computation, it can be shown that

$$\text{Res}_0^{N-k} = N - k \qquad \text{for } 0 \leqslant k < N$$

Thus,

$$\overset{\circ}{\varphi}_x(k) = N - |k| \qquad \text{for } 0 \leqslant |k| < N$$

2.4.9 Convolution

The z-transform of a convolution also leads to a simple product. If $y(k)$ is the convolution of two signals $x(k)$ and $g(k)$, then

$$y(k) = \sum_{l=-\infty}^{+\infty} x(l)g(k-l) \tag{2.54}$$

The z-transform of this signal is given by

$$Y(z) = \sum_{k=-\infty}^{+\infty} \sum_{l=-\infty}^{+\infty} x(l)g(k-l)z^{-k} = \sum_{l=-\infty}^{+\infty} x(l) \sum_{k=-\infty}^{+\infty} g(k-l)z^{-k} \tag{2.55}$$

With the change of variable $m = k - l$, we obtain the following expression:

$$Y(z) = \sum_{l=-\infty}^{+\infty} x(l)z^{-l} \cdot \sum_{m=-\infty}^{+\infty} g(m)z^{-m} \tag{2.56}$$

Thus, for the values of z that are within the intersection of the regions of convergence of $X(z)$ and $G(z)$, the following equation is derived:

$$Y(z) = X(z)G(z) \tag{2.57}$$

The region of convergence of $Y(z)$ can be larger than the intersection of the regions of convergence of $X(z)$ and $G(z)$ if the zeros of one of the transforms balance the poles of the other. Equation (2.57) is the more general form of the similar equation (1.85), established with the Fourier transform.

2.4.10 Example

Let us consider a linear shift invariant system characterized by the first-order difference equation:

$$y(k) - a y(k-1) = x(k) \quad \text{with } |a| < 1$$

with the initial condition:

$$y(k) = 0 \qquad \text{for } k < 0$$

The impulse response of this system is obtained by considering the unit impulse $x(k) = d(k)$ as the excitation. Then,

$$g(k) = a^k \epsilon(k)$$

Let us also consider an input signal $x(k) = b^k \epsilon(k)$ with $|b| < 1$. These two signals are shown in figure 2.5. The corresponding z-transforms are given by

$$G(z) = \sum_{k=0}^{+\infty} a^k z^{-k} = \frac{1}{1 - az^{-1}} \qquad \text{for } |z| > |a|$$

$$X(z) = \sum_{k=0}^{+\infty} b^k z^{-k} = \frac{1}{1 - bz^{-1}} \qquad \text{for } |z| > |b|$$

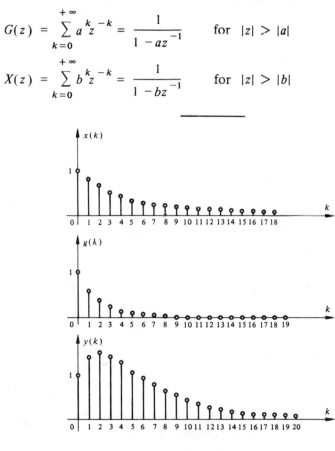

Fig. 2.5

Moreover, let us assume that $|a| < |b|$:

$$Y(z) = G(z) \cdot X(z) = \frac{1}{1 - az^{-1}} \frac{1}{1 - bz^{-1}}$$

$$Y(z) = \frac{z^2}{(z-a)(z-b)} \qquad \text{for } |z| > |b|$$

$Y(z)$ can be decomposed into a partial fraction leading to the following result:

$$Y(z) = \left(\frac{a}{a-b}\right)\frac{1}{1-az^{-1}} + \left(\frac{b}{b-a}\right)\frac{1}{1-bz^{-1}} \qquad \text{for } |z| > |b|$$

The response of the system to this excitation is the inverse z-transform of $Y(z)$. From example 2.2.5, it is a sum of two exponential signals:

$$y(k) = \frac{a}{a-b}a^k\epsilon(k) + \frac{b}{b-a}b^k\epsilon(k)$$

This signal is shown in figure 2.5 for $a = 0.6$ and $b = 0.8$.

2.4.11 Theorem of complex convolution

In the previous chapter (sect. 1.6.9), we showed that a simple product of the Fourier transforms in the frequency domain corresponds to a convolution in the time domain. The same equation was proved (sect. 2.4.9) with the z-transform. However, additionally, we saw that a continuous and periodic convolution in the frequency domain corresponds to a simple product in the time domain (sect. 1.3.10). In this section, the corresponding equation will be established using the z-transform. This equation is known as the *complex convolution theorem*.

Let $v(k)$ be a signal obtained as the product of two signals $x(k)$ and $y(k)$:

$$v(k) = x(k)y(k) \tag{2.58}$$

The z-transform of this signal is given by

$$V(z) = \sum_{k=-\infty}^{+\infty} x(k)y(k)z^{-k} \tag{2.59}$$

The signal $x(k)$ can be expressed in terms of its inverse z-transform, leading to the following result:

$$V(z) = \sum_{k=-\infty}^{+\infty} \frac{1}{2\pi j} \oint_\Gamma X(w)w^{k-1}dw\, y(k)z^{-k}$$

$$= \frac{1}{2\pi j} \oint_\Gamma \left[\sum_{k=-\infty}^{+\infty} y(k)\left(\frac{z}{w}\right)^{-k}\right] w^{-1}X(w)dw \tag{2.60}$$

However, the expression in brackets is nothing other than the transform $Y(z/w)$. Consequently,

$$V(z) = \frac{1}{2\pi j} \oint_\Gamma X(w) Y\left(\frac{z}{w}\right) w^{-1} dw \tag{2.61}$$

If $y(k)$ is expressed in (2.59) in terms of its inverse z-transform, the following expression is obtained exactly as in (2.61):

$$V(z) = \frac{1}{2\pi j} \oint_\Gamma X\left(\frac{z}{w}\right) Y(w) w^{-1} dw \tag{2.62}$$

In expressions (2.61) and (2.62), the integration contour Γ must be chosen within the intersection of the regions of convergence of $X(w)$ and $Y(z/w)$, or of $X(z/w)$ and $Y(w)$, respectively.

If $[R_{x-}, R_{x*}]$ and $[R_{y-}, R_{y*}]$ are the regions of convergence of $X(z)$ and $Y(z)$, respectively, the following condition must be satisfied when $x(k)$ is substituted in (2.59):

$$R_{x-} < |w| < R_{x+} \tag{2.63}$$

Likewise, in the transform:

$$Y\left(\frac{z}{w}\right) - \sum_{k=-\infty}^{+\infty} y(k) \left(\frac{z}{w}\right)^{-k} \tag{2.64}$$

we must have

$$R_{y-} < |z/w| < R_{y+} \tag{2.65}$$

The intersection of the regions of convergence is obtained by combining the two equations (2.63) and (2.65):

$$R_{x-} R_{y-} < |z| < R_{x+} R_{y+} \tag{2.66}$$

To extract the convolution in (2.61) or (2.62), the variables w and z must be represented in polar coordinates in the complex plane. With $w = a \exp(jb)$ and $z = \alpha \exp(j\beta)$:

$$V(\alpha e^{j\beta}) = \frac{1}{2\pi} \int_{-\pi}^{\pi} X(a e^{jb}) \cdot Y\left(\frac{\alpha}{a} e^{j(\beta-b)}\right) db \tag{2.67}$$

From this expression, we can see that this is a continuous and periodic (of period 2π) convolution of the functions $X[a \exp(jb)]$ and $Y[\alpha \exp(j\beta)]$.

2.4.12 Theorem of the initial value

With the initial value theorem, the value at the origin for a signal of known z-transform can be derived relatively simply. However, it is only valid for causal signals, as indicated by its name.

The z-transform of the causal signal $x(k)$ is given by

$$X(z) = \sum_{k=0}^{+\infty} x(k)z^{-k}$$

$$= x(0) + \frac{x(1)}{z} + \frac{x(2)}{z^2} + \dots + \frac{x(k)}{z^k} + \dots \qquad (2.68)$$

If the limit of $X(z)$ for $z \to \infty$ is taken, all the terms of the series, except the first, tend toward zero. Consequently,

$$x(0) = \lim_{z \to \infty} X(z) \qquad (2.69)$$

This is the expression for the initial value theorem.

2.4.13 Example

Let us consider the signal $y(k)$ studied in section 2.4.10:

$$y(k) = \frac{a}{a-b} a^k \epsilon(k) + \frac{b}{b-a} b^k \epsilon(k) \qquad (2.70)$$

with

$$Y(z) = \frac{z^2}{(z-a)(z-b)} = \frac{z^2}{z^2 - (a+b)z + ab}$$

Application of the initial value theorem leads to

$$y(0) = \lim_{z \to \infty} \frac{z^2}{z^2 - (a+b)z + ab} = 1$$

This is exactly the result obtained if $k = 0$ is set in (2.70).

2.4.14 Remark

Other properties of the z-transform can be found in references [7] and [9]. A review of the properties studied here is given in table 2.6.

Table 2.6 Principal properties of the z-transform

$x(k) = \dfrac{1}{2\pi j} \oint_\Gamma X(z)\, z^{k-1} dz$	$X(z) = \sum\limits_{k=-\infty}^{+\infty} x(k)\, z^{-k}$ for $R_{x-} < |z| < R_{x+}$
$a\, x_1(k) + b\, x_2(k)$	$a\, X_1(z) + b\, X_2(z)$ for $\max\,[R_{x_1-}, R_{x_2-}] < |z| < \min\,[R_{x_1+}, R_{x_2+}]$
$x(k - k_0)$	$z^{-k_0} X(z)$ for $R_{x-} < |z| < R_{x+}$
$a^k x(k)$	$X\left(\dfrac{z}{a}\right)$ for $|a|\, R_{x-} < |z| < |a|\, R_{x+}$
$k\, x(k)$	$-z\, \dfrac{\mathrm{d}X(z)}{\mathrm{d}z}$ for $R_{x-} < |z| < R_{x+}$
$\overset{\circ}{\varphi}_{xy}(k) = \sum\limits_{l=-\infty}^{+\infty} x(l)\, y(l+k)$	$\overset{\circ}{\Phi}_{xy}(z) = X\left(\dfrac{1}{z}\right) Y(z)$ for $\max\left[\dfrac{1}{R_{x+}}, R_{y-}\right] < |z| < \min\left[\dfrac{1}{R_{x-}}, R_{y+}\right]$
$y(k) = \sum\limits_{l=-\infty}^{+\infty} x(l)\, g(k-l)$	$Y(z) = X(z)\, G(z)$ for $\max\,[R_{x-}, R_{g-}] < |z| < \min\,[R_{x+}, R_{g+}]$
$x(k)\, y(k)$	$\dfrac{1}{2\pi j} \oint_{\Gamma'} X(w)\, Y\left(\dfrac{z}{w}\right) w^{-1}\, dw$ for $R_{x-} R_{y-} < |z| < R_{x+} R_{y+}$
$x(0)$ (causal)	$\lim\limits_{z \to \infty} X(z)$

2.5 RELATIONSHIP TO THE FOURIER AND LAPLACE TRANSFORMS

2.5.1 Relationship to the Fourier transform

To establish the relationship between the Fourier transform and the z-transform, let us consider definition (2.1):

$$X(z) = \sum_{k=-\infty}^{+\infty} x(k) z^{-k} \tag{2.71}$$

The complex variable z can be represented by the polar coordinates in the complex plane:

$$z = r \exp(j\theta) \tag{2.72}$$

When this relation is substituted in definition (2.1), the following result is obtained:

$$X(z) = \sum_{k=-\infty}^{+\infty} x(k)(r e^{j\theta})^{-k} = \sum_{k=-\infty}^{+\infty} x(k) r^{-k} e^{-j\theta k} \tag{2.73}$$

or, with $\theta = 2\pi f$:

$$X(z) = \sum_{k=-\infty}^{+\infty} x(k) r^{-k} e^{-j2\pi fk} \qquad (2.74)$$

When this equation is compared with the definition of the Fourier transform (1.36), the z-transform of a signal $x(k)$ can be considered as the Fourier transform of the product of this signal and an exponential signal. In particular, for $r = 1$, that is, for $|z| = 1$, the z-transform and the Fourier transform are identical:

$$X(z)\Big|_{|z|=1} = \sum_{k=-\infty}^{+\infty} x(k) e^{-j2\pi fk} = X(f) \qquad (2.75)$$

In other words, the Fourier transform is the z-transform evaluated on the unit circle $|z| = 1$.

2.5.2 Comments

Comparing (1.36) and (2.74) indicates that even if the Fourier transform of a given signal does not converge (condition (1.37) unsatisfied), the z-transform of the same signal can converge, due to the presence of the real exponential signal r^{-k}.

The role played by the unit circle in the study of the z-transform is of fundamental importance. We will show that if all the poles of a z-transform are within the unit circle, then the inverse transform corresponds to the impulse response of a stable linear system. A large number of ideal systems, such as the ideal low-pass filter or the ideal differentiator, have a z-transform that converges only on the unit circle. The Fourier transform (frequency response) exists for these systems, but the z-transform does not. The reason is that the corresponding Fourier transform is not continuous. Consequently, it is not an analytical function. The region of convergence for the corresponding z-transform is reduced to zero.

2.5.3 Relationship to the Laplace transform

Let us consider the two-sided Laplace transform $X_a(s)$ of an analog signal $x_a(t)$ (vol. IV, sect. 8.1.1):

$$X_a(s) = \int_{-\infty}^{+\infty} x_a(t) e^{-st} dt \qquad (2.76)$$

If the signal $x_a(t)$ is periodically sampled, with a period Δt and in accordance with the sampling theorem (sect. 1.7.2), it can be represented by

$$x_e(t) = x_a(t)e(t) \qquad (2.77)$$

where

$$e(t) = \sum_{k=-\infty}^{+\infty} \delta(t - k\Delta t) \qquad (2.78)$$

and where $\delta(t)$ is the Dirac impulse.

The Laplace transform of the sampled signal is then given by

$$
\begin{aligned}
X_e(s) &= \int_{-\infty}^{+\infty} \sum_{k=-\infty}^{+\infty} x_a(t)\delta(t - k\Delta t)e^{-st}dt \\
&= \sum_{k=-\infty}^{+\infty} \int_{-\infty}^{+\infty} x_a(t)\delta(t - k\Delta t)e^{-st}dt \\
&= \sum_{k=-\infty}^{+\infty} x_a(k\Delta t)e^{-ks\Delta t} \qquad (2.79)
\end{aligned}
$$

Comparing this result with the z-transform of a digital signal $x(k) = x_a(k\Delta t)$, we can see that the Laplace transform of a sampled signal is the z-transform of the corresponding digital signal computed for $z = \exp(s\Delta t)$. Thus,

$$X(z)\Big|_{z = \exp(s\Delta t)} = X_e(s) \qquad (2.80)$$

2.5.4 Definition

The complex logarithm is defined by

$$\ln[z] = \ln|z| + j\,\arg(z) \qquad (2.81)$$

2.5.5 Mapping from the s plane to the z plane

The equation:

$$z = \exp(s\Delta t) \qquad (2.82)$$

is the mapping of the complex s plane on the complex z plane and *vice versa*. However, this mapping is not one-to-one. Indeed, z, like any complex number, can be represented as follows:

$$z = |z|\, e^{j\, \arg(z)}$$

$$= e^{\ln|z|}\, e^{j\, \arg(z)} = e^{\ln|z| + j\, \arg(z)} \tag{2.83}$$

The inverse equation of (2.82) then becomes

$$s = \frac{1}{\Delta t} \ln[z] = \frac{1}{\Delta t} \ln|z| + j\, \frac{\arg(z)}{\Delta t} \tag{2.84}$$

Usually, s is written as $s = \sigma + j\omega$ or $s = \sigma + j2\pi f$. Comparing the real and imaginary parts of this representation with those in (2.84), we obtain

$$\sigma = \frac{1}{\Delta t} \ln|z| \tag{2.85}$$

$$f = \frac{\arg(z)}{2\pi \Delta t} \tag{2.86}$$

However, the argument of z is determined modulo 2π (a complete rotation in the complex plane), so (2.86) shows that each interval of width $1/\Delta t$ of the variable f leads to the same argument of z:

$$\arg(z) = 2\pi f \Delta t \tag{2.87}$$

In other words, $\arg(z)$ is periodic in f with period $1/\Delta t$. This shows that an infinite number of points in the s plane are mapped onto the same point in the z plane.

2.5.6 Interpretation

Equation (2.87) indicates that the imaginary axis of the s plane, represented by the variable f, is cut into slices of length $1/\Delta t$. The image of each slice in the z plane is the unit circle, as can be seen in (2.85) by setting $\sigma = 0$. We obtain $|z| = 1$.

The horizontal bands determined by these cuts are mapped, from (2.82), one over the other, in the z plane. Thus, the z plane can be considered as a Rieman surface.

The left half of each band in the s plane is mapped within the unit circle in the z plane. This can be observed directly from (2.85). Because σ is negative in this case, the modulus of z is necessarily less than one.

The right half of each band in the s plane is mapped onto the z plane outside the unit circle. This interpretation is summarized in figure 2.7.

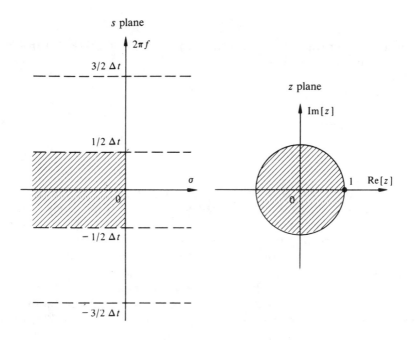

Fig. 2.7

2.5.7 Comment

The correspondence between the s plane and the z plane shows in a more general manner the results of the periodic sampling of an analog signal. In fact, the integral Fourier transform (1.34) is nothing other than the Laplace transform computed along the imaginary axis in the s plane. As we saw in section 1.7.4, the Fourier transform of a sampled signal is obtained by the periodic repetition of the Fourier transform of the analog signal (1.112).

According to the previous discussion, it can be proved that, more generally, the Laplace transform of the sampled signal is given by

$$X_e(s) = \frac{1}{\Delta t} \sum_{n=-\infty}^{+\infty} X_a\left(s - j\frac{2\pi n}{\Delta t}\right) \tag{2.88}$$

2.6 REPRESENTATION OF A SIGNAL BY ITS POLES AND ZEROS

2.6.1 Principle

Let us consider the z-transform $X(\dot{z})$ of a signal $x(k)$.

The poles of $X(z)$ have been defined as the values of z for which the z-transform $X(z)$ tends toward infinity. The zeros of $X(z)$ have been defined as the values of z for which the transform $X(z)$ is zero. If $X(z)$ has M zeros z_m and N poles p_n, it can be written as

$$X(z) = A \frac{\prod\limits_{m=1}^{M} (z - z_m)}{\prod\limits_{n=1}^{N} (z - p_n)} \tag{2.89}$$

Expression (2.89) is equivalent to writing $X(z)$ as the quotient of two polynomials of z, $P(z)$ and $Q(z)$, of degree M and N, respectively. The representation of the poles and the zeros in the complex plane gives an attractive description of the corresponding signal. In practice, it is always possible to write a z-transform in the form of (2.89) and thereby represent the corresponding signal by its poles and zeros.

For real signals, the coefficients of the polynomials $P(z)$ and $Q(z)$ are real numbers. The roots of the polynomials of this type are either real or come in complex conjugate pairs. Thus, the complex poles or zeros of $X(z)$ obtained from a real signal $x(k)$ are of the form $\alpha \pm j\beta$.

2.6.2 Example

Let us consider the following signal:

$$x(k) = a^k \epsilon(k) \quad \text{with } |a| < 1 \tag{2.90}$$

Its z-transform is given by

$$X(z) = \sum_{k=0}^{+\infty} a^k z^{-k} = \frac{1}{1 - az^{-1}} = \frac{z}{z - a} \quad \text{for } |z| > |a| \tag{2.91}$$

This transform has one zero $z_1 = 0$ and one pole $p_1 = a$. They are shown in figure 2.8 for the case where a is real and positive.

2.6.3 Geometrical interpretation of the Fourier transform

The representation of a signal by its poles and its zeros also gives information about the form of its Fourier transform. The Fourier transform can be obtained by evaluating the z-transform on the unit circle. In the case of the previous example 2.6.2, we must set $z = \exp(j2\pi f_0)$ in (2.91) to obtain the value

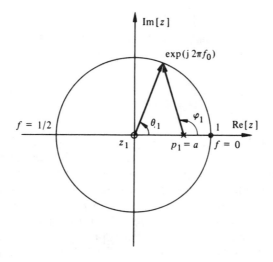

Fig. 2.8

of the Fourier transform $X(f)$ for a given frequency f_0. Thus, the following result is obtained:

$$X(f_0) = \frac{\exp(j\,2\pi f_0)}{\exp(j\,2\pi f_0) - a} \qquad (2.92)$$

The numerator can be considered as a vector joining the origin and the point $z = \exp(j2\pi f_0)$ on the unit circle. The denominator can be considered as the difference of two vectors (Fig. 2.8). The modulus of the Fourier transform at the frequency f_0 is simply the ratio of the moduli of the vectors which represent the numerator and the denominator in (2.92). For the same value of the frequency, the phase is the difference of the angles between these vectors and the real axis. This stems directly from the division of two complex numbers. The modulus is the ratio of the moduli and the argument is the difference of the arguments. To determine the Fourier transform for other frequencies, the same operations are repeated with respect to another point along the unit circle.

2.6.4 Comment

Because the Fourier transform of a digital signal is periodic of period 1 (or $1/\Delta t$, where Δt is the sampling period), one round trip about the unit circle is enough to determine the Fourier transform completely. The round trip starting at the point $z = 1$ in the positive direction (counterclockwise) leads to a frequency representation in the interval $[0, 1]$ or $[0, 1/\Delta t]$. To obtain a two-sided representation, it is necessary to start at point $z = -1$ ($f = -1/2$ or $f = -1/2\Delta t$) and to follow the circle in the positive direction, ending at the point $f = 1/2$ or $f = 1/2\Delta t$.

2.6.5 Remarks

The representation of a signal by its poles and zeros indicates that if the point on the unit circle is close to a pole, then the corresponding magnitude spectrum of the signal has a peak for the value of the frequency taken on the unit circle. For example, in the case of figure 2.8, the point on the unit circle closest to the pole $p_1 = a$ is the point $z = 1$ or $f = 0$. Moving counterclockwise around the circle, the modulus of the numerator of (2.92) does not change. However, the modulus of the vector that represents the denominator increases progressively and has its maximum for $f = 1/2$. The corresponding magnitude spectrum decreases continuously between these two diametrically opposed points. It is obvious that the closer the pole $p_1 = a$ is to the unit circle, the higher is the peak around the origin of the frequencies.

If the moving point on the unit circle is close to a zero, the magnitude spectrum has a minimum for the corresponding value of the frequency. In the example considered (Fig. 2.8), this fact is not obvious because the only zero is at the center of the unit circle. The magnitude spectra obtained from (2.92) for various real values of a are shown in figure 2.9 with a linear scale in decibels (dB). The corresponding phase spectra are also shown in the same figure. It must not be forgotten that these spectra are periodic of period 1. Moreover, because the signal $x(k)$ is real, the magnitude spectrum is an odd function and the phase spectrum is even.

2.6.6 Generalization

It is possible to determine the shape of the Fourier transform in the general case where $X(z)$ is given by (2.89). This case is more complicated than the simple case of only one zero and one pole, but, with a little practice, we can sketch the shape of the magnitude and the phase spectra of a signal represented by its poles and its zeros.

To achieve this, relation (2.89) can be solved for the moduli and arguments, leading to

$$|X(z)| = |A| \frac{\displaystyle\prod_{m=1}^{M} |z - z_m|}{\displaystyle\prod_{n=1}^{N} |z - p_n|} \qquad (2.93)$$

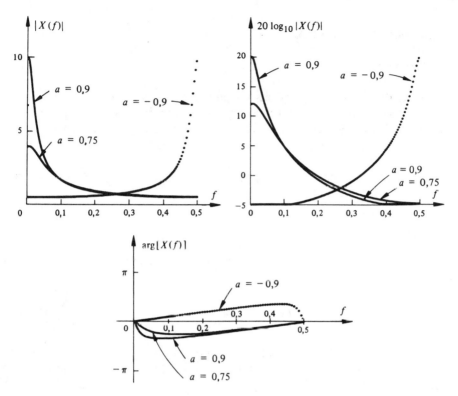

Fig. 2.9

and

$$\arg[X(z)] = \arg(A) + \sum_{m=1}^{M} \arg(z - z_m) - \sum_{n=1}^{N} \arg(z - p_n) \qquad (2.94)$$

For the geometrical interpretation of the Fourier transform from the *z*-transform, we must construct the vectors connecting all the poles and zeros of $X(z)$ with a given point of the unit circle. These vectors are sometimes called *pole vectors* and *zero vectors*. The modulus of the Fourier transform (magnitude spectrum) is the quotient of the product of the moduli of the zero vectors and the product of the moduli of the pole vectors. Moving along the unit circle, the magnitude spectrum will have a peak in the vicinity of a pole near the circle and a minimum in the vicinity of a zero near the circle. The phase spectrum is the difference between the sum of the arguments of the zero vectors and the sum of the arguments of the pole vectors. It is clear, that, in general, the factor *A* introduced in (2.93) and (2.94) must also be taken into account.

2.7 TRANSFER FUNCTION

2.7.1 Introduction

As mentioned in the previous chapter, a linear system can be described by its impulse response or its frequency (or harmonic) response by using the Fourier transform. We also saw that the harmonic response of a linear shift invariant system described by a difference equation is a quotient of polymomials in $\exp(j2\pi f)$.

In this section, a more general description of these problems is presented, using the z-transform.

2.7.2 Definition

We have seen that the output signal of a linear shift invariant system is given by the convolution:

$$y(k) = \sum_{l=-\infty}^{+\infty} x(l)g(k-l) \tag{2.95}$$

where $x(k)$ is the input signal and $g(k)$ is the impulse response of the system. Using the convolution property of the z-transform (sect. 2.4.9), the transformation of (2.95) leads to

$$Y(z) = G(z)X(z) \tag{2.96}$$

The function $G(z)$ is called the *transfer function*. If the transfer function is computed on the unit circle for $|z| = 1$, the frequency response $G(f)$ is obtained. If the transfer function of a system is known, the method described in the previous section can be used to obtain the shape of the frequency response quickly.

2.7.3 Transfer function of a causal system

The transfer function $G(z)$ is the z-transform of the impulse response:

$$G(z) = \sum_{k=-\infty}^{+\infty} g(k)z^{-k} \quad \text{for } R_{g-} < |z| < R_{g+} \tag{2.97}$$

If the system is causal, the impulse response is zero for negative time values. Thus,

$$G(z) = \sum_{k=0}^{+\infty} g(k)z^{-k} \tag{2.98}$$

Using the Cauchy criterion (sect. 2.2.3), the series converges if

$$|z| > R_{g-} \qquad (2.99)$$

where

$$R_{g-} = \lim_{k \to \infty} |g(k)|^{1/k} \qquad (2.100)$$

As shown in section 2.2.7, the z-transform of a causal system converges outside a circle of radius R_{g-}. All the poles of $G(z)$ must be within this circle because a z-transform never converges at a pole.

2.7.4 Transfer function of a stable system

If the system is stable (not necessary causal), its impulse response is absolutely integrable (sect. 1.6.20):

$$\sum_{k = -\infty}^{+\infty} |g(k)| < \infty \qquad (2.101)$$

The transfer function of a stable system has the general form (2.97), which converges inside a ring in the z plane. However, because condition (2.101) is satisfied for a stable system, the transfer function consequently converges for $|z| = 1$, that is, on the unit circle. Thus, the ring of convergence necessarily includes the unit circle. This result can also be proved by using the Fourier transform. Because condition (2.101) is satisfied, the Fourier transform of the impulse response — that is, the frequency response — exists. Because the Fourier transform is obtained by evaluating the z-transform on the unit circle, the region of convergence of the transfer function contains the unit circle.

2.7.5 The transfer function of a stable causal system

If a linear shift invariant system is causal and stable, the unit circle must be included in the region of convergence (2.99) of a causal system. Consequently, for a stable causal system, the transfer function converges outside a circle whose radius, in any case, must be less than one. We must have $R_{g-} < 1$. Thus, *the poles of the transfer function of a stable and causal linear shift invariant system must be within the unit circle.*

This shows the importance of the unit circle in the representation of the poles and zeros in the z plane. We can immediately determine whether the system is stable from the position of the poles in this plane.

2.7.6 Remarks

The fact that the poles of the transfer function of a stable causal system must be within the unit circle is also consistent with the analog equivalent of digital systems. The transfer function of an analog system is the Laplace transform of the analog impulse response.

If the system is causal and stable, the poles of the transfer function must be in the left half of the s plane. The mapping defined by (2.82) maps this part of the s plane onto the unit circle in the z plane. This is very important in the case of digital systems used for the simulation of analog signal processing. The stability is ensured if the poles are within the unit circle, but not if they are in the left half of the z plane.

Clearly, the preceding remarks are also valid for signals because the impulse response of a linear system is itself a signal: the output signal of a system excited by a unit impulse.

In the general case, we speak of absolutely integrable signals in the same terms, since absolute integrability is merely the generalization of stability. Thus, the region of convergence of the z-transform $X(z)$ for an absolutely integrable causal signal is outside of a circle whose radius is less than one. The poles of $X(z)$ are also within the unit circle.

If the zeros of $X(z)$ are also inside the unit circle, the corresponding signal is then called a *minimum phase signal*. This is studied in more detail in chapter 5.

Using the property for a change of complex scale (sect. 2.4.4), a unstable system can be modified into a stable system (or, equivalently, a signal not absolutely integrable can be modified into an absolutely integrable signal). Its impulse response must be multiplied by an exponential signal of the form a^k. For small values of a, the poles can be moved inside the unit circle.

2.7.7 The transfer function of a system governed by a difference equation

Let us consider the class of systems characterized by an Nth-order difference equation:

$$\sum_{n=0}^{N} a_n y(k-n) = \sum_{m=0}^{M} b_m x(k-m) \tag{2.102}$$

where $y(k)$ and $x(k)$ are the output signal (response) and the input signal (excitation), respectively. If the $M + N$ initial conditions have been applied and the response of the system has reached its stationary state, the z-transforms of the two terms of (2.102) can be computed:

$$\sum_{k=-\infty}^{+\infty} \left[\sum_{n=0}^{N} a_n y(k-n) \right] z^{-k} = \sum_{k=-\infty}^{+\infty} \left[\sum_{m=0}^{M} b_m x(k-m) \right] z^{-k} \tag{2.103}$$

Using the properties of linearity and shift of the z-transform, this equation can be written as follows:

$$Y(z) \sum_{n=0}^{N} a_n z^{-n} = X(z) \sum_{m=0}^{M} b_m z^{-m} \tag{2.104}$$

where $X(z)$ and $Y(z)$ are the z-transforms of $x(k)$ and $y(k)$. Comparing (2.96), which is unique for a given system, with (2.104), we deduce the transfer function:

$$G(z) = \frac{\displaystyle\sum_{m=0}^{M} b_m z^{-m}}{\displaystyle\sum_{n=0}^{N} a_n z^{-n}} \tag{2.105}$$

Thus, the transfer function of a system characterized by a difference equation is the quotient of two polynomials in z. The coefficients of these polynomials are the coefficients of the difference equation. In general, these are polynomials of degrees M and N, with M roots $z = z_m$ and N roots $z = p_n$. Consequently, the transfer function $G(z)$ can be written as follows:

$$G(z) = G_0 \frac{\displaystyle\prod_{m=1}^{M} (1 - z_m z^{-1})}{\displaystyle\prod_{n=1}^{N} (1 - p_n z^{-1})} = \frac{P(z)}{Q(z)} \tag{2.106}$$

If this shape of the transfer function is established, the system can be characterized by the poles p_n and the zeros z_m of $G(z)$. Furthermore, the frequency response of the system can be derived from the position of the poles and the zeros by using the method described in section 2.6.3.

2.7.8 Remarks

As we will see in chapter 5, the difference equation (2.102) and the particular forms of convolution (2.95), which are written for causal signals and systems, are used directly in the realization of a large number of processing operations. With the Fourier transform, it is possible to obtain the samples of the impulse response for the system from the coefficients of the difference equation.

However, the inverse of this transformation is complicated because of the difficulty in writing a complex function $G(f)$ of the real variable f as the quotient of two polynomials in $\exp(j2\pi f)$. The bridge between the impulse response of a system and the coefficients of its difference equation is established more easily with the z-transform. From the coefficients of the difference equation (2.102), we can easily deduce the form of the transfer function (2.105). The impulse response is then the inverse z-transform of $G(z)$.

For example, inversion by expansion into partial fractions can be used advantageously if the poles of $G(z)$ are known or can be easily computed.

Working the other way, the transfer function is computed by taking the z-transform of the impulse response. If the result is already the quotient of two polynomials, the coefficients a_n and b_m are obtained by identification. Otherwise, the poles and zeros of $G(z)$ must be sought, and $G(z)$ must be written in the form (2.106). The polynomials $P(z)$ and $Q(z)$ are then obtained by calculating the products in the numerator and in the denominator $G(z)$. The coefficients a_n and b_m are again obtained by identification.

Because of (2.96) and (2.105), the z-transform can be used to solve a difference equation of the form (2.102). From this point of view, the z-transform is a powerful tool, like the Laplace transform, for solving differential equations.

The region of convergence of the transfer function $G(z)$, obtained from the difference equation, is not specified. With different choices for the region of convergence, many different impulse responses can be derived from the same transfer function (2.105) because there is a family of solutions to equation (2.102). All of these impulse responses correspond to the same difference equation. In general, the chosen region of convergence must be a ring containing the poles of $G(z)$.

For example, if the system is stable, the region of convergence must include the unit circle. If the system is required to be causal, the region of convergence must be chosen outside a circle surrounding all the poles of $G(z)$. In addition, if the stability condition is required, the poles of $G(z)$ must be within the unit circle. These problems are studied in more detail in chapter 5, which is devoted to digital filtering.

2.7.9 First-order difference equation

The first-order difference equation is obtained by setting $N = 1$ in (2.102). For simplicity, let us assume that $M = 0$.

With $a = -a_1$ and the simplifying hypothesis $a_0 = b_0 = 1$, (2.102) takes the following form:

$$y(k) - a y(k-1) = x(k) \tag{2.107}$$

The z-transform of this equation is given by

$$Y(z)(1 - a z^{-1}) = X(z) \tag{2.108}$$

If the excitation $x(k)$ is the unit impulse $d(k)$, the z-transform of the corresponding response is the transfer function $G(z)$. Thus, with $X(z) = 1$ and $Y(z) = G(z)$, we obtain

$$G(z) = \frac{1}{1 - az^{-1}} = \frac{z}{z - a} \qquad (2.109)$$

This transfer function has one zero $z_1 = 0$ and one pole $p_1 = a$ (sect. 2.6.2 and Fig. 2.8).

2.7.10 Causality and stability

If a stable system is required, the region of convergence of $G(z)$ must include the unit circle. The system is stable if $|a| \neq 1$ because the poles must be outside the region of convergence.

If this condition is satisfied, any region of convergence that contains the unit circle but excludes the pole $p_1 = a$ leads to a stable system.

If a causal system is required, the region of convergence must be outside the circle of radius $|a|$. The system is noncausal if the region of convergence is chosen as the interior of this circle.

2.7.11 Impulse response

When such a region is chosen, the inverse z-transform, that is, the impulse response, can be computed. With a contour Γ, which is a circle of radius less than a, the following result is obtained:

$$g(k) = \frac{1}{2\pi j} \oint_\Gamma \frac{z^k}{z - a} \, dz \qquad (2.110)$$

For $k \geq 0$, the function to be integrated has no poles inside the contour Γ. The theorem of residues leads directly to $g(k) = 0$ for $k \geq 0$. For $k < 0$, there is a kth-order pole at the origin. The analytical form of the response $g(k)$ can be obtained by applying the theorem of the residues. However, direct integration of the form (2.110) is very long because the order of the pole changes for each value of k. A more practical method is to make the change of variable $z = w^{-1}$. Equation (2.110) then becomes

$$g(k) = -\frac{1}{2\pi j} \oint_{\Gamma'} \frac{w^{-k}}{w^{-1} - a} \, w^{-2} \, dw \qquad (2.111)$$

In this expression, the contour is in the negative direction. The direction of integration can be reversed by multiplying the expression by -1. Then,

$$g(k) = \frac{1}{2\pi j} \oint_{\Gamma'} \frac{w^{-k-1}}{1-aw} \, dw \qquad \text{for } |w| > \frac{1}{|a|} \tag{2.112}$$

With the change of variable, the region of convergence $|z| < |a|$ is changed outside a circle of radius $1/|a|$. The contour of integration is thus a circle whose radius is greater than $1/|a|$. For $k < 0$, there is only one simple pole $p_1 = 1/a$ inside this contour. The residue at this pole is given by

$$\operatorname{Res}_{1/a} \lim_{w \to 1/a} \frac{(aw-1)}{a} \frac{w^{-k-1}}{(1-aw)} = -a^k \tag{2.113}$$

The impulse response of the noncausal system is then given by

$$g(k) = -a^k \epsilon(-k-1) \tag{2.114}$$

For another type of system, let us return to expression (2.109). If a causal system is desired, the region of convergence must be outside the circle of radius $|a|$. It was shown previously (sect. 2.6.2) that the corresponding impulse response has the following form:

$$g(k) = a^k \epsilon(k) \tag{2.115}$$

Furthermore, if the system is stable, the condition $|a| < 1$ must be satisfied in order to include the unit circle in the region of convergence. In this case, the response (2.115) is a decreasing exponential. The frequency responses corresponding to the response (2.115) for various values of a were given in figure 2.9.

2.7.12 Second-order difference equation

By setting $N = 2$ in (2.102), we obtain a second-order difference equation. Again, for reasons of simplicity, we shall assume that the coefficients are real numbers, and that $M = 0$ and $a_0 = b_0 = 1$. Thus,

$$y(k) + a_1 y(k-1) + a_2 y(k-2) = x(k) \tag{2.116}$$

The z-transform for both sides of this equation is given by

$$Y(z)(1 + a_1 z^{-1} + a_2 z^{-2}) = X(z) \tag{2.117}$$

The transfer function of a second-order linear shift invariant system can be derived from this expression by using the unit impulse $d(k)$ as an input signal. In this case, $X(z) = 1$ and $Y(z) = G(z)$. This leads to

$$G(z) = \frac{1}{1 + a_1 z^{-1} + a_2 z^{-2}} = \frac{z^2}{z^2 + a_1 z + a_2} \tag{2.118}$$

To find the poles of $G(z)$, we must compute the roots of the denominator. They are given by

$$p_1 = \frac{-a_1 - \sqrt{a_1^2 - 4 a_2}}{2} \tag{2.119}$$

and

$$p_2 = \frac{-a_1 + \sqrt{a_1^2 - 4 a_2}}{2} \tag{2.120}$$

2.7.13 Stability and causality

The two poles of $G(z)$ must be within the unit circle to ensure the stability and the causality of the system. Thus, the following two conditions must be simultaneously verified:

$$|p_1| < 1 \quad \text{and} \quad |p_2| < 1 \tag{2.121}$$

These conditions impose some constraints on the possible choice of the coefficients a_1 and a_2. Thus, to ensure the stability and the causality of the system, it is necessary to define a region in the (a_1, a_2) plane, inside which conditions (2.121) are satisfied. Applying these conditions to (2.119) and (2.120) leads to

$$\left| -a_1 - \sqrt{a_1^2 - 4 a_2} \right| < 2 \tag{2.122}$$

$$\left| -a_1 + \sqrt{a_1^2 - 4 a_2} \right| < 2 \tag{2.123}$$

Two cases can be distinguished: complex poles or real poles. If $4a_2 < a_1^2$, the poles are real numbers. The parabola $a_2 = a_1^2/4$, shown in figure 2.10, defines the region of real poles (outside the parabola) and that of complex poles (inside the parabola).

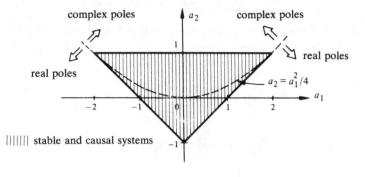

Fig. 2.10

2.7.14 Real poles

In the case of real poles, (2.122) and (2.123) are written as follows:

$$-2 < -a_1 - \sqrt{a_1^2 - 4a_2} < 2 \tag{2.124}$$

$$-2 < -a_1 + \sqrt{a_1^2 - 4a_2} < 2 \tag{2.125}$$

The inequality on the right-hand side of (2.124) gives

$$-a_1 - \sqrt{a_1^2 - 4a_2} < 2$$
$$-\sqrt{a_1^2 - 4a_2} < 2 + a_1 \tag{2.126}$$
$$a_2 > -(1 + a_1)$$

The inequality on the left-hand side of (2.124) leads to

$$a_2 > a_1 - 1 \tag{2.127}$$

Inequalities (2.126) and (2.127) define two straight lines and delineate a v-shaped region whose base is at the point $a_z = -1$ (see Fig. 2.10). Conditions (2.121) are satisfied inside the v. The two inequalities of (2.125) lead to the same results (2.126) and (2.127).

2.7.15 Complex poles

In the case of complex poles, (2.119) and (2.120) can be written as follows:

$$p_1 = \frac{-a_1 - j\sqrt{4a_2 - a_1^2}}{2} \qquad p_2 = \frac{-a_1 + j\sqrt{4a_2 - a_1^2}}{2} \tag{2.128}$$

The moduli are given by

$$|p_1| = |p_2| = \frac{\sqrt{a_1^2 + 4a_2} - a_1^2}{2} \qquad (2.129)$$

The conditions (2.121) imply that

$$\sqrt{a_2} < 1$$

or

$$|a_2| < 1$$

This condition, together with inequalities (2.126) and (2.127), leads to a triangular region in the (a_1, a_2) plane, which is shown in figure 2.10. If the values of a_1 and a_2 are chosen inside this region, the system whose transfer function is given by (2.118) is necessarily stable. An unstable system is obtained for values of a_1 and a_2 which are outside this region.

2.7.16 Impulse response

In order to compute the impulse response of a second-order linear shift invariant system by inversion of the transfer function (2.118), it is necessary to define a region of convergence. Three different regions of convergence can be defined with the two poles, p_1 and p_2, and with the hypothesis that these two poles are distinct. Let us assume, for example,

$$|p_1| > |p_2| \qquad (2.130)$$

A first region of convergence can be defined as the outside of a circle of radius $|p_1|$ (fig. 2.11). In this case, the impulse response obtained corresponds to that of a causal system. The corresponding region of convergence is given by $|z| > R_{g-}^{11} = |p_1|$, where the superscript 11 refers to figure 2.11. The second region of convergence may be the ring defined by the circles of radius $|p_1|$ and $|p_2|$ (Fig. 2.12). This is the region where

$$R_{g-}^{12} = |p_2| < |z| < R_{g+}^{12} = |p_1| \qquad (2.131)$$

The corresponding impulse response in this case is the entire real axis k. Finally, the third possible choice is the interior of the circle of radius $|p_2|$ with (Fig. 2.13):

$$|z| < R_{g+}^{13} = |p_2| \qquad (2.132)$$

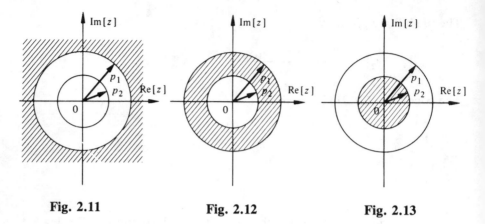

Fig. 2.11   Fig. 2.12   Fig. 2.13

The impulse response obtained corresponds to that of a noncausal system. These three impulse responses can be easily computed by using the different inversion methods presented in section 2.3.

The following results are obtained:

- For $|z| > R_{g-}^{11} = |p_1|$

$$g_a(k) = \left(\frac{p_1}{p_1 - p_2} \, p_1^k + \frac{p_2}{p_2 - p_1} \, p_2^k \right) \epsilon(k) \qquad (2.133)$$

- For $R_{g-}^{12} = |p_2| < |z| < R_{g+}^{12} = |p_1|$

$$g_b(k) = \begin{cases} \dfrac{p_2}{p_2 - p_1} \, p_2^k & \text{for } k \geqslant -1 \\[2mm] \dfrac{p_1}{p_2 - p_1} \, p_1^k & \text{for } k \leqslant -2 \end{cases} \qquad (2.134)$$

- For $\quad g_c(k) = \left(\dfrac{p_1}{p_2 - p_1} \, p_1^k + \dfrac{p_2}{p_1 - p_2} \, p_2^k \right) \epsilon(-k-2) \qquad (2.135)$

2.7.17 Remarks

The only feasible system is the one whose impulse response is given by (2.133). In the case of real poles $(a_1^2/4)$, this response is simply the superposition of two exponential signals. Additionally, if the system is stable, these signals are decreasing functions of k (thus, conditions (2.129) are satisfied). In the case of complex poles ($a_2 > a_1^2/4$), (2.133) can be written in a more simple form. If θ is the argument of p_1, since the poles are the complex conjugates of each other, we can write

$$g_a(k) = \frac{p^{k+1}}{\sqrt{a_1^2 - 4a_2}} \left[e^{j(k+1)\theta} - e^{-j(k+1)\theta} \right] \epsilon(k) \qquad (2.136)$$

where

$$p = |p_1| = |p_2| = \sqrt{a_2}$$

and

$$\tan \theta = \frac{\sqrt{4a_2 - a_1^2}}{a_1} = \frac{\sqrt{1 - a_1^2/4a_2}}{a_1/2\sqrt{a_2}} \qquad (2.137)$$

but

$$\tan \theta = \frac{\sin \theta}{\cos \theta} = \pm \frac{\sqrt{1 - \cos^2 \theta}}{\cos \theta} \qquad (2.138)$$

If this result is identified with (2.137), we obtain

$$\cos \theta = \frac{a_1}{2\sqrt{a_2}} \qquad (2.139)$$

Thus, finally,

$$g_a(k) = \pm \left[\frac{1}{\sin \theta} \sqrt{a_2}^k \sin(k\theta + \theta) \right] \epsilon(k) \qquad (2.140)$$

Expression (2.140) shows that the impulse response of a second-order system with two complex poles is a sinusoidal signal modulated in amplitude by an exponential signal. Furthermore, since the poles are complex numbers, we have $a_2 > a_1^2/4$, which implies that $a_2 > 0$. The impulse response is, therefore, a real signal. If the value of a_2 is chosen in the region of stability ($a_1 < 1$) the impulse response is a damped sine function.

2.7.18 Frequency response

The frequency response of a second-order system can be obtained by computing the z-transform (2.118) over the unit circle $|z| = 1$, either geometrically or analytically. In the latter case, the following result is obtained:

$$G(f) = \frac{1}{1 + a_1 e^{-j2\pi f} + a_2 e^{-j4\pi f}} \qquad (2.141)$$

The modulus and the phase of this response are shown in figure 2.14 for various values of a_1 and a_2 chosen in the region of stability. Based on the previous remarks (sect. 2.7.17) and these curves, we can see that, for two complex poles, the second-order linear system is a digital oscillator.

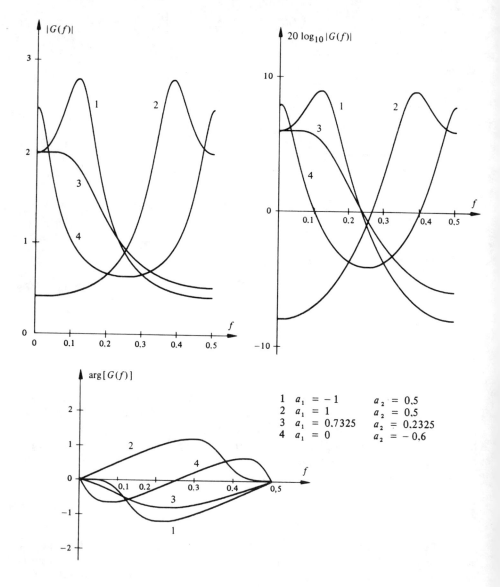

1	$a_1 = -1$	$a_2 = 0.5$
2	$a_1 = 1$	$a_2 = 0.5$
3	$a_1 = 0.7325$	$a_2 = 0.2325$
4	$a_1 = 0$	$a_2 = -0.6$

Fig. 2.14

2.8 EXERCISES

2.8.1 Determine the z-transform and its region of convergence for the following signals:

- $x(k) = (1/2)^k \epsilon(k)$
- $x(k) = d(k-1)$
- $x(k) = (1/2)^k \epsilon(-k)$
- $x(k) = d(k)$
- $x(k) = d(k-k_0), k_0 = \text{constant}$
- $x(k) = -(1/2)^k \epsilon(-k-1)$

2.8.2 Compute the z-transform of the signal $x(k) = \text{rect}_K(k)$:
- by directly applying the definition;
- by using the signal $\epsilon(k)$ and the shift theorem.

Compare the two results.

2.8.3 Using the expansion into partial fractions, determine the analytical form of the signal $x(k)$ whose z-transform is given by

$$X(z) = \frac{z^2}{z^2 - 3z + 2} \quad \text{with } |z| > 2$$

Compare this result with that of example 2.3.11.

2.8.4 Let us consider a signal $x(k)$, which can be complex, of finite duration K with $x(k) = 0$ for $k < 0$ or for $k \geqslant K$. How many such signals with the same magnitude spectrum $|X(f)|$ can be found? Justify your answer.
Use exercise 1.8.13.

2.8.5 Let $G(z)$ be the transfer function of a causal linear shift invariant system with

$$G(z) = (az - 1)/(z - a) \quad \text{where } a \text{ is a real number}$$

Determine the values of a for which $G(z)$ corresponds to a stable system. Determine the poles, the zeros, and the region of convergence for one of these values. Compute $|G(f)|$.

2.8.6 Prove equations (2.133), (2.134), and (2.135).

2.8.7 Let us consider a linear shift invariant system defined by the difference equation:

$$y(k) = 3y(k-1) - 2y(k-2) + x(k)$$

Determine the transfer function of this system. What can be said about the stability and the causality of this system? Compute the impulse response.

2.8.8 Compute the z-transform of the signal $y(k) = k^2(k)$ in terms of the z-transform $X(z)$ of the signal $x(k)$.

2.8.9 Prove that the z-transform is unique, that is, if

$$x(k) \longleftrightarrow X(z)$$
$$y(k) \longleftrightarrow Y(z)$$

with $X(z) \equiv Y(z)$, then $x(k) \equiv y(k)$ for any k.

2.8.10 Let $g(k)$ be the real, even impulse response of duration $2K + 1$ of a linear shift invariant system. Show that if $G(z) = 0$ for $z = a \exp(jb)$, then $G(z) = 0$ for $z = (1/a) \exp(jb)$.

2.8.11 Let us consider a linear shift invariant system defined by the difference equation:

$$y(k) = 0.3y(k-1) + 0.3y(k+1) - 0.3x(k)$$

Determine its impulse response and its transfer function. Study the causality and the stability of the system.

Chapter 3

The Discrete Fourier Transform

3.1 INTRODUCTION

3.1.1 Introductory remarks

The Fourier transform of digital signals, as defined and studied in chapter 1, is inappropriate for use in digital processing. There are two reasons for this. First, the frequency is represented by a continuous (or analog) variable, and, second, an infinite number of samples is needed to characterize the signal.

Given the importance of this transform in signal processing, we must find a more practical way to express it. In this new form, it is called *the discrete Fourier transform* and will be denoted hereafter by the abbreviation DFT.

3.1.2 Organization of the chapter

In this chapter, we present the modifications required to obtain the discrete version of the Fourier transform. In particular, we will study the conditions under which a digital signal can be completely characterized by a set of samples taken from its Fourier transform. This leads to the study of the discrete Fourier transform (DFT) for periodic signals and real signals. The principal properties of the DFT and its use in the evaluation of correlation functions and convolutions are examined next. Also presented is a detailed study of the means for limiting the length of infinite-length signals and of the rules for use of the DFT for this type of signal.

3.1.3 Comments

The DFT is attractive due to the existence of a particularly fast and efficient algorithm for its computation. This algorithm, presented in detail in the next chapter, also enables the DFT to be used for analog signals. For this reason, the tools and the rules for using the DFT for analog signals are studied as well. Some of these rules are very different from those given for digital signals. Consequently, to avoid severe estimation errors, the rules must be followed carefully. The DFT obviously can be used to evaluate analog correlation functions and convolutions because the digital representation of an analog signal by periodic sampling does not alter the correlation and convolution.

3.2 DISCRETE FOURIER TRANSFORM

3.2.1 Review

The Fourier transform of a signal $x(k)$ is defined by (sect. 1.2.3):

$$X(f) = \sum_{k=-\infty}^{+\infty} x(k) e^{-j2\pi fk} \tag{3.1}$$

The function $X(f)$ is periodic of period 1 and, in general, is a complex function of the real variable f. The kth sample of the signal is given by the inverse equation:

$$x(k) = \int_{-1/2}^{1/2} X(f) e^{j2\pi fk} df \tag{3.2}$$

Because $X(f)$ is a periodic function of period 1, this integral can, in fact, be computed over any period $[f_0, f_0 + 1]$. The limits chosen in (3.2) allow for a two-sided representation of the function $X(f)$ about the frequency $f = 0$ within the main interval.

3.2.2 Remarks

There are two difficulties associated with (3.1) and (3.2). First, f is a continuous variable and, thus, cannot be handled by a digital processing system. Second, it is impossible, in practice, to process an infinite number of samples of the signal $x(k)$.

The solutions to these two problems are straightforward. The continuous variable f must be replaced by a discrete variable and the length of the signal $x(k)$ must be limited. However, even if the solutions are simple, their use and the interpretation of the results after modifications require a careful and detailed study to avoid being incorrectly applied.

3.2.3 Discretization of the frequency

The replacement of the continuous variable f by a discrete variable n can be written as follows:

$$f = n \Delta f \tag{3.3}$$

where Δf is the increment used on the frequency axis. The discrete frequencies $f_n = n\Delta f$ are called the *harmonic frequencies of the DFT*.

Because $X(f)$ is periodic of period 1, it is sufficient to use substitution (3.3) over a single period. One period can be divided into N intervals:

$$\Delta f = 1/N \tag{3.4}$$

If the period chosen is between $-1/2$ and $1/2$, the N values of the discrete variable are given by

$$n = -N/2, \ -N/2 + 1, \ .... \ , N/2 - 1 \tag{3.5}$$

3.2.4 Effects of the discretization of the frequency

After the change of variable, the integral equation (3.2) is approximated by a sum, which is written in the form:

$$x(k) \cong \frac{1}{N} \sum_{n=-N/2}^{N/2-1} X(n) \exp\left(j \, 2\pi \, \frac{nk}{N} \right) \tag{3.6}$$

The exact value of this sum will be denoted by $x_p(k)$:

$$x_p(k) = \frac{1}{N} \sum_{n=-N/2}^{N/2-1} X(n) \exp\left(j \, 2\pi \, \frac{nk}{N} \right) \tag{3.7}$$

Thus,

$$x(k) \cong x_p(k) \tag{3.8}$$

We must determine the quality of this approximation and the conditions for which (3.8) becomes an identity. For this purpose, let us consider the properties of the term:

$$\exp\left(j \, 2\pi \, \frac{nk}{N} \right) \tag{3.9}$$

which represents a complex exponential digital signal. The notation almost unanimously agreed upon to express the Nth root of unity is

$$W_N = \exp\left(j\,\frac{2\pi}{N}\right) \tag{3.10}$$

Thus, the complex signal (3.9) is denoted by

$$W_N^{nk} = \exp\left(j\,2\pi\,\frac{nk}{N}\right) \tag{3.11}$$

3.2.5 Property: Separability

The property of separability is written as follows:

$$W_N^{k+l} = \exp\left[j\,\frac{2\pi}{N}(k+l)\right] = \exp\left(j\,\frac{2\pi}{N}k\right)\exp\left(j\,\frac{2\pi}{N}l\right) \tag{3.12}$$
$$= W_N^k \cdot W_N^l$$

3.2.6 Property: Periodicity

The property of periodicity is written as follows:

$$W_N^{k+lN} = \exp\left(j\,\frac{2\pi}{N}k\right)\cdot \exp\left(j\,2\pi l\right) = W_N^k \tag{3.13}$$

where l is an integer. Thus:

$$W_N^k = W_N^{k \bmod N} \tag{3.14}$$

3.2.7 Property: Special values

For any integer l, we have

$$W_N^{lN} = \exp(j\,2\pi l) = 1 \tag{3.15}$$
$$W_N^{N/2} = \exp(j\,\pi) = -1 \tag{3.16}$$

$$W_N^{k+N/2} = \exp\left(j \frac{2\pi}{N}k\right) \exp(j\pi) = -W_N^k$$

(3.17)

$$W_N^2 = \exp\left(2j \frac{2\pi}{N}\right) = \exp\left(j\frac{2\pi}{N/2}\right) = W_{N/2}$$

(3.18)

3.2.8 Property: Orthonormality

The orthonormality of a signal is written in the form:

$$\frac{1}{N} \sum_{n=0}^{N-1} W_N^{nk} = \begin{cases} 1 & \text{for } k = lN \text{ with } l \text{ integer} \\ 0 & \text{otherwise} \end{cases}$$

(3.19)

3.2.9 Proof

For $k = lN$ all the terms of this sum are equal to 1 (3.15). The sum of N terms divided by N gives 1. For other values of k, the vectorial sum of the N Nth roots of unity is always zero because these roots are separated from each other by an angular increment of $2\pi/N$ radians.

3.2.10 Quality of the discrete approximation

From (3.13), we can see that W_N^{nk} is periodic in k of period N. Thus, the signal $x_p(k)$ given by (3.7) is a periodic signal of period N. This means that the approximation of the initial signal $x(k)$, obtained by the discretization of its Fourier transform $X(f)$, is periodic, even if the original signal is not. This is a fundamental difference for approximation (3.8). To find the relationship between the signals $x_p(k)$ and $x(k)$, relation (3.1), modified by the change (3.3), can be substituted into relation (3.7). This leads to

$$x_p(k) = \frac{1}{N} \sum_{n=-N/2}^{N/2-1} \left[\sum_{l=-\infty}^{+\infty} x(l)\exp\left(-j\,2\pi \frac{nl}{N}\right) \right] \exp\left(j\,2\pi \frac{nk}{N}\right)$$

(3.20)

Reversing the order of the summations gives

$$x_p(k) = \sum_{l=-\infty}^{+\infty} x(l)\left[\frac{1}{N} \sum_{n=-N/2}^{N/2-1} \exp\left(-j\,\frac{2\pi n}{N}(l-k)\right) \right]$$

(3.21)

Equation (3.1) shows that the expression in brackets is 1 if $l - k = iN$, where i is an integer. It is zero for all other values of $l - k$. Thus, $l = iN + k$. Or

$$x_p(k) = \sum_{i=-\infty}^{+\infty} x(iN + k) \qquad k = -\infty, ..., +\infty \qquad (3.22)$$

Equation (3.22) indicates that the periodic signal $x_p(k)$ is obtained by a periodic repetition (of period N) of the signal $x(k)$.

3.2.11 Remarks

The duality between the time and frequency domains related by the Fourier transform is well known. In section 1.7, we saw that the sampling of an analog signal introduces a periodic repetition of the Fourier transform of this signal. By duality, the change of variable (3.3), which is nothing other than the sampling of the Fourier transform $X(f)$ used in (3.20), introduces the periodic repetition shown in (3.22). This points out a fundamental difference between the Fourier transforms of analog and digital signals. The Fourier transform, as well as the inverse transform, of a sampled function is a periodic function whose period is the inverse of the sampling period. Forgetting this fact can lead to serious errors.

Equation (3.22) indicates that each period of the signal $x_p(k)$ is an exact replica of $x(k)$ if the length of the nonperiodic signal $x(k)$ is limited to N. If this length is greater than N, an overlapping (or aliasing) occurs, as discussed in section 1.7.4. Consequently, in this case, $x(k)$ cannot be exactly extracted from a period of $x_p(k)$. In other words, the approximate equation (3.8) becomes an exact identity only for signals $x(k)$ of finite length. It is clear that signals of length M less than N can be considered as signals of length N by attaching $N - M$ zero samples.

3.2.12 Definitions

For nonperiodic signals of finite length N, (3.1) becomes

$$X(f) = \sum_{k=k_0}^{k_0+N-1} x(k)e^{-j2\pi fk} \qquad (3.23)$$

Taking into account (3.3) and (3.4), and using notation (3.10), this equation can be written as follows:

$$X(n) = \sum_{k=k_0}^{k_0+N-1} x(k)W_N^{-nk} \qquad (3.24)$$

with

$$n = -N/2, ..., N/2 - 1$$

The inverse transform is given by

$$x(k) = \frac{1}{N} \sum_{n=-N/2}^{N/2-1} X(n) W_N^{nk} \qquad (3.25)$$

for

$$k = k_0, ..., k_0 + N - 1$$

Equations (3.24) and (3.25) define the *discrete Fourier transform (DFT) for a nonperiodic signal of finite length N*. Equation (3.24) is referred to as the direct transform, or analysis, of the signal, while (3.25) is the inverse transform, or synthesis, of the signal.

If $X(n)$ is written in polar coordinates in the complex plane, then

$$X(n) = |X(n)| \exp[j\theta_x(n)] \qquad (3.26)$$

with

$$\theta_x(n) = \arg[X(n)]$$

The terms $|X(n)|$ and $\theta_x(n)$ respectively represent the samples of the magnitude and phase spectra. They are called the *discrete magnitude spectrum* and the *discrete phase spectrum*.

3.2.13 Comments

It is clear that the transform $X(n)$ is periodic of period N and can be computed for any integer value of n. The period chosen in (3.24) leads to the usual two-sided representation. If N is an odd number, the boundaries of N in (3.24) are obtained from the integer part of the division $N/2$, since N must be an integer. For example, for $N = 15$, n lies in the interval $[-7, 7]$. If N is an even number, we can complete the symmetry of the bilateral representation by using the equation $X(N/2) = X(-N/2)$. For reasons which will become apparent, N is very often chosen to be even.

The finite-length signal $x(k)$, whose nonzero values are assumed to lie within the interval $[k_0, k_0 + N - 1]$, can be reconstructed exactly by using (3.25), if the values of k are chosen from the interval $[k_0, k_0 + N - 1]$. If (3.25) is computed for other values of k, the following periodic signal is obtained:

$$x_p(k) = \sum_{i=-\infty}^{+\infty} x(k + iN) \qquad (3.27)$$

3.2.14 Reconstruction of the signal

Signal (3.27) is the periodic repetition (of period N) of the desired signal $x(k)$. This signal can be extracted from the period of $x_p(k)$ corresponding to the interval where $x(k)$ is nonzero. Therefore, the rectangular signal $\text{rect}_N(k)$, or any shifted version of this signal, can be used, leading to

$$x(k) = x_p(k)\,\text{rect}_N(k - k_0) \tag{3.28}$$

This operation is only required if (3.25) was computed for values of k outside the defined interval of $x(k)$. Since this interval is known *a priori*, we can write

$$x(k) = \begin{cases} x_p(k) & \text{for } k_0 \leqslant k \leqslant k_0 + N - 1 \\ 0 & \text{otherwise} \end{cases} \tag{3.29}$$

3.2.15 Ambiguity of the discrete Fourier transform

If only the coefficients $X(n)$ for $n = -N/2, ..., N/2 - 1$ are known with no other information about the signal from which they come, we cannot determine whether the signal is periodic or of finite length.

If the signal is assumed to be of finite length, the interval $[k_0, k_0 + N - 1]$, which includes the nonzero samples of $x(k)$, must also be known. If the parameter k_0, which characterizes this interval, is not known, then (3.25) leads to any period of $x_p(k)$ depending on the value of k_0 used. The form of the desired signal $x(k)$ can be found only after cyclic permutations of the samples over this period.

3.3 DISCRETE FOURIER TRANSFORM OF PERIODIC AND REAL SIGNALS

3.3.1 Discrete Fourier transform of periodic signals

The discrete Fourier transform of a periodic signal $x_p(k)$ of period N can be deduced directly from the previous development. Such a signal is not absolutely integrable in the sense of (1.37). The infinite sum in (3.1) must, therefore, be reduced to a period of $x_p(k)$. Considering the period $[0, N - 1]$, we obtain

$$X_p(f) = \sum_{k=0}^{N-1} x_p(k)e^{-j2\pi fk} \tag{3.30}$$

It can be easily verified that the function $X_p(f)$ is periodic of period 1. Using the change of variable (3.3), it can be also written as

$$
] \quad X_p(n) = \sum_{k=0}^{N-1} x_p(k)\exp\left(-j\,2\pi\,\frac{nk}{N}\right) \tag{3.31}
$$

Thus, the inverse transform becomes

$$
x_p(k) = \frac{1}{N} \sum_{n=-N/2}^{N/2-1} X_p(n)\exp\left(j\,\frac{2\pi nk}{N}\right) \tag{3.32}
$$

Using notation (3.10), expressions (3.31) and (3.32) can be written as follows:

$$
X_p(n) = \sum_{k=0}^{N-1} x_p(k)W_N^{-nk} \quad \text{for } n = -N/2, ..., N/2 - 1 \tag{3.33}
$$

and

$$
x_p(k) = \frac{1}{N} \sum_{n=-N/2}^{N/2-1} X_p(n)W_N^{nk} \quad \text{for } k = 0, ..., N-1 \tag{3.34}
$$

These equations define *the discrete Fourier transform (DFT) for periodic signals* of period N.

3.3.2 Comments

The transformations (3.33) and (3.34) are identical to relations (3.24) and (3.25) for nonperiodic signals of finite length. However, the principal difference between (3.34) and (3.25) is the limitation of the possible values of k. The periodic signal $x_p(k)$ is defined for all values of k. Furthermore, aliasing cannot occur in the exact equation (3.34) because the signal $x_p(k)$ is periodic.

In contrast to the case of periodic analog signals, a periodic digital signal of period N is represented by only N complex exponential signals, the frequencies of which are integer multiples of a fundamental frequency $1/N$. This can be deduced from the periodicity discussed in section 3.2.6.

3.3.3 Remarks

The important role played by the DFT in digital signal processing is greatly emphasized by the existence of a fast computation algorithm (chap. 4). In a digital processing system, the computation time is proportional to the number of operations that are to be performed. Therefore, redundant computations must be reduced as much as possible. In (3.24) and (3.25), or (3.33) and (3.34), the signal $x(k)$ is not assumed to be real. These equations are valid for both real and complex signals. However, in the case of real signals, the symmetry of the Fourier transform can be used to avoid redundant computations.

3.3.4 Property: Symmetry of the DFT

Let us consider the Fourier transform of a real signal. We have established that its real part is an even function and its imaginary part is an odd function (sect. 1.3.8). Thus, we have

$$X(-f) = X^*(f) \tag{3.35}$$

This equation is also valid for the samples $X(n)$ taken from $X(f)$ if the symmetry of these samples about origin $f = 0$ is conserved. This is the case in (3.24) because, for $n = 0$, $X(n)$ is the value of $X(f)$ at the origin [see (3.3)]. In this case (3.24) can be written as follows:

$$X(n) = \sum_{k=0}^{N-1} x(k) W_N^{-nk} \tag{3.36}$$

with

$$n = 0, ..., N/2 \quad \text{et} \quad X(-n) = X^*(n)$$

Thus, half of the computation in (3.24) can be avoided if the signal $x(k)$ is real.

3.3.5 Case of real signals

Equation (3.24) can be used to compute the DFT of two real signals. If $x_1(k)$ and $x_2(k)$ are two real signals, a complex signal $x_3(k)$ can be constructed in the following way:

$$x_3(k) = x_1(k) + j x_2(k) \tag{3.37}$$

Using the linearity of the Fourier transform (sect. 1.8.4), we have

$$X_3(f) = X_1(f) + j X_2(f) \tag{3.38}$$

Because $X_1(f)$ and $X_2(f)$ are generally complex transformations, we can also write

$$\begin{aligned} X_3(f) &= \text{Re}\,[X_1(f)] + j\,\text{Im}\,[X_1(f)] + j\,\{\text{Re}\,[X_2(f)] + j\,\text{Im}\,[X_2(f)]\} \\ &= \text{Re}\,[X_1(f)] - \text{Im}\,[X_2(f)] + j\,\{\text{Im}\,[X_1(f)] + \text{Re}\,[X_2(f)]\} \end{aligned} \tag{3.39}$$

Thus, the real and imaginary parts of $X_3(f)$ are given by

$$\mathrm{Re}\,[X_3(f)] = \mathrm{Re}\,[X_1(f)] - \mathrm{Im}\,[X_2(f)] \qquad (3.40)$$

$$\mathrm{Im}\,[X_3(f)] = \mathrm{Im}\,[X_1(f)] + \mathrm{Re}\,[X_2(f)] \qquad (3.41)$$

However, any function can be decomposed into a sum of an odd function and an even function (see vol. IV, sect. 7.2.7):

$$a(t) = a_{\mathrm{ev}}(t) + a_{\mathrm{odd}}(t) \qquad (3.42)$$

where $a_{\mathrm{ev}}(t)$ and $a_{\mathrm{odd}}(t)$ are even and odd functions, respectively. The two functions can be derived from $a(t)$ by using the following equations:

$$a_{\mathrm{ev}}(t) = [a(t) + a(-t)]/2 \qquad (3.43)$$

$$a_{\mathrm{odd}}(t) = [a(t) - a(-t)]/2 \qquad (3.44)$$

Because the signals $x_1(k)$ and $x_2(k)$ are assumed to be real, the real parts of their Fourier transform are even functions and the imaginary parts are odd functions. From the real and imaginary parts of the Fourier transform of the complex signal $x_3(k)$, the following equations can be deduced:

$$\mathrm{Re}\,[X_1(f)] = \{\,\mathrm{Re}\,[X_3(f)] + \mathrm{Re}\,[X_3(-f)]\}\,/2 \qquad (3.45)$$

$$\mathrm{Im}\,[X_1(f)] = \{\,\mathrm{Im}\,[X_3(f)] - \mathrm{Im}\,[X_3(-f)]\}\,/2 \qquad (3.46)$$

Thus, (3.45) and (3.46) give the Fourier transform of the real signal $x_1(k)$. Similarly, the following equations can be written for the Fourier transform of the real signal $x_2(k)$:

$$\mathrm{Re}\,[X_2(f)] = \{\,\mathrm{Im}\,[X_3(f)] + \mathrm{Im}\,[X_3(-f))\}\,/2 \qquad (3.47)$$

$$\mathrm{Im}\,[X_2(f)] = -\,\{\,\mathrm{Re}\,[X_3(f)] - \mathrm{Re}\,[X_3(-f)]\}\,/2 \qquad (3.48)$$

3.3.6 Case of a discrete frequency

Equations (3.45) to (3.48) remain valid for the samples $X_1(n)$ and $X_2(n)$ taken on $X_1(f)$ and $X_2(f)$ if the symmetry of these samples with respect to the origin $f = 0$ is conserved. This is always the case when (3.3) is used for the sampling of the Fourier transform. Thus,

$$\mathrm{Re}\,[X_1(n)] = \{\,\mathrm{Re}\,[X_3(n)] + \mathrm{Re}\,[X_3(-n)]\}\,/2 \qquad (3.49)$$

for $n = 0, \ldots, N/2$

and

$$\mathrm{Re}\,[X_1(-n)] = \mathrm{Re}\,[X_1(n)] \qquad (3.50)$$

For the imaginary part, we obtain

$$\text{Im}\,[X_1(n)] = \{\text{Im}\,[X_3(n)] - \text{Im}\,[X_3(-n)]\}\,/2 \tag{3.51}$$

for $n = 0, \ldots, N/2$

and

$$\text{Im}\,[X_1(-n)] = -\text{Im}\,[X_1(n)] \tag{3.52}$$

Similar equations can be deduced for the DFT $X_2(n)$. The number of operations involved in (3.49) and (3.51) is small compared with that required for computing a DFT. This dissimilarity is even more pronounced when the number of samples N is large. The advantages and the importance of these equations will be shown in section 3.6.4.

3.4 RECONSTRUCTION OF THE z-TRANSFORM AND THE FOURIER TRANSFORM

3.4.1 Preliminary remarks

We have seen (sect. 3.2.11) that a nonperiodic signal of finite length N can be completely reconstructed with the N samples of its Fourier transform. However, for such a signal, the conditions for the existance of its z and Fourier transforms are satisfied if and only if the signal does not contain infinite samples. The z-transform of a finite length signal converges in the entire complex z plane, except possibly at the points $z = 0$ and $z \to \infty$. Because the unit circle is included in the region of convergence, the Fourier transform can be obtained from the value of the z-transform on this circle. It is obvious that if the signal can be completely reconstructed, its z-transform and, consequently, its Fourier transform can also be reconstructed from N samples $X(n)$.

3.4.2 Reconstruction of the z-transform

The z-transform of a finite-length signal $x(k)$ is given by (2.1):

$$X(z) = \sum_{k=0}^{N-1} x(k)z^{-k} \tag{3.53}$$

substituting (3.25) into this expression, we obtain:

$$X(z) = \sum_{k=0}^{N-1} \left[\frac{1}{N} \sum_{n=-N/2}^{N/2-1} X(n)W_N^{nk} \right] z^{-k}$$

$$= \frac{1}{N} \sum_{n=-N/2}^{N/2-1} X(n) \sum_{k=0}^{N-1} W_N^{nk} z^{-k} \tag{3.54}$$

However, the sum over k in this equation is a geometric series in $W_N^n z^{-1}$, which leads to

$$X(z) = \frac{1}{N} \sum_{n=-N/2}^{N/2-1} X(n) \frac{W_N^{nN} z^{-N} - 1}{W_N^n z^{-1} - 1} \tag{3.55}$$

Finally, using property (3.15), we have

$$X(z) = \frac{z^{-N} - 1}{N} \sum_{n=-N/2}^{N/2-1} X(n) \frac{1}{W_N^n z^{-1} - 1} \tag{3.56}$$

3.4.3 Remarks

Thus, the z-transform of a finite length signal can be obtained from the N values $X(n)$ of its Fourier transform. Because the values $X(n)$ are the samples of $X(z)$ computed on the unit circle, we may consider (3.56) to be an interpolation formula in the complex z plane. In chapter 5, this equation will be directly involved in the recursive implementation of filters with finite-length impulse responses.

3.4.4 Reconstruction of the Fourier transform

The coefficients $X(n)$ of the DFT can also be considered as samples of the z-transform taken at N regularly spaced points, each separated by $2\pi/N$ radians, on the unit circle. Making the substitution:

$$z = \exp(j2\pi f) \tag{3.57}$$

we thus obtain the Fourier transform:

$$X(f) = \frac{e^{-j2\pi fN} - 1}{N} \sum_{n=-N/2}^{N/2-1} X(n) \frac{1}{e^{j2\pi n/N} e^{-j2\pi f} - 1} \tag{3.58}$$

Using the following equation:

$$\begin{aligned} e^{jx} - 1 &= e^{jx/2}(e^{jx/2} - \bar{e}^{jx/2}) \\ &= 2j\, e^{jx/2} \sin x/2 \end{aligned} \tag{3.59}$$

statement (3.58) can be written more compactly as

$$X(f) = \frac{1}{N} \sum_{n=-N/2}^{N/2-1} X(n) \frac{\sin \pi fN}{\sin \pi(f - n/N)} \exp\{j\pi[f(1-N) - n/N]\} \tag{3.60}$$

Equation (3.60) enables us to obtain the Fourier transform by interpolation between the values $X(n)$.

3.4.5 Example

To illustrate the notions just developed, let us consider the following signal (Fig. 3.1):

$$x(k) = a^k \operatorname{rect}_N(k) \qquad \text{with } a \text{ real} < 1$$

Its DFT is given by

$$X(n) = \sum_{k=0}^{N-1} a^k W_N^{-nk} = \frac{1 - a^N W_N^{-nN}}{1 - a W_N^{-n}} = \frac{1 - a^N}{1 - a W_N^{-n}} \tag{3.61}$$

The discrete amplitude and phase spectra are obtained by computing the modulus and the argument of $X(n)$, respectively. After replacing W_n by its expression (3.10), the following result is obtained:

$$X(n) = \frac{1 - a^N}{1 - a\cos(2\pi n/N) + ja\sin(2\pi n/N)} \tag{3.62}$$

Thus,

$$|X(n)| = \frac{1 - a^N}{\sqrt{1 + a^2 - 2a\cos(2\pi n/N)}} \tag{3.63}$$

and

$$\theta_x(n) = \arg[X(n)] = \arctan\left(\frac{a\sin(2\pi n/N)}{a\cos(2\pi n/N) - 1}\right) \tag{3.64}$$

Expressions (3.63) and (3.64) are shown in figure 3.1. It is easy to prove that these values are the samples of the Fourier transform $X(f)$ taken from the given signal. Thus,

$$X(f) = \sum_{k=0}^{N-1} a^k e^{-j2\pi fk} = \frac{1 - a^N e^{-j2\pi fN}}{1 - a e^{-j2\pi f}} \tag{3.65}$$

The amplitude and phase spectra deduced from this expression are also shown in figure 3.1.

Figure 3.1 shows that expressions (3.63) and (3.64) correspond to samples of the amplitude and phase spectra taken within the period $[-1/2, 1/2]$, using an increment $\Delta f = 1/N$. The signal $X_p(k)$, obtained by calculating the inverse DFT from (3.61) for all values of k, is also shown in figure 3.1. In particular, if these values are in the interval $[k_0, k_0 + N - 1]$ with $k_0 = 2$, we obtain the signal $x'(k)$ (Fig. 3.1). The form of the original signal can be obtained by cyclic permutation of two samples. This shows the importance of having some prior knowledge about the signal (that it is a finite-length nonperiodic signal, and the

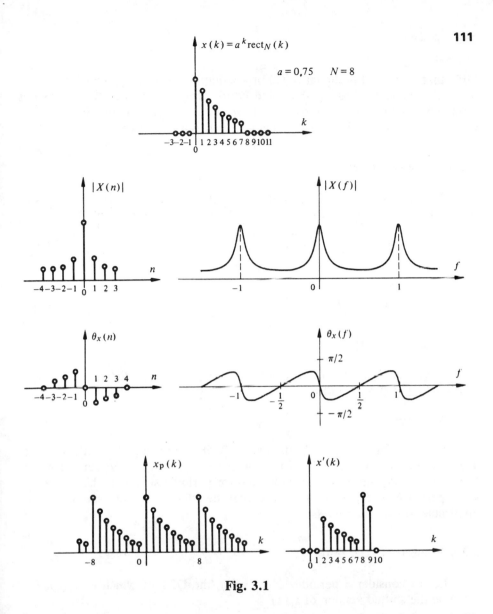

Fig. 3.1

value of k_0). It is clear that expression (3.62) is also the DFT of the periodic signal $x_p(k)$, which has the same discrete amplitude and phase spectra.

3.5 MAIN PROPERTIES OF THE DISCRETE FOURIER TRANSFORM

3.5.1 Introduction

For fruitful and efficient use of the DFT, it is important to know some of its main properties. In this section, we will cover the main properties of the

DFT that are useful in digital signal processing. Most of the properties studied here are similar to those of the z and Fourier transforms because the DFT is obtained by sampling the Fourier transform or the z-transform on the unit circle. Consequently, when the proof is obvious, it shall not be given.

3.5.2 Property: Linearity

The DFT is a linear transformation. If $X_1(n)$ and $X_2(n)$ are the DFTs of two finite-length signals, $x_1(k)$ and $x_2(k)$, then the DFT of the signal:

$$x(k) = a x_1(k) + b x_2(k) \tag{3.66}$$

is

$$X(n) = a X_1(n) + b X_2(n) \tag{3.67}$$

3.5.3 Comments

If N_1 and N_2 are the respective lengths of $x_1(k)$ and $x_2(k)$, then the length of the signal $x(k)$ is given by

$$N = \max [N_1, N_2] \tag{3.68}$$

If, for example, $N_1 > N_2$, the two DFT must be computed with $N = N_1$. Thus, $X_2(n)$ will be the DFT of the signal $x_2(k)$ plus $N_1 - N_2$ zero samples. Equations (3.66) and (3.67) remain valid for periodic signals if they have the same period $N = N_1 = N_2$. It is clear that the addition of zero samples is not applicable to periodic signals.

3.5.4 Property: Cyclic shift

Let us consider a periodic signal $x_p(k)$, the DFT of which is $X_p(n)$. Let $y_p(k)$ be the shifted version of $x_p(k)$:

$$y_p(k) = x_p(k - k_0) \tag{3.69}$$

It is clear that if k_0 is greater than the period N of $X_p(n)$, this shift and the shift of $k_0' = k_0 - lN$, where l is an integer, lead to identical results. Consequently, we consider only k_0 modulo N. The DFT of $y_p(k)$ is given by

$$Y_p(n) = W_N^{-n k_0} X_p(n) \tag{3.70}$$

If the period of a periodic signal $x_p(k)$ is associated with a finite-length signal $x(k)$, the length of which is less than or equal to the period N of $x_p(k)$, then a shift of k_0 in the signal $x_p(k)$ corresponds to a cyclic or circular shift of the samples of $x(k)$ for a given period. This is equivalent to observing the periodic signal within an observation window of fixed width N. What we see is the cyclic shift of the samples of the signal $x(k)$. The DFT of a finite-length signal $x(k)$, circulary shifted by k_0, is given by

$$Y(n) = W_N^{-nk_0} X(n) \tag{3.71}$$

where $Y(n)$ is the DFT of the signal $y(k)$ defined by

$$y(k) = x_p(k - k_0) \operatorname{rect}_N(k) \tag{3.72}$$

Similarly, if the DFT of a signal, periodic or not, is shifted by n_0, that is, if

$$Y(n) = X(n - n_0) \tag{3.73}$$

then, the corresponding signal is given by

$$y(k) = W_N^{+n_0 k} x(k) \tag{3.74}$$

This equation is valid for periodic signals of period N as well as for finite-length signals of length N.

3.5.5 Correlation functions

Let us consider two real periodic signals $x_p(k)$ and $y_p(k)$, both of period N. The crosscorrelation function of these two signals, evaluated over one period, is given by

$$\varphi_{xy}(k) = \frac{1}{N} \sum_{l=0}^{N-1} x_p(l) y_p(l + k) \tag{3.75}$$

The correlation function of two periodic signals of period N is itself a periodic signal of period N. In fact, when the shift in the product $x_p(l) y_p(l + k)$ reaches N, we obtain the same configuration as for $k = 0$. The DFT of (3.74) is

$$\Phi_{xy}(n) = \frac{1}{N} X_p^*(n) Y_p(n) \tag{3.76}$$

Figure 3.2 shows two periodic signals of period N and their periodic correlation function $\varphi_{xy}(k)$. It is obvious that this result can be obtained either directly from (3.75) or by the inverse DFT of $\Phi_{xy}(n)$.

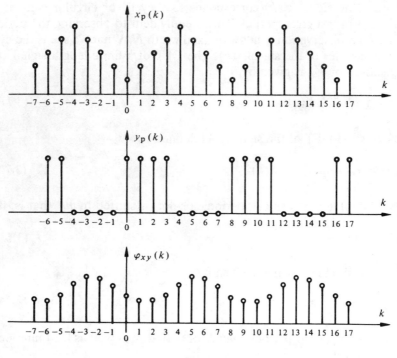

Fig. 3.2

3.5.6 Case of finite-length signals

The crosscorrelation function of two finite-length signals is given by

$$\overset{\circ}{\varphi}_{xy}(k) = \sum_{l=0}^{N-1} x(l)y(l+k) \tag{3.77}$$

where N is the length of $x(k)$. The DFT $X(n)$ can be computed with N samples (sect. 3.2.11) because the length of $x(k)$ is limited to N. The DFT for both sides of this equation leads to the following expression:

$$\overset{\circ}{\Phi}_{xy}(n) = X^*(n)\ Y(n) \tag{3.78}$$

Instead of computing (3.77) directly, we may use expression (3.78), for which the DFTs of $x(k)$ and $y(k)$ must be evaluated. We obtain $\overset{\circ}{\varphi}_{xy}(k)$ as the inverse DFT of $\overset{\circ}{\Phi}_{xy}(n)$. However, this result must be interpreted carefully. To illustrate the use of (3.78) for finite-length signals, let us consider the signals $x(k)$ and $y(k)$, obtained by taking one period of their periodic versions $x_p(k)$ and $y_p(k)$, shown in figure 3.2. Using the methods of section 1.4.1, we can compute the crosscorrelation function $\overset{\circ}{\varphi}_{xy}(k)$.

3.5.7 Remarks

The finite-length signals $x(k)$ and $y(k)$, and their correlation function $\mathring{\varphi}_{xy}(k)$, are shown in figure 3.3. It is clear that $\mathring{\varphi}_{xy}(k)$ is not periodic. It is a finite-length signal. If N_x and N_y are the respective lengths of $x(k)$ and $y(k)$, then the length $\mathring{\varphi}_{xy}(k)$ is $N_x + N_y - 1$. This is obtained directly by looking for the number of unit shifts for which the product $x(l)y(l + k)$ is nonzero. However, if (3.78) is used, the signal obtained by taking one period of the inverse DFT of $\Phi_{xy}(n)$ is of length $N = N_x$. Thus, aliasing occurs, which is shown in figures 3.2 and 3.3. This aliasing of $\mathring{\varphi}_{xy}(k)$ is simply a consequence of the underestimated sampling of $\mathring{\Phi}_{xy}(f)$. Because the length of $\mathring{\varphi}_{xy}(k)$ is $N_x + N_y - 1$, $\mathring{\Phi}_{xy}(f)$ should be sampled at a rate of $1/(N_x + N_y - 1)$, not $1/N_x$.

To avoid such aliasing, zero samples must be added to both signals in order to give them the same length $N = N_x + N_y - 1$. The modified signals and the corresponding result obtained using the inverse DFT of (3.78) are shown in figure 3.4. Only in this case can a period of the new result $\varphi'_{ry}(k)$ be considered as the crosscorrelation function of two finite-length signals computed without aliasing.

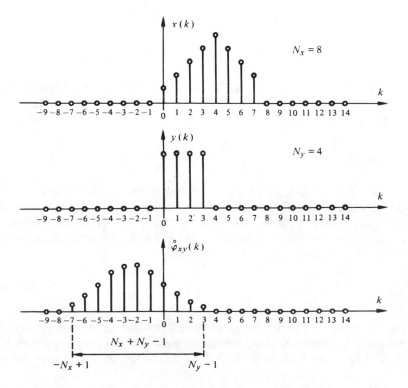

Fig. 3.3

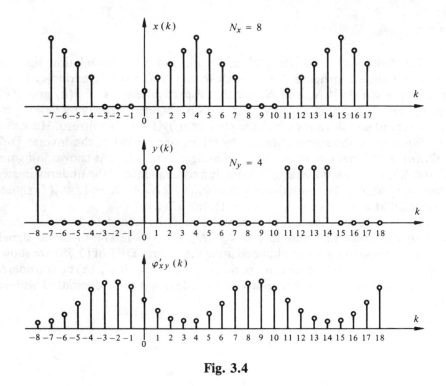

Fig. 3.4

Despite the similarity between equations (3.75) and (3.77) and those for the analog case (see vol. VI, chap. 4), the aliasing effect illustrated here occurs only in digital processing.

3.5.8 Convolution

Given that the convolution of two signals is the crosscorrelation function of these signals with one of them being reversed about the origin, the convolution has properties similar to those established in the three preceding sections. The convolution of two periodic real signals, $x_p(k)$ and $g_p(k)$, of period N, is a periodic signal of period N, which is given by

$$y_p(k) = \sum_{l=0}^{N-1} x_p(l)g_p(k-l) \tag{3.79}$$

Because these are not finite-energy signals, the summation must be limited to a single period, which, for simplicity, is chosen here to be between zero and $N - 1$. The DFT for both sides of (3.79) leads to the following expression:

$$Y_p(n) = X_p(n)\, G_p(n) \tag{3.80}$$

Figure 3.5 shows two periodic signals of period N and their convolution. Because the periodic signals are completely represented by their DFTs, the result $y_p(k)$ can be obtained either directly from (3.79) or indirectly by computing the inverse transform of the product of the two DFTs in (3.80). Comparing figures 3.2 and 3.5 shows once more the strong relationship between a correlation function and a convolution.

3.5.9 Remarks

Equations (3.79) and (3.80) can be used for finite-length signals if the precautions mentioned in section 3.5.7 are taken into account for the correlation function. The convolution of finite-length signals obtained by taking one period of the periodic signals, $x_p(k)$ and $g_p(k)$, is shown in figure 3.6.

As with the crosscorrelation function, the convolution of two finite-length signals is a finite-length signal of length $N_x + N_g - 1$, where N_x and N_g are the lengths of $x(k)$ and $g(k)$, respectively. Consequently, a DFT taken over a length N_x cannot be used to represent the convolution of two signals of lengths N_x and N_g without aliasing effects.

The overlaps can be avoided if zero samples are added to the signals in

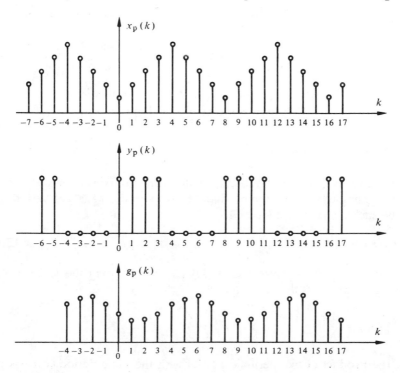

Fig. 3.5

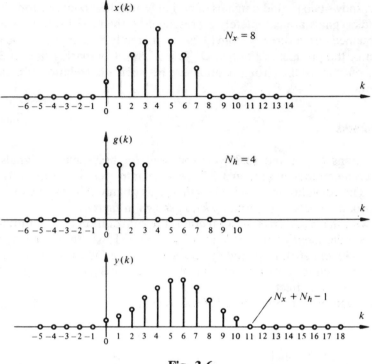

Fig. 3.6

order to make them of length $N = N_x + N_g - 1$. In this case, a period of the inverse DFT of (3.80) represents the results without aliasing.

3.5.10 Frequency convolution

The duality between the frequency and time domains allows us to use the same reasoning to consider the product of two periodic signals of period N. Let $y_p(k)$ be such a signal. We have

$$y_p(k) = x_p(k)\, g_p(k) \tag{3.81}$$

After expressing the signals by their DFTs and grouping the terms W_N^{nk}, the following statement is derived:

$$Y_p(n) = \frac{1}{N} \sum_{m=0}^{N-1} X_p(m) G_p(n-m) \tag{3.82}$$

Thus, the product of two periodic signals with the same period corresponds in the spectral domain to the convolution of their DFTs.

3.5.11 Geometric interpretation

It was previously shown that the DFT is a sampling of the z-transform at points which are uniformly spaced on the unit circle. Coefficients of the DFT are then plotted on the cylinder generated by the unit circle, the intersection of which with the complex surface $X(z)$ is the Fourier transform $X(f)$. To compute (3.82), the two cylinders representing the DFTs $X_p(n)$ and $G_p(n)$ must be fit together and the product of the superposed coefficients must be taken two by two. The sum of these products gives a value $Y_p(n)$ in (3.82). Then, for another value of N, one cylinder must be rotated by $2\pi/N$ radians, and the

Table 3.7 Principal properties of the discrete Fourier transform.

$$x(k) = \frac{1}{N} \sum_{n=-N/2}^{N/2-1} X(n) W_N^{nk} \qquad X(n) = \sum_{k=0}^{N-1} x(k) W_N^{-nk}$$
$$n = -N/2, \ldots, N/2 - 1$$

$$a x_1(k) + b x_2(k) \qquad a X_1(n) + b X_2(n)$$

$$x(k - k_0) \text{ cyclic} \qquad W_N^{-nk_0} X(n)$$

$$x_p(k - k_0) \qquad W_N^{-nk_0} X_p(n)$$

$$W_N^{n_0 k} x(k) \qquad X(n - n_0)$$

$$\varphi_{xy}(k) = \frac{1}{N} \sum_{l=0}^{N-1} x_p(l) y_p(l+k) \qquad \Phi_{xy}(n) = \frac{1}{N} X_p^*(n) Y_p(n)$$
$$\text{DFT with } N \text{ samples}$$

$$\overset{\circ}{\varphi}_{xy}(k) = \sum_{l=0}^{N-1} x(l) y(l+k) \qquad \overset{\circ}{\Phi}_{xy}(n) = X^*(n) Y(n)$$
$$\text{DFT with } N_x + N_y - 1 \text{ samples}$$

$$y_p(k) = \sum_{l=0}^{N-1} x_p(l) g_p(k-l) \qquad Y_p(n) = X_p(n) G_p(n)$$
$$\text{DFT with } N \text{ samples}$$

$$y(k) = \sum_{l=0}^{N-1} x(l) g(k-l) \qquad Y(n) = X(n) G(n)$$
$$\text{DFT with } N_x + N_g - 1 \text{ samples}$$

$$y_p(k) = x_p(k) g_p(k) \qquad Y_p(n) = \frac{1}{N} \sum_{m=0}^{N-1} X_p(m) G_p(n-m)$$

$$y(k) = x(k) g(k) \qquad Y(n) = \frac{1}{N} \sum_{m=0}^{N-1} X(m) G(n-m)$$

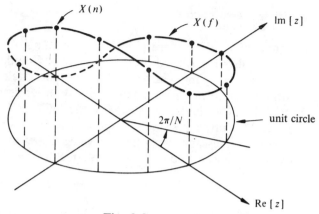

Fig. 3.8

multiplications and summation of the coefficients performed again. After completing one turn, all the values of $Y_p(n)$ are obtained.

It can be shown that (3.82) is also valid for the product of two finite-length signals. The DFT of the product is the periodic convolution of the two DFTs.

3.6 SPLIT CORRELATION AND CONVOLUTION

3.6.1 Introduction

Situations where the two signals to be correlated or convolved are of different lengths often occur and, in general, one is much longer than the other. For example, if we want to detect an impulse of known form and length that is drowned in noise (for instance, in the case of radar echos), we evaluate the crosscorrelation function of the received signal and of the impulse with the much shorter length. Similar cases also occur in filtering problems. Quite often, a very long signal must be filtered by a system with a very short impulse response compared to that of the signal to be filtered. A typical example is when the long signal representing voice must be filtered. The existence of a particularly efficient and fast algorithm for the computation of the DFT enables us to consider the indirect method, which consists of computing two DFTs, taking their product, and computing the inverse DFT [see (3.78) and (3.80)]. The analysis presented in section 4.7.6 shows that the indirect method using the DFT is more efficient than the direct computation of (3.77) or (3.79) when the lengths of the signals are greater than about $N_0 = 30$, which is commonly used. Furthermore, the processed signals are generally nonperiodic. The DFT must then be computed over an interval greater than $N_1 + N_2 - 1$, where N_1 and N_2 are the lengths of the two signals to be correlated or convolved. Let us assume, for example, that N_1 is much greater than N_2.

However, some problems arise when computing the DFT over a very long interval. In theory, this operation is possible, but, in practice, large amounts of memory and computation time are required. Furthermore, computations must be completed before the first sample of the result can be obtained. Therefore, the methods developed by Stockham [10] are preferred. They consist of dividing the signal into several parts, correlating or convolving them separately, then recombining the partial result to obtain the total signal.

These methods are shown only in the case of convolution because of the strong relationship between correlation and convolution. Their application to correlation is straightforward.

3.6.2 First method: Superposed addition

Let us consider a signal $x(k)$ of great length N_x, which must be filtered by a filter with an impulse response $g(k)$ of length N_g. If the signal $x(k)$ is divided into K sections, each of length M, a signal $x_m(k)$, corresponding to each section, can be defined as follows:

$$x_m(k) = \begin{cases} x(k) & \text{for } mM \leqslant k \leqslant (m+1)M-1 \\ 0 \end{cases} \tag{3.83}$$

The signals $x(k)$, $g(k)$, and $x_m(k)$ are shown in figure 3.9. The original signal is

$$x(k) = \sum_{m=m_0}^{m_0+K-1} x_m(k) \tag{3.84}$$

Using notation (1.78) and property (1.81), the entire filtered signal can be written as follows:

$$y(k) = \sum_{m=m_0}^{m_0+K-1} y_m(k) \tag{3.85}$$

where

$$y_m(k) = x_m(k) * g(k) \tag{3.86}$$

Because the length of $g(k)$ is N_g, *the DFT must be computed with $M + N_g - 1$ samples* for each partial convolution in order to avoid aliasing. Since the length of each section is M, each partial result overlaps the first $N_g - 1$ samples of the following partial result in the sum (3.85). This is shown in figure 3.9.

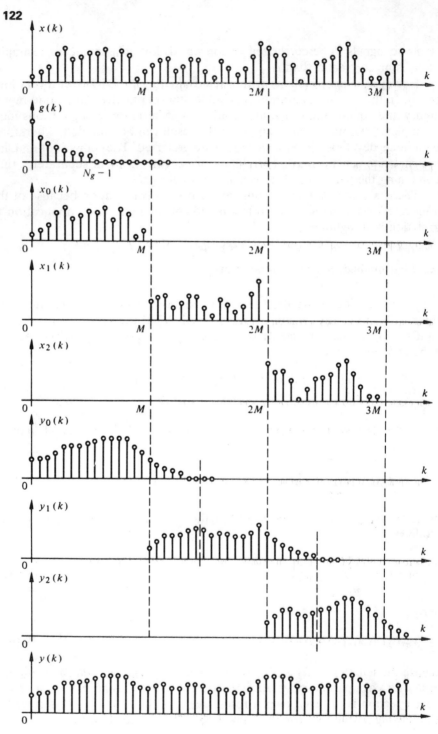

Fig. 3.9

3.6.3 Second method: Juxtaposition

A variation of the first method can be obtained by computing the convolution of the DFT for each part of the signal with M samples (M is assumed to be greater than N_g). The $N_g - 1$ first samples of each partial result will be aliased, and, therefore, incorrect.

To remove these samples from the final result, the sections of $x(k)$ must be chosen so that $N_g - 1$ samples overlap. The final result is then obtained by the juxtaposition of the partial results from which the $N_g - 1$ first values were removed. This method is illustrated in figure 3.10.

3.6.4 Remarks

If these methods are used for real signals, we can benefit from the properties of the DFT of real signals (sect. 3.3.5). Equation (3.80) is valid even for complex signals. Equation (3.78), representing the DFT of the correlation, is also valid for complex signals if their correlation is defined by

$$\overset{\circ}{\varphi}_{xy}(k) = \sum_{l=0}^{N-1} x^*(l)y(l+k) \tag{3.87}$$

In fact, (3.78) and (3.80) are equivalent to four real correlations or convolutions. To prove this, let us again consider the case of convolution. The results obtained can be easily transposed to the case of correlation. The convolution of two complex signals $x(k)$ and $g(k)$ is

$$y(k) = x(k) * g(k) \tag{3.88}$$

If each signal is expressed by its real and imaginary parts in this equation, then

$$y(k) = \text{Re}\,[x(k)] * \text{Re}\,[g(k)] - \text{Im}\,[x(k)] * \text{Im}\,[g(k)]$$
$$+ j\,\{\text{Re}\,[x(k)] * \text{Im}\,[g(k)] + \text{Im}\,[x(k)] * \text{Re}\,[g(k)]\} \tag{3.89}$$

Thus, complex convolution consists of four real convolutions. If the signal $g(k)$ is real, then (3.89) reduces to two distinct real convolutions:

$$y(k) = \text{Re}\,[x(k)] * g(k) + j\,\text{Im}\,[x(k)] * g(k) \tag{3.90}$$

The DFT of (3.90) is always given by expression (3.80). By examining (3.90), we can see that two real signals, represented here by the real and the imaginary parts of $x(k)$, can be convolved in parallel by a simple real signal $g(k)$. For example, these two signals may be two neighboring sections of a long real signal. Thus, it is possible to obtain a convolution of two real signals with half the operations needed when the indirect method is used with the DFT.

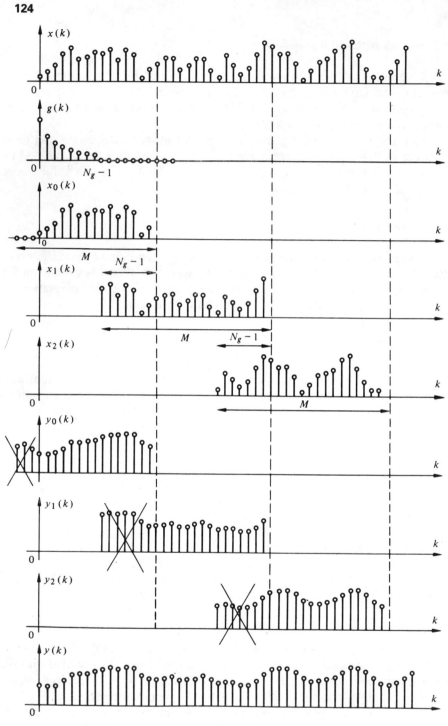

Fig. 3.10

3.7 DISCRETE FOURIER TRANSFORM FOR FINITE-LENGTH SIGNALS

3.7.1 Preliminary remarks

In the beginning of this chapter, we saw that a signal can be entirely represented by N samples of its Fourier transform if and only if its length is limited to N. The DFT of an infinite-length signal cannot be defined exactly. For such a signal, the DFT is defined, approximately, by limiting the length of the signal by some appropriate means. In this section, the principal ways to limit the length of a signal and the quality of the approximation associated with the DFT are presented.

The Fourier transform $X(f)$ of a digital signal $x(k)$ is a periodic function. The samples of this signal are, in fact, the coefficients of the Fourier series decomposition of the periodic function $X(f)$.

In general, the Fourier series representation of a periodic function requires an infinite number of coefficients. This is particularly the case for all discontinuous periodic functions. Thus, the problem of limiting the length of a digital signal is similar to the problem of convergence of a Fourier series.

3.7.2 Rectangular window function

The brute force method for limiting the length of a signal consists of multiplying it by a rectangular signal $\text{rect}_N(k)$, which has N unit samples. Such a signal is often called the *temporal window function,* or, more simply, *temporal window.* From the start, the multiplication of the signal by the window $\text{rect}_N(k)$ presents two problems. The first is the location of this window with respect to the signal and the second is the choice of N. To locate the window, the particular signal must be known. However, in general, the position of the window is chosen in such a way as to keep the important samples of the signal and neglect those of comparatively small amplitude. For example, for an exponential signal of the form:

$$x(k) = a^{|k|} \quad \text{with } |a| < 1 \tag{3.91}$$

the samples of large amplitude are in the vicinity of the origin. Consequently, the window is located about the origin. In this section, the windows studied will be located symmetrically about the origin in order to simplify the calculations, although this is not a general rule. The result can be easily modified for other locations by using the shift theorem.

3.7.3 Limitation of the length in the time domain

The finite-length version of a signal of infinite length $x(k)$ is

$$x_N(k) = x(k) \, \text{rect}_N(k + N/2) \tag{3.92}$$

or

$$x_N(k) = \begin{cases} x(k) & \text{for } |k| \leqslant N/2 \\ 0 & \text{otherwise} \end{cases} \tag{3.93}$$

3.7.4 Effects of limitation of the length

The product (3.92) in the time domain corresponds to a periodic convolution in the frequency domain. Thus,

$$X_N'(f) = \int_{g_0}^{g_0+1} X(g) W_R(f-g) \, dg \tag{3.94}$$

where $X_N(f)$ and $W_R(f)$ are the Fourier transforms of the signals $x_N(k)$, $x(k)$ and $w_R(k) = \text{rect}_N(k + N/2)$, respectively.

The function $W_R(f)$, and more generally the Fourier transform of any time window, is called a *spectral window*.

The DFT can be defined for the signal $x_N(k)$ because by definition its length is bounded by N. The coefficients $X_N(n)$ of this DFT approximately represent the samples taken of $X(f)$. The convolution (3.94) must be analyzed in order to study the quality of this approximation. The approximation depends on the shape of the function $W_R(f)$, which can be easily computed:

$$W_R(f) = \sum_{k=-N/2}^{N/2} e^{-j\,2\pi f k} = \frac{\sin \pi f N}{\sin \pi f} \tag{3.95}$$

This function is shown in figure 3.11 for $N = 9$. On the principal interval $[-1/2, 1/2]$, this function includes a main lobe with a base of $2/N$, and secondary sidelobes, which decrease at the boundaries of this interval. The zeros of this function are at frequencies $f_i = i/N$, with $i \neq mN$, where i and m are integers.

3.7.5 Property

The ratio of the amplitudes of the main lobe and the principal sidelobes of $W_R(f)$ varies little as a function of N. As long as N stays finite, it can be increased without any significant change of this ratio. For example, for $N = 9$ (the case of Fig. 3.11), we have

$$\left| \frac{W_R(0)}{W_R(1,5/N)} \right| = 4.5 \tag{3.96}$$

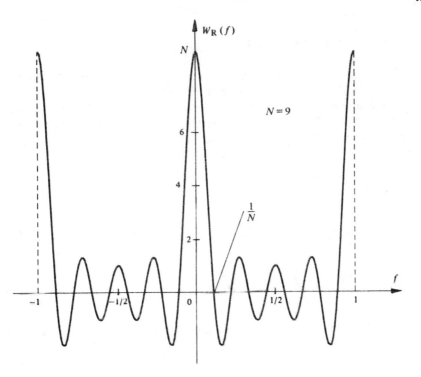

Fig. 3.11

For $N = 50$ and $N = 100$, the ratio is 4.705 and 4.711, respectively. The threshold value for large N is $3\pi/2 = 4.712$.

3.7.6 Characterization of spectral windows

In general, the global form of the spectral windows used is similar to that of $W_R(f)$. The spectral windows include, on the principal interval $[-1/2, 1/2]$, a main lobe and the sidelobes. Two principal parameters enable us to characterize the windows. The first is the width of the base of the main lobe and the second is the ratio of the amplitudes of the main lobe and the principal sidelobes. This ratio is expressed in decibels as

$$\lambda_i = 20 \log_{10} \left| \frac{W_i(f_s)}{W_i(0)} \right| \tag{3.97}$$

where f_s is the middle frequency of the sidelobe of the window function $W_i(f)$. For example, in the case of the function $W_R(f)$, for $N = 9$, we have

$$\lambda_R = 20 \log_{10} \frac{1}{4.5} \cong -13 \quad dB$$

3.7.7 Example of time limitation

The approximation of a Fourier transform $X(f)$ of an infinite-length signal by the function $X_N(f)$ obtained after time limitation makes ripples appear around the discontinuities of $X(f)$. This is known as the Gibbs phenomenon (vol. IV, sect. 7.3.36). To illustrate this phenomenon, let us consider the response of the ideal low-pass filter. This signal (sect. 1.6.13) has the following form:

$$g(k) = \frac{\sin 2\pi f_c k}{\pi k} \qquad (3.98)$$

The corresponding harmonic response is given on the principal period $[-1/2, 1/2]$ by

$$G(f) = \begin{cases} 1 & |f| < f_c \\ 0 & \text{otherwise} \end{cases} \qquad (3.99)$$

If the length of $g(k)$ is limited by the window function $W_R(k) = \text{rect}_N(k - N/2)$ the following result is obtained:

$$g_N(k) = g(k)\, w_R(k) \qquad (3.100)$$

The Fourier transform of this finite-length signal is given by

$$G_N(f) = \sum_{k=-N/2}^{N/2} \frac{\sin 2\pi f_c k}{\pi k} e^{-j2\pi f k} \qquad (3.101)$$

Using (3.94), it can also be written:

$$G_N(f) = \int_{g_0}^{g_0+1} G(g)\, \frac{\sin \pi N(f-g)}{\sin \pi(f-g)}\, dg \qquad (3.102)$$

On the principal period, this convolution is simply

$$G_N(f) = \int_{-f_c}^{f_c} \frac{\sin \pi N(f-g)}{\sin \pi(f-g)}\, dg \qquad (3.103)$$

Figure 3.12 shows the Fourier transform $G(f)$ and its approximations $G_N(f)$ obtained for different values of N. The ripples around the discontinuities $f = \pm f_c$ (Gibbs phenomenon) are obvious. The errors in the approximation

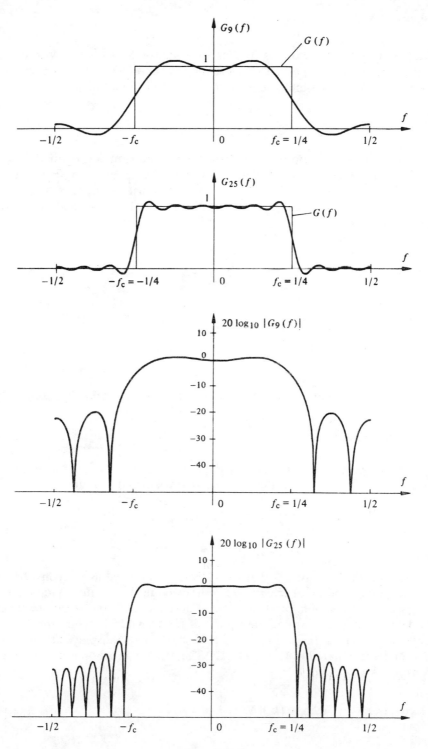

Fig. 3.12

of $G(f)$ by $G_N(f)$ are more easily seen by using the logarithmic representation of the amplitudes. In particular, increasing N increases the frequency of the ripples, but does not simultaneously decrease their amplitudes.

3.7.8 Generalization

It is possible to improve the above-mentioned situation by modifying the shape of the window function while keeping the same length N. However, the integral over one period of the spectral window must be equal to one in order to avoid changing the scale in the convolution (3.94). This condition may be written as follows:

$$\int_{-1/2}^{1/2} W(f)\mathrm{d}f = 1 \tag{3.104}$$

This integral is the value of the corresponding signal at the origin. In fact, using the definition of the Fourier transform, we have

$$w(0) = \int_{-1/2}^{1/2} W(f)\mathrm{d}f = 1 \tag{3.105}$$

This condition is implicitly satisfied for the rectangular window function $w_R(k)$.

3.7.9 Triangular window

To decrease the amplitude of the sidelobes of $W_R(f)$, another window function may be chosen having the following triangular shape:

$$w_T(k) = \begin{cases} 1 - \dfrac{2\,|k|}{N} & \text{for } |k| \leqslant N/2 \\ 0 & \text{otherwise} \end{cases} \tag{3.106}$$

The reason for this choice is that $w_T(k)$ can be expressed as the convolution of a rectangular window of length $N/2$ with itself. In the frequency domain, this corresponds to a simple product of two Fourier transforms of this rectangular window, or, in other words, the square of the transform. When the square of (3.95) is computed, the ratio λ_T of the function $W_T(f)$ displays an attenuation of -24 dB, compared to $\lambda_R = -13$ dB. The window function is mathematically expressed by

$$w_T(k) = \frac{2}{N}\ \mathrm{rect}_{N/2}\,(k+N/4) * \mathrm{rect}_{N/2}\,(k+N/4) \tag{3.107}$$

The factor $2/N$ comes from condition (3.104). In the frequency domain, we have

$$W_T(f) = \frac{2}{N}\left(\frac{\sin \pi f N/2}{\sin \pi f}\right)^2 \tag{3.108}$$

Only one restriction of little importance appears in (3.107) and (3.108). Because $N/2$ represents the number of samples of the rectangular function $\text{rect}N/2(k)$, this number must be an integer. Therefore, N must be an even number in (3.107) and (3.108). Nevertheless, the symmetry of the window $w_T(k)$ with respect to the origin is not destroyed. Figure 3.13 shows convolution (3.107) and the function $W_T(f)$ for $N = 10$.

The effective length of $w_T(k)$ is obviously $N - 1$. The unused sample $w_T(N/2)$ due to the restriction on N is considered to be zero in the Fourier transform of $w_T(k)$:

$$W_T(f) = \sum_{k=-(N/2-1)}^{N/2-1} w_T(k)e^{-j2\pi f k} \tag{3.109}$$

Comparing the functions $W_R(f)$ and $W_T(f)$, represented with the same scale in Figures 3.11 and 3.13, respectively, clearly shows the attenuation of the sidelobes. This, however, is obtained only at the price of increasing the width of the main lobe. In fact, the base width of the main lobe of the function $W_T(f)$

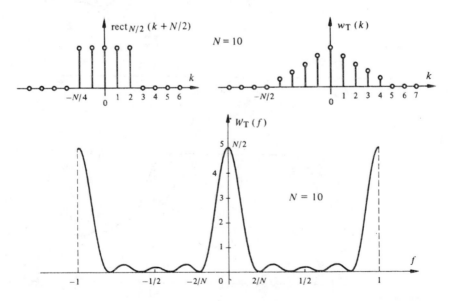

Fig. 3.13

is $4/N$. The zeros of this function are at the frequencies $f_i = 2i/N$, with $i \neq mN/2$, where i and m are integers and N is an even number.

3.7.10 Example

The results obtained by approximating the Fourier transform of the infinite-length signal $g(k)$ given by (3.98), using the window function $w_T(k)$, are shown in figure 3.14 for $N = 10$ and $N = 26$. Comparing these results to the corresponding results of figure 3.12 shows the improvement obtained in the attenuation of the ripples around the discontinuities, as well as the enlargement of the transition region around the frequencies $\pm f_c$.

3.7.11 Parabolic window

The amplitude of the ripples can be further decreased by using a parabolic window function $w_P(k)$ obtained by the convolution of a rectangular window

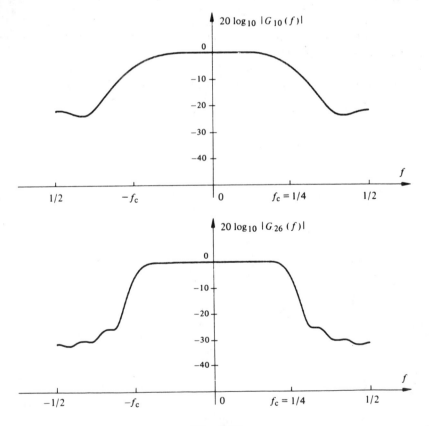

Fig. 3.14

with a triangular one. It can also be obtained by the double convolution of a rectangular window of length $N/3$ with itself. The Fourier transform of this window has the following form:

$$W_p(f) = \frac{1}{A}\left(\frac{\sin \pi f N/3}{\sin \pi f}\right)^3 \tag{3.110}$$

where the coefficient A is chosen to satisfy condition (3.104). Its expression is

$$A = (N/3)^2 - \text{int}\,(N/6)(1 + \text{int}\,(N/6)) \tag{3.111}$$

In this equation, int represents the integer or truncated part. The length of the parabolic window must be a multiple of three. Once again, the attenuation of the sidelobes is obtained only at the price of an enlargement of the main lobe, which is in this case $6/N$.

3.7.12 Polynomial windows

More generally, it is possible to construct a window whose Fourier transform is of the form:

$$W_G(f) = \frac{1}{B}\left(\frac{\sin \pi f N/m}{\sin \pi f}\right)^m \tag{3.112}$$

by convolving a rectangular window of length N/m with itself $m - 1$ times. For large values of the integer m, considerable attenuation of the secondary sidelobes is obtained. However, the width of the main lobe of $W_G(f)$ is $2m/N$, which increases in proportion to m. The consequence of this is an enlargement of the regions of discontinuity. The ideal value of m depends on the particular problem, the most commonly used values being $m = 2, 3$, and 4.

3.7.13 Cosine window

Polynomial windows are not the only possibilities. Another method is to construct a window that attenuates the sidelobes by superposition. To see this, let us examine again the rectangular window $w_R(k)$. The sign of the sidelobes of the function $W_R(f)$ changes every $1/N$ (Fig. 3.11). The same alternation is observed around the discontinuities after the convolution with $W_R(f)$ (Fig. 3.12). Moreover, the global form varies little from one sidelobe to the next. If we superpose a Fourier transform $G_N(f)$ onto this function, shifted by the width $(1/N)$ of a sidelobe, we can expect considerable attenuation of these lobes due to the alternation of the signs. The logical and precise way to perform this superposition is to compute the sum of two identical transforms, the first shifted toward positive frequencies by $1/2N$ and the second shifted toward negative frequencies by the same amount. This sum is then multiplied by 1/2 to satisfy condition (3.104). This enables us to conserve the position of the Fourier transform on the frequency axis. Using the shift theorem in the frequency domain (sect. 1.8.14), the window corresponding to this superposition can be expressed as follows:

$$w_c(k) = \begin{cases} \dfrac{1}{2}\left[\exp\left(j\,\dfrac{\pi k}{N}\right) + \exp\left(-j\,\dfrac{\pi k}{N}\right)\right] = \cos\left(\dfrac{\pi k}{N}\right) & \text{for } |k| < N/2 \\ 0 & \text{otherwise} \end{cases}$$

(3.113)

Thus, the window $w_c(k)$ is a cosine taken over a half period about the origin $k = 0$. It is shown in figure 3.15.

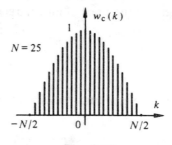

Fig. 3.15

The Fourier transform of this window can be easily computed by expressing it as the product of an infinite-length cosine signal and a rectangular window function of length N. If the corresponding periodic convolution is performed in the frequency domain, the following result is obtained:

$$W_c(f) = \frac{1}{2}\,\frac{\sin \pi N(f - 1/2N)}{\sin \pi(f - 1/2N)} + \frac{1}{2}\,\frac{\sin \pi N(f + 1/2N)}{\sin \pi(f + 1/2N)}$$

(3.114)

We can see from this expression that the function $w_c(f)$ is obtained by the superposition of two functions $W_R(f)$ shifted by $\pm 1/2N$. This superposition leads to a main lobe in $W_c(f)$ of width $3/N$. The attenuation of the sidelobes can be measured by the ratio λ_c, which is -24 dB for $N = 9$. Comparing this window function to the triangular window, we obtain the same ratio $\lambda_c \cong \lambda_T$. The advantage of the cosine window lies in the width of the main lobe, which is $3/N$ instead of $4/N$ for the triangular window.

3.7.14 Example

The effect of the use of the window function $w_c(k)$ in the approximation of the Fourier transform $G(f)$ given by (3.99) is represented in figure 3.16.

Comparing these results with those of figure 3.12 shows the attenuation of the ripples around the discontinuities. However, just as for the window function $w_T(k)$, this attenuation is obtained at the price of an enlargement of the transition regions around the frequencies $\pm f_c$, which in this case is less pronounced than in the case of the function $w_T(k)$.

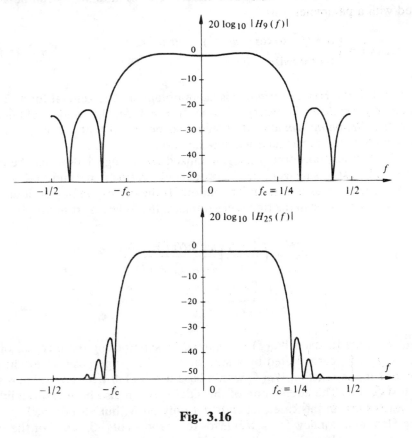

Fig. 3.16

3.7.15 Hamming window

The next stage in the refinement of the attenuation of the sidelobes by superposition is to superpose three copies of the same transform: two shifted by $1/N$ and the third unshifted, but given a double weight in the superposition. The window function corresponding to this superposition in the frequency domain is given by

$$W_h(k) = \frac{1}{4}\left[\exp\left(j\,\frac{2\pi k}{N}\right) + 2 + \exp\left(-j\,\frac{2\pi k}{N}\right)\right]$$

$$= \frac{1}{2}\left(1 + \cos\frac{2\pi k}{N}\right) \quad \text{for } |k| \leqslant -N/2 \tag{3.115}$$

As before $w_\mathrm{h}(k)$ is equal to zero outside the interval $[-N/2, N/2]$. A window function of this form is known by the name of *Hanning* [11]. In the superposition of three copies of a transform, the weights 1/4, 1/2, 1/4 can be modified to obtain other attenuation effects. The window $w_\mathrm{h}(k)$ can be generalized with a parameter α as

$$
w_\mathrm{H}(k) = \begin{cases} \alpha + (1 - \alpha) \cos (2\pi k/N) & \text{for } |k| \leqslant N/2 \\ 0 \text{ otherwise} \end{cases}
\tag{3.116}
$$

For $\alpha = 1/2$, the Hanning window is again obtained. The general form (3.116) is called *a generalized Hamming window*. If $\alpha = 0.54$, the window obtained is called the *Hamming window*. It should also be noted that the value $\alpha = 1$ corresponds to the rectangular window function.

The Fourier transform of the generalized Hamming window may be easily computed by expressing $w_\mathrm{H}(k)$ as the product of a cosine signal and a rectangular window function plus a rectangular window. If the corresponding periodic convolution is performed in the frequency domain, the following result is obtained:

$$
W_\mathrm{H}(f) = \alpha \, \frac{\sin \pi f N}{\sin \pi f} + \left(\frac{1 - \alpha}{2} \right) \frac{\sin \pi N(f - 1/N)}{\sin \pi(f - 1/N)}
$$
$$
+ \left(\frac{1 - \alpha}{2} \right) \frac{\sin \pi N(f + 1/N)}{\sin \pi(f + 1/N)}
\tag{3.117}
$$

For $\alpha \neq 1$, the function $W_\mathrm{H}(f)$ is obtained by superposing three copies of the function $W_R(f)$, each shifted by a step of $1/N$. This superposition produces a main lobe in the function $W_\mathrm{H}(f)$ with a base width of $4/N$ for any value α such that $0 \leqslant \alpha < 1$. The attenuation of the sidelobes can also be measured by the ratio λ_H, which, in this case, depends not only on N, but also on α. For $N = 9$, the Hamming window ($\alpha = 1/2$) leads to a ratio λ_h of -32 dB. For the same value of N, the Hamming window with $\alpha = 0.54$ leads to $\lambda_\mathrm{H} = -90$ dB. This shows the dramatic increase in the attenuation of the principal sidelobe obtained by using the Hamming window ($\alpha = 0.54$) instead of the Hanning window ($\alpha = 0.5$). However, it must not be forgotten that the ratio λ was defined for the principal sidelobe. The weight $\alpha = 0.54$ enables us to attenuate this sidelobe, but the other sidelobes are not attenuated as much. They are at about -43 dB from the main lobe.

3.7.16 Blackman window

It is also possible to generalize the window construction by increasing the number of functions superposed from 3 to M. The M copies of the original function, shifted by $1/N$, are superposed with carefully chosen weights. To conserve the symmetry of the shifted copies about the origin, M must be an odd integer. In the time domain, the general form of the window corresponding to these M superpositions is

$$
w_B(k) = \begin{cases} a_0 + 2 \sum_{l=1}^{L} a_l \cos \dfrac{2\pi k l}{N} & \text{for } |k| \leqslant N/2 \\ 0 \quad \text{otherwise} \end{cases}
\tag{3.118}
$$

where $L = (M - 1)/2$. The coefficients a_l must satisfy the condition:

$$
a_0 + 2 \sum_{l=1}^{L} a_l = 1
\tag{3.119}
$$

to avoid a scale factor in the estimation. The Fourier transform of the window $w_B(k)$ may be computed as in the case of the window $w_H(k)$. The following result is obtained:

$$
W_B(f) = a_0 \frac{\sin \pi f N}{\sin \pi f} + \sum_{l=1}^{L} a_l \frac{\sin \pi N(f - l/N)}{\sin \pi(f - l/N)} + \sum_{l=1}^{L} a_l \frac{\sin \pi N(f + l/N)}{\sin \pi(f + l/N)}
\tag{3.120}
$$

The values of the coefficients a_l are obtained by optimization according to a given criterion. In principle, the criterion is to minimize the energy of the signal $w_B(k)$ outside some given frequency interval. This interval generally corresponds to the base width of the main lobe of $W_B(f)$. The window function $w_B(k)$ is called the *generalized Blackman window function* [11]. The window function obtained in the particular case $L = 2$ is called the *Blackman window*. The Blackman window is characterized by the weights $a_0 = 0.42$, $a_1 = 0.25$, and $a_2 = 0.04$. The Fourier transform of the Blackman window ($L = 2$) has a main lobe with a width of $6/N$. The corresponding ratio λ_B is about -59 dB.

3.7.17 Kaiser window

Another family of window functions was proposed by Kaiser [12]. The Kaiser window makes it possible, according to the value of a parameter β, to specify in the frequency domain the compromise between the width of the main lobe and the amplitude of the sidelobes. An important characterization of this refined family of windows is that it is possible to obtain large attenuation of the sidelobes while preserving a minimal width for the main lobe. The general form of this window is the following:

$$
w_K(k) = \begin{cases} \dfrac{I_0[\beta \sqrt{N^2 - 4k^2}]}{I_0(\beta N)} & \text{for } |k| \leqslant N/2 \\ 0 \quad \text{otherwise} \end{cases}
\tag{3.121}
$$

where I_0 is modified *zero*th-order Bessel function of first kind, and β is the parameter characterizing the energy exchange between the main lobe and sidelobes. For best performance, the values of the product βN must be chosen within the interval [4, 9]. The width of the principal sidelobe of $W_K(f)$ increases with β. Due to the Bessel function I_0, the computation of the analytic form of the Fourier transform $W_K(f)$ is very complicated and will not be considered here. We will instead refer the reader to the graphical representation of $W_K(f)$ obtained by digital computation.

The different window functions studied in this section and their Fourier transforms are summarized in figure 3.17.

3.7.18 Choice of the length N in the time domain

Up to this point, the effect of the shape of the window function on the quality of the approximation of a Fourier transform of an infinite-length signal for a given length N has been considered. In particular, it was shown that, with a nonrectangular window, the ripples of the Gibbs phenomenon can be dramatically decreased, but at the cost of a less accurate approximation of fast transitions. The choice of the value of N is of prime importance. Because the sampling of the Fourier transform $X(f)$ with a step of $1/N$ corresponds to the periodic repetition, of period N, of the signal $x(k)$, N must be chosen so that a period of the periodic signal obtained $x_p(k)$ is a valid approximation of the signal $x(k)$. We have

$$x_p(k) = \sum_{i=-\infty}^{+\infty} x(k+iN) \tag{3.122}$$

Because the length of $x(k)$ is unlimited, unavoidable aliasing occurs in this equation. In general, aliasing can be made negligible, or at least tolerable, if N is chosen to be large. Thus, we have, approximately

$$x_p(k) \cong x(k) \quad \text{pour } i = 0 \tag{3.123}$$

The shape of the signal $x(k)$ must be known, more or less, to make a choice of N on this basis. Only with this information can we determine the minimum value of N for which the approximation (3.123) can be considered valid.

3.7.19 Choice of the length N in the frequency domain

If the shape of the signal $x(k)$ is unknown, or hardly known, an alternative is to use spectral information in the frequency domain. Because $X(f)$ is sampled with a step of $1/N$, the resolution with which we wish to analyse the DFT of the signal provides the information needed to choose N.

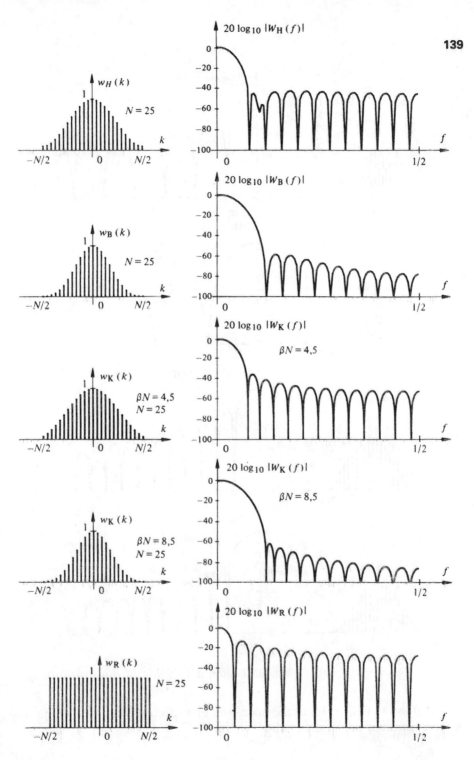

Fig. 3.17

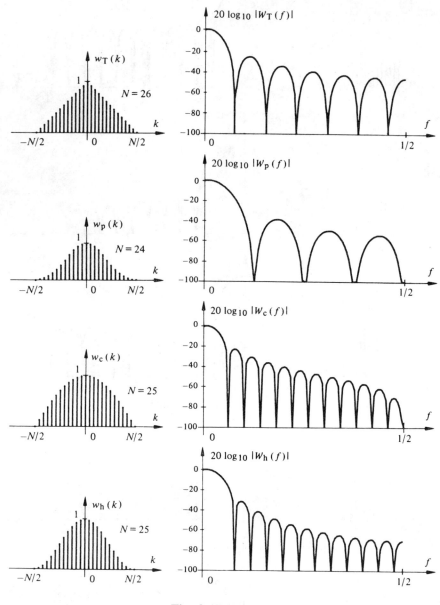

Fig. 3.17 (cont'd.)

If B_i is the base width of the main lobe of a spectral window $W_i(f)$ for the window functions previously studied, we have

$$B_i = \gamma_i / N \tag{3.124}$$

where γ_i is a characteristic function $W_i(f)$. For example, for the rectangular window $\gamma_R = 2$. If the required frequency resolution is Δf, N must be chosen to satisfy the condition:

$$B_i \leqslant \Delta f \tag{3.125}$$

or, in terms of N:

$$N \geqslant \gamma_i / \Delta f \tag{3.126}$$

3.7.20 Example

The poor resolution which results when condition (3.126) is violated is shown in figure 3.18 for the case of the signal:

$$x(k) = \cos 2\pi f_1 k + \cos 2\pi f_2 k$$

and the Hamming window function $w_H(k)$. The DFT is obtained by sampling the convolution (3.94). For a window function $w_i(k)$, we have

$$X_N(n) = \int_{g_0}^{g_0+1} X(g) W_i\left(\frac{n}{N} - g\right) dg \tag{3.127}$$

When condition (3.126) is satisfied, the resolution becomes sufficient to discriminate between the two sidelobes (Fig. 3.18).

3.7.21 Distribution phenomenon

From equation (3.127), we can see that each coefficient $X_N(n)$ of the DFT computed at frequency $f_n = n/N$ is the weighted mean of the desired Fourier transform $X(f)$. The window $W_i(f)$ distributes the energy, or the power, at a given frequency over the entire period of $X_N(f)$. This phenomenon is clearly seen in figure 3.18. It is worsened by the sampling of $X_N(f)$ for the DFT. In effect, a line in $X(f)$ observed at the frequency f_n may in fact be at a frequency f_n such that

$$f_n - 1/(2N) \leqslant f_i \leqslant f_n + 1/(2N) \tag{3.128}$$

The estimation of the frequency of this line is, therefore, false. This error depends only on N. An example is shown in figure 3.19. There are two different ways to decrease this error. The first is to increase the length N to decrease the frequency interval $\Delta f = 1/N$. The second, used in the case where only a fixed

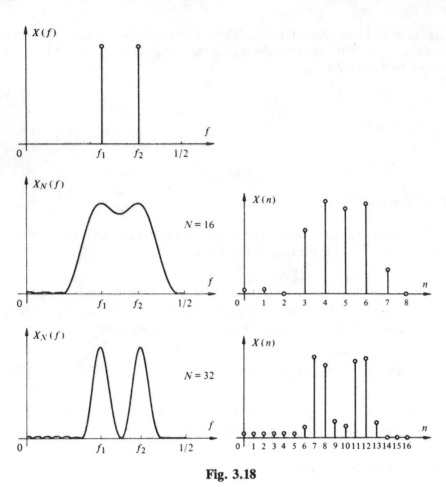

Fig. 3.18

number of samples of the signal $x(k)$ is available, consists of considering a pseudolength N' by defining a new finite-length signal $x_{N'}(k)$ as

$$x_{N'}(k) = \begin{cases} x_N(k) & \text{for } k_0 \leqslant k \leqslant k_0 + N-1 \\ 0 & \text{for } k_0 + N \leqslant k \leqslant k_0 + N'-1 \\ & \text{with } N' > N \end{cases} \qquad (3.129)$$

The new interval of frequency $\Delta f'$ is $1/N'$. It is smaller than the previous one, which means that the sampling of $X_{N'}(f)$ will be finer.

The distribution phenomenon will be covered in chapter 6 for practical cases.

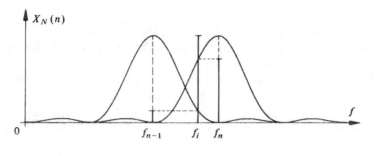

Fig. 3.19

3.7.22 General indications

The value of N can be chosen independently of, or in relationship to, the choice of the window function $W_i(k)$, depending on the case. This depends on the different constraints placed on the allowable error. For example, a function $W_i(f)$ with a large main lobe can be chosen because of the severe constraints placed on the sidelobes of $W_i(f)$. If, in addition, other constraints require a good approximation of the fast transitions, the value of N must be increased by using one of the two above-mentioned methods.

The choice of the particular shape of a window function mainly depends on the width B_i of the principal sidelobe of $W_i(f)$ and the amplitude of the secondary sidelobes with respect to the amplitude of the principal sidelobe. If the principal sidelobe is wide, the rapid transitions of the Fourier transform of the original signal are approximated very poorly. If the relative amplitude of the secondary sidelobes is large, the approximation obtained has inconvenient ripples, especially around the discontinuities. The compromise, which is often hard to find at first glance, depends on the particular signal and on the allowable error. Specific examples will be studied in section 6.7.

3.8 APPROXIMATION OF THE FOURIER TRANSFORM OF ANALOG SIGNALS

3.8.1 Introduction

The existence of a particularly fast algorithm to compute the DFT makes the idea of using it as an approximation for the Fourier transform of analog signals very attractive. This is even more so when the analog signals being considered are very-low-frequency signals. In fact, tremendous difficulties are involved in the application of low-frequency analog spectrum analyzers. These systems are best used for high-frequency analog signals. Digital spectrum analyzers, despite their frequency limitation, on the other hand, have two advantages, which are typical of any digital system. Their time stability is guaranteed and their precision is simpler to control and to improve.

In this section, we cover the use of the DFT as an approximation for the integral Fourier transform and the Fourier series for periodic analog signals. However, the reader is expected to be familiar with the elementary concepts of analog signal processing. If this is not the case, texts such as those by de Coulon (vol. VI) and Papoulis [13] are excellent references. Moreover, only deterministic signals are studied here. Chapter 6 is devoted to the spectral analysis of random signals.

The development of the discrete equations will also make it possible to express the DFT in a form where the sampling period Δt appears as in the notation $x(k\Delta t)$, instead of the notation $x(k)$. This is done only to allow us to rescale the frequency axis according to the sampling frequency.

3.8.2 Periodic signals

The coefficients of the Fourier series expansion of a periodic analog signal of period T are given by (vol. IV, sect. 7.4.6; vol. VI, chap. 6):

$$X_a(n) = \frac{1}{T} \int_{t_0}^{t_0+T} x_a(t) \exp(-j2\pi f_n t)\,dt \tag{3.130}$$

with $f_n = n\Delta f$ and $\Delta f = 1/T$

The Fourier series decomposition of the periodic signal $x_a(t)$ is

$$x_a(t) = \sum_{n=-\infty}^{+\infty} X_a(n) \exp(j2\pi f_n t) \tag{3.131}$$

If the coefficients $X_a(n)$ are zero, or can be considered as such for $|n| \geq N/2$, then the signal $x_a(t)$ can be periodically sampled with a period Δt. Conforming to the sampling theorem, Δt is given by (sect. 1.7.2):

$$\Delta t = 1/f_N = 1/(N\Delta f) \tag{3.132}$$

Since $\Delta f = 1/T$, (3.132) leads to

$$T = N\Delta t \tag{3.133}.$$

If N samples are taken over one period of the signal $x_a(t)$, (3.130) can be written in the approximate form:

$$\hat{X}_a(n) = \frac{1}{T} \sum_{k=k_0}^{k_0+N-1} x_a(t_k) \exp(-j 2\pi f_n t_k) \Delta t$$

$$= \frac{1}{N} \sum_{k=k_0}^{k_0+N-1} x_a(t_k) \exp\left(-j 2\pi \frac{nk}{N}\right) \qquad (3.134)$$

with $n = -N/2, ..., N/2 - 1$

We can easily see that the series of the coefficient $X_a(N)$ given by (3.134) is rendered periodic by sampling with a period N. Moreover, expression (3.134) is an approximation of integral (3.130) using juxtaposed rectangles. Thus, the approximation becomes better when the width of the rectangles Δt becomes smaller. The samples of the signal $x_a(t)$ are given by

$$\hat{x}_a(t_k) = \sum_{n=-N/2}^{N/2-1} \hat{X}_a(n) \exp\left(j 2\pi \frac{nk}{N}\right) \qquad (3.135)$$

with $k = k_0, ..., k_0 + N - 1$

Comparing (3.134) and (3.135) with (3.24) and (3.25), we note that the DFT can be easily modified for use in the approximate Fourier series representation of periodic analog signals. The difference is only a multiplicative factor.

3.8.3 Comments

The spectra of periodic signals are line spectra. They can be easily obtained digitally if the period T of the signal is known. In this case, *the samples must be taken over a length corresponding exactly to one or several periods.* If this condition is violated, the coefficients $X_a(f)$ are no longer computed at harmonic frequencies $f_n = n\Delta f = n/T$. An error similar to that shown in figure 3.19 occurs. In this case, the Fourier transform of the window function is of the form (3.95) because of the rectangular window implicitly contained in the data acquisition. When the period T of the signal is not known in advance, the observation time of the signal must be large with respect to the presumed period T. This ensures sufficient resolution for the estimation of the harmonic frequencies and the coefficients $X_a(n)$.

3.8.4 Digital integration

To avoid an overestimation or underestimation of an integral obtained digitally, the samples must always be chosen in the middle of the approximation intervals. This is also the case for (3.134). These errors are shown in figure 3.20. Thus, if the periodic signal to be analyzed includes a discontinuity, for example, in its known analytic form, the samples must not be taken over the discontinuities in order to avoid a delay or an advance of $\Delta t/2$. This appears only in the phase spectra. The rule must also be observed in the case of finite-energy signals, which shall be studied later. An example which illustrates the $\Delta t/2$ delay will also be presented.

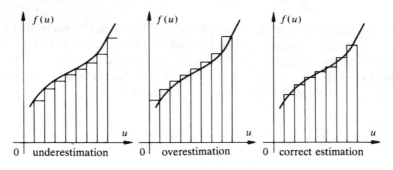

Fig. 3.20

3.8.5 Signals with finite energy

The Fourier transform of a finite-energy analog signal $x_a(t)$ is given by (vol. IV, sect. 7.3.1; vol. VI, sect. 4.1):

$$X_a(f) = \int_{-\infty}^{+\infty} x_a(t)\exp(-j\,2\pi ft)dt \tag{3.136}$$

The inverse Fourier transform is given by

$$x_a(t) = \int_{-\infty}^{+\infty} X_a(f)\exp(j\,2\pi ft)df \tag{3.137}$$

If the signal $x_a(t)$ is sampled periodically at times t_k with a sampling period Δt chosen according to the sampling theorem, then (3.136) becomes

$$X_a(f) \cong \sum_{k=-\infty}^{+\infty} x_a(t_k)\exp(-j\,2\pi ft_k)\Delta t \tag{3.138}$$

The complex function $X_a(f)$ can be considered a signal and sampled periodically with the period Δf. Δf must be chosen to satisfy as well as possible the converse sampling theorem. Equation (3.138) then becomes

$$X_a(f_n) \cong \sum_{k=-\infty}^{+\infty} x_a(t_k)\exp(-j\,2\pi f_n t_k)\Delta t \tag{3.139}$$
$$\text{with } n = -\infty, ..., +\infty \qquad f_n = n\,\Delta f$$

The inverse transform is given by the following approximate equation:

$$x_a(t_k) \cong \sum_{n=-\infty}^{+\infty} X_a(f_n)\exp(j\,2\pi f_n t_k)\Delta f \tag{3.140}$$
$$\text{with } k = -\infty, ..., +\infty$$

Also, in this case, (3.139) and (3.140) are approximations of the integrals (3.136) and (3.137), respectively. Moreover, an unavoidable aliasing error is included in the coefficient $X_a(f_n)$, or in the samples of the signal, because it is well known that the sampling theorem cannot be simultaneously satisfied by a finite-energy signal $x_a(t)$ and its Fourier transform $X_a(f)$ (vol. VI, sect. 4.6).

3.8.6 Properties

If the discrete variables f_n and t_k are replaced by $n\Delta f$ and $k\Delta t$, respectively, in (3.139) and (3.140), we can see that the sequences of coefficients $X_a(n\Delta f)$ and the samples of $x_a(k\Delta t)$ are periodic, with the same period $(\Delta t\Delta f)^{-1}$, in n and in k, respectively. Since n and k are integers, this common period must also be an integer. If this number is denoted by N, then the following result is obtained:

$$N = 1/(\Delta t\,\Delta f) \tag{3.141}$$

from which, we derive

$$\Delta f = 1/(N\,\Delta t) \tag{3.142}$$

The product $N\Delta t$ is a length, which will be denoted hereafter by T and called the *observation length*. Equation (3.142) then becomes

$$\Delta f = 1/T \tag{3.143}$$

3.8.7 Remarks

Even if the original signal $X_a(t)$ and its Fourier transform $X_a(f)$ are non-periodic, their sampled versions are periodic. Since the coefficients $X_a(n\Delta f)$ and the digital signal $x_a(k\Delta t)$ are periodic, they can be entirely represented by all the values of one period. In principle, any period can be chosen to represent the signal. However, it is useful to choose the more significant period of $x_a(t)$ so as to preserve some parallel between the two representations. The period for the transform $X_a(n\Delta f)$ is generally chosen as the main period $[-N/2, N/2 - 1]$. For the variable f, it corresponds to the period $[-1/2\Delta t, 1/2\Delta t]$.

The equivalent forms of (3.139) and (3.140) are, then,

$$X_a(n\Delta f) = \sum_{k=k_0}^{k_0+N-1} x_a(k\Delta t)\exp\left(-j\frac{2\pi nk}{N}\right)\Delta t \qquad (3.144)$$

and

$$x_a(k\Delta t) = \sum_{n=-N/2}^{N/2-1} X_a(n\Delta f)\exp\left(j\frac{2\pi nk}{N}\right)\Delta f \qquad (3.145)$$

These expressions are the approximations of the direct and inverse integral Fourier transforms of a finite-energy analog signal $x_a(t)$. Once again, when these expressions are compared with (3.24) and (3.25), we see that the DFT can be used to compute them. Only the factors Δt, Δf, and N must be modified.

The DFT $Y(n)$ of a digital signal $y(k\Delta t)$ is always periodic in n of period N. However, its Fourier transform $Y(f)$ is periodic, but the period is $1/\Delta t$ rather than 1. The principal period $[-1/2, 1/2]$ becomes $[-1/2\Delta t, 1/2\Delta t]$ using the notation $y(k\Delta t)$.

3.8.8 Difference between analog and digital signals

When using (3.144) and (3.145), the following important point must be taken into account. In the beginning of this chapter, it was shown that *a digital signal of finite length N is entirely represented by N samples of its Fourier transform.* This is not at all true for an analog signal. The length T of the observation must be greater than the effective length of the signal for an accurate approximation of the integral Fourier transform of a finite-length analog signal by the coefficients $X_a(n\Delta f)$. The larger is this length, the closer the frequencies $f_n = n\Delta f = n/T$ are to each other. The small separation between the frequencies f_n allows for an adequate description of the desired Fourier transform. This is directly linked to the problem of frequency resolution, which was studied in section 3.7.21.

3.8.9 Example

Let $x_a(t)$ be a rectangular analog signal defined by

$$x_a(t) = \begin{cases} 1 & \text{for } |t| < 0.5 \text{ s} \\ 0 & \text{otherwise} \end{cases} \tag{3.146}$$

For this signal, the condition of the sampling theorem is never satisfied. In addition, the sampling frequency was chosen small enough (8 Hz) to demonstrate the aliasing error. The chosen frequency resolution is also small (0.125 Hz). This leads to an observation time $T = 1/0.125 = 8$ s, which is large compared to the effective length of the signal. The number of samples is given by $N = 8.8 = 64$. The results obtained under these conditions are shown in figure 3.21. The aliasing errors are apparent on the magnitude spectrum for $|f| > 2$ Hz. This error can be decreased if the sampling frequency is increased, that is, if Δt

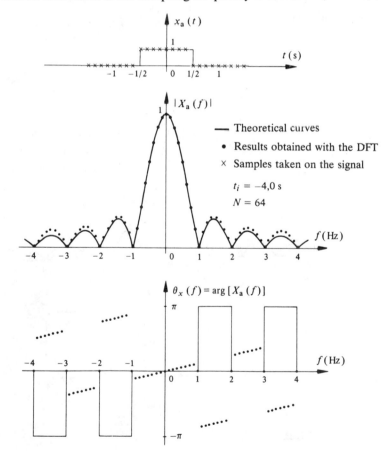

Fig. 3.21

is decreased. If the frequency resolution must be increased with a constant sampling frequency (Δt constant), the number of samples N, and hence the observation time T, must be increased. A tremendous error is also observed in the phase spectrum as a result of the poor choice of the sampling instants. The first nonzero sample of the signal is taken on a discontinuity. This error can be corrected if all the sampling times are shifted by $+\Delta t/2$. The results obtained with this correction are shown in figure 3.22. Despite the aliasing error in the modulus of $X_a(f)$, its argument is correctly computed.

3.9 EXERCISES

3.9.1 Let $x_p(k)$ be a periodic signal of period N. Show that the DFT of this signal delayed by k_0 is given by

$$W_N^{-nk_0} X_p(n)$$

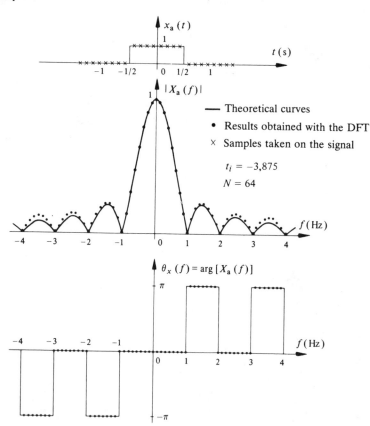

Fig. 3.22

Is there any ambiguity for different values of k_0?

3.9.2 Show that the DFT of the correlation of two periodic signals with the same period is given by

$$\Phi_{xy}(n) = (1/N) X_p^*(n) Y_p(n)$$

3.9.3 Let $X(n)$ be the DFT of a real signal $x(k)$. Among the N coefficients $X(n)$, how many are real numbers? Discuss your result for N odd or even. Is this result also valid for the number of imaginary coefficients $X(n)$? Justify your response.

3.9.4 Let $x_p(k)$ be a periodic signal of period N and $X_{1p}(n)$ its DFT. We may also consider $x_p(k)$ as a periodic signal of period $2N$. Let $X_{2p}(n)$ be its DFT when it is considered as periodic of period $2N$. Determine $X_{2p}(n)$ as a function of $X_{1p}(n)$.

3.9.5 An analog signal is sampled at a frequency of 10 kHz. A DFT is computed on $N = 1024$ samples of this signals. What is the frequency interval between the two coefficients $X(n)$ and $X(n + 1)$? Justify your response.

3.9.6 Show that the transform $X_p(f)$ given by (3.30) is periodic in f of period 1.

3.9.7 Let $x(k)$ be a finite-length signal of length $N = 8$. Its DFT has the following form:

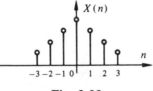

Fig. 3.23

A new signal $y(k)$ of finite length $N = 16$ is formed from $x(k)$ in the following manner:

$$y(k) = \begin{cases} x(k/2) & \text{for } k \text{ even} \\ 0 & \text{for } k \text{ odd} \end{cases}$$

Sketch the form of $Y(n)$ and justify your response.

3.9.8 Let $x(k)$ and $y(k)$ be two finite-length signals of lengths $N_x = 8$ and $N_y = 20$, respectively. The two DFTs $X(n)$ and $Y(n)$ are computed with $N =$

20, the product $X(n)$ $Y(n)$ is taken, and the inverse DFT of this product is computed with $N = 20$. Let $z(k)$ be the inverse DFT of $X(n)$ $Y(n)$. Determine the samples of $z(k)$ which could be obtained by the direct computation of the linear convolution.

3.9.9 Let $X(f)$ be the Fourier transform of $x(k) = (1/2)^k \epsilon(k)$. Let $Y(n)$ be the DFT with $N = 10$ samples of a finite-length signal $y(k)$. We set

$$Y(n) = X(f)\Big|_{f=n/10} \qquad n = -5, ..., 4$$

Determine $y(k)$.

3.9.10 Which type of window function and which length N must be used to compute the DFT of a signal with a frequency resolution better than $\Delta f = 0.01$ (on the normalized scale) and oscillations in the separation band of less than $\lambda_i = 30$ dB?

3.9.11 Let $X(n)$ be the DFT of a finite-length signal $x(k)$ of length N. Considering $X(n)$ as a complex signal, its DFT $y(k)$ can be computed. Express $y(k)$ in terms of $x(k)$.

3.9.12 Show that (3.120) is the Fourier transform of the generalized Blackman window.

3.9.13 What is the DFT of the signal $x(k) = \cos 2\pi f_0 k$, where $f_0 = 1.28125$ is computed with a rectangular window and $N = 32$ samples. Is the result subject to the distribution phenomenon?

Chapter 4

Fast Unitary Transformations

4.1 INTRODUCTION

4.1.1 Preamble

Since its rediscovery in 1965 by Cooley and Tukey [14], the algorithm for the fast Fourier transform (FFT), thanks to its efficiency, has revolutionized digital signal analysis. The discrete Fourier transform can be considered as the product of a matrix, called the transformation matrix, with a vector composed of the samples of a signal. The efficiency of the FFT comes from the redundancy of the elements in the transformation matrix. This redundancy is not arbitrary and has a well determined structure. This structure can be used advantageously to synthesize other transformation matrices with similar redundancy, and thereby to generalize the algorithm for the fast computation of the FFT.

This structure, obtained by using the Kronecker product (or tensorial product), seems to have been used for the first time by Sylvester [15] in 1867. Since then, the method has been optimized and generalized so that today it constitutes an efficient tool in digital signal processing. The basis of the FFT was established in 1939 by Runge, Konig, and Strumpff [16, 17]. Most of the present computation algorithms were thus in existance at a time when there were neither computers nor efficient computation algorithms.

4.1.2 Organization of the chapter

The general theory presented in this chapter uses matrix computations. The mathematical developments are given, in most cases, under two complementary forms; the first using matrix notation and representation, and the second using series expansions. After the general theory is presented, particular transformations, and especially the FFT, are studied in detail. Finally, the different applications in signal processing are studied.

4.2 SIGNALS AND VECTOR SPACES

4.2.1 Review: Definitions

In a vector space, the vector obtained by summing N vectors, each one multiplied by a coefficient, is called a *linear combination*. The corresponding expression is

$$x = \sum_{i=1}^{N} \alpha_i x_i \tag{4.1}$$

A set of N vectors E_i is called *linearly independent* if the equation:

$$\sum_{i=1}^{N} \beta_i E_i = 0 \tag{4.2}$$

is satisfied, if and only if each coefficient β_i is zero. The space ϵ_N of the linear combination of N linearly independent vectors is an N-dimensional space. The set of the vectors $\{E_i\}$ is called the *basis* of ϵ_N. Each vector in ϵ_N is a unique linear combination of the E_i. Any set of N linearly independent vectors can be a basis of ϵ_N. Any vector x in ϵ_N can be represented by

$$x = x_1 E_1 + x_2 E_2 + ... + x_N E_N$$

$$= \sum_{i=1}^{N} x_i E_i \tag{4.3}$$

The set of the coefficients x_i is called the *representation of the vector x* in the space ϵ_N with respect to the basis $\{E_i\}$.

The *scalar product* of two vectors x and y in ϵ_N is defined by

$$(x \cdot y) = \sum_{i=1}^{N} x_i y_i^* = (y \cdot x)^* \tag{4.4}$$

If the vector x corresponds to a signal, the space ϵ_N is called the *signal space*.

4.2.2 Relationship between a signal and its representation

The scalar product of the two terms of equation (4.3) with the vector E_j leads to

$$(x \cdot E_j) = \sum_{i=1}^{N} (E_i \cdot E_j) x_i \qquad \text{with } j = 1, ..., N \tag{4.5}$$

This is a set of N equations with N unknowns x_i, which can be solved. A more convenient way to obtain the x_i is to establish an other basis $\{F_i\}$ of ϵ_N, whose vectors are orthogonal by twos to those of the basis $\{E_i\}$:

$$(E_i \cdot F_j) = \delta_{ij} \qquad \text{with } i, j = 1, ..., N \qquad (4.6)$$

where δ_{ij} is the *Kronecker symbol* defined by

$$\delta_{ij} = \begin{cases} 0 & \text{for } i \neq j \\ 1 & \text{for } i = j \end{cases} \qquad (4.7)$$

The notation $\delta(i, j)$ will be also used in the case of indices with a complicated form. If the two terms of (4.3) are multiplied by F_j, the following expression is obtained:

$$(x \cdot F_j) = \sum_{i=1}^{N} (E_i \cdot F_j) x_i \qquad (4.8)$$

Using (4.6), the result is

$$(x \cdot F_j) = x_j \qquad (4.9)$$

Thus, each coefficient x_j can be determined independently of one another. The orthogonality introduced by (4.6) corresponds to the decomposition into a series of orthogonal functions, which is well known in analog signal processing (vol. VI, chap. 3). Bases satisfying (4.6) are called *reciprocal bases*.

4.2.3 Orthonormal bases

A basis $\{E_i\}$ is called *orthonormal* if it is its own reciprocal:

$$(E_i \cdot E_j) = \delta_{ij} \qquad (4.10)$$

In this case, we have

$$x_i = (x \cdot E_i) \qquad \text{with } i = 1, ..., N \qquad (4.11)$$

This set of equations enables us to obtain the representation x_i of the vector x relative to the original basis $\{E_i\}$.

If $\{F_i\}$ is another basis in ϵ_N, then it is possible to write

$$x = \sum_{i=1}^{N} x_i E_i = \sum_{i=1}^{N} X_i F_i \qquad (4.12)$$

where X_i is the representation of x using the basis $\{F_i\}$.

4.2.4 Relationship between two representations

Each vector E_i is a vector of ϵ_N, and, consequently, in the basis $\{F_i\}$, we have the representation:

$$E_i = \sum_{j=1}^{N} a_{ji} F_j \qquad \text{with } i = 1, ..., N \qquad (4.13)$$

Substituting this equation into (4.12) and identifying the coefficients of each F_i, the following result is obtained:

$$X_1 = a_{11} x_1 + a_{12} x_2 + ... + a_{1N} x_N$$
$$\vdots \qquad\qquad\qquad\qquad\qquad (4.14)$$
$$X_N = a_{N1} x_1 + a_{N2} x_2 + ... + a_{NN} x_N$$

Using matrix notation, this equation can be written in the following form:

$$\begin{pmatrix} X_1 \\ \vdots \\ X_N \end{pmatrix} = \begin{pmatrix} a_{11} & a_{12} & ...a_{1N} \\ \vdots & & \\ a_{N1} & a_{N2} & ...a_{NN} \end{pmatrix} \begin{pmatrix} x_1 \\ \vdots \\ x_N \end{pmatrix} \qquad (4.15)$$

or, more compactly,

$$X = Ax \qquad (4.16)$$

4.2.5 Interpretation

The elements a_{kl} of the square matrix A can be obtained by multiplying both sides of (4.13) by G_k, where $\{G_k\}$ is a basis of ϵ_N. The following equations are obtained:

$$(E_i \cdot G_k) = \sum_{j=1}^{N} a_{ji} (F_j \cdot G_k) \qquad \text{with } i, k = 1, ..., N \qquad (4.17)$$

This is a set of N^2 equations with N^2 unknowns, which can, in principle, be solved to obtain the a_{kl}. If $\{G_k\}$ is the reciprocal basis of $\{F_i\}$, using definition (4.6), we obtain

$$a_{ki} = (E_i \cdot G_k) = (G_k \cdot E_i)^* \qquad \text{with } i, k = 1, ..., N \qquad (4.18)$$

Equation (4.15) or (4.16) can be considered as a *change of basis* in the space ϵ_N. This is a *coordinate transformation,* or *rotation,* in the space ϵ_N. If the bases $\{E_i\}$ and $\{F_j\}$ are orthonormal, it can be shown that

$$a_{jk} = (E_k \cdot F_j) = (F_j \cdot E_k)^* \qquad \text{with } j, k = 1, ..., N \qquad (4.19)$$

and that

$$\sum_{i=1}^{N} a_{ik} a_{ij}^* = \delta_{kj} \qquad (4.20)$$

4.2.6 Definitions

Equation (4.20) indicates that the columns of the matrix A are normalized and mutually orthogonal. Using matrix notation, it can be written in the following form:

$$A^H A = I \qquad (4.21)$$

where A^H is the *Hermitian transpose* of A defined by $A^H = (A^T)^*$ and I is the *unitary matrix.* According to the definition of the *inverse matrix,* we have

$$A A^{-1} = A^{-1} A = I \qquad (4.22)$$

4.2.7 Property

By comparing the two relations (4.21) and (4.22), we may remark that the Hermitian transpose of A is equal to its inverse:

$$A^H = A^{-1} \qquad (4.23)$$

Thus, the following equality must also be true:

$$A A^H = I \qquad (4.24)$$

This can be expressed by a series as:

$$\sum_{i=1}^{N} a_{ki} a_{ji}^* = \delta_{kj} \qquad (4.25)$$

Equation (4.25) indicates that the rows of the matrix A are normalized and mutually orthogonal.

4.2.8 Definition

A matrix representing the coordinate transformation between two orthonormal bases is called a *unitary matrix*.

4.2.9 Properties

If A and B are two unitary matrices, then the matrix $C = AB$ is also a unitary matrix. For A and B, we can verify the following equations:

$$A^H A = I = A A^H$$
$$B^H B = I = B B^H$$
$$C^H C = (AB)^H AB = B^H A^H AB = B^H (A^H A) B = B^H B = I \qquad (4.26)$$

We can also show that $CC^H = I$.

If the elements a_{ij} are real numbers, then

$$A^T A = A A^T = I \qquad (4.27)$$

In this case, the matrix A is called *orthonormal*. Such a matrix is identical to the inverse of its transpose. Moreover, if the matrix A is symmetric, it is equal to its transpose. Thus, a symmetric and orthonormal matrix is identical to its inverse.

On the other hand, it can be proved that the determinant of a unitary matrix is 1. This ensures the existence of the inverse matrix, which allows us to compute the inverse transformation of (4.16):

$$x = A^{-1} X \qquad (4.28)$$

4.2.10 Application to signals

The vector x can be constructed with the samples of a signal $x(k)$ with $x_k = x(k)$ and $k = 1, ..., N$. *A transformation* of type (4.16) *makes it possible to point out some properties of the signal $x(k)$ that are not obvious in the series of numbers $x(k)$.*

The transformation (4.16) is one-dimensional because it is applied to a one-dimensional signal $x(k)$. This transformation can be written as a series in the following manner:

$$X(l) = \sum_{k=1}^{N} a_{kl} x(k) \qquad \text{with } l = 1, ..., N \qquad (4.29)$$

4.2.11 Generalization

The general form of a transformation that can be applied to a p-dimensional signal is given by

$$X(l_1, l_2, l_3, \ldots, l_p) = \sum_{k_1} \sum_{k_2} \cdots \sum_{k_p} a_{k_1 k_2 \ldots k_p l_1 l_2 \ldots l_p} x(k_1, k_2, k_3, \ldots, k_p)$$

$$(4.30)$$

The most interesting cases are those of two- and three-dimensional transformations. Three-dimensional transformations are used in TV image processing. The three dimensions consist of two spatial dimensions and time. Two-dimensional transformations, which are more commonly utilized, are used in image processing and pattern recognition.

For $p = 2$, (4.30) becomes

$$X(l_1, l_2) = \sum_{k_1} \sum_{k_2} a_{k_1 k_2 l_1 l_2} x(k_1, k_2) \tag{4.31}$$

where $a_{k_1 k_2 l_1 l_2}$ is the general element of a table $A[l_1, l_2, k_1, k_2]$. The table is, by definition, the *kernel* of the two-dimensional transformations.

4.2.12 Properties

The kernel is called *separable* if it can be written in the following form:

$$A[l_1, l_2, k_1, k_2] = A'[l_1, k_1] A''[l_2, k_2] \tag{4.32}$$

A two-dimensional separable transformation can be computed in two steps. First, a one-dimensional transformation is taken for each row of the matrix $x(k_1, k_2)$ which leads to

$$X_{k_2}(l_1, l_2) = \sum_{k_1} a'_{l_1 k_1} x(k_1, k_2) \tag{4.33}$$

Then, a second one-dimensional transformation is computed for each column of the matrix $X(l_1, k_2)$, which gives the following result:

$$X(l_1, l_2) = \sum_{k_2} a''_{l_2 k_2} X_{k_2}(l_1, l_2) \tag{4.34}$$

The kernel of a two-dimensional transformation is called *separable* and *symmetric* if the following is true:

$$A[l_1, l_2, k_1, k_2] = A'[l_1 \ k_1] A'[l_2, k_2] \tag{4.35}$$

If x is a matrix representing a two-dimensional signal and A is the matrix of a symmetric and separable kernel, a two-dimensional transformation can be written as follows:

$$X = A x A^H \tag{4.36}$$

where X is the matrix representing the two-dimensional transform of the matrix x. The inverse transformation is obtained by multiplying by A^{-1} from the left and by $(A^{-1})^H$ from the right, thusly,

$$A^{-1} X (A^{-1})^H = A^{-1} A x A^H (A^{-1})^H = x \tag{4.37}$$

Moreover, if A is unitary and hermitian, then

$$A^H = A = A^{-1} \tag{4.38}$$

which is a very convenient form for practical applications.

4.3 SYNTHESIS OF MATRICES WITH REDUNDANT ELEMENTS

4.3.1 The Kronecker product

Let us consider two matrices A and B of order $m \times n$ and $p \times q$, respectively. The indices m and p represent the number of rows, and n and q denote the number of columns. The *Kronecker product* of A and B, denoted by $A \otimes B$ is a new matrix C of order $mp \times nq$, which is obtained by replacing each element a_{ij} of the matrix A by the following matrix $a_{ij}B$:

$$\begin{pmatrix} a_{ij} b_{11} & a_{ij} b_{12} & \cdots & a_{ij} b_{1q} \\ \vdots & & & \\ a_{ij} b_{p1} & a_{ij} b_{p2} & \cdots & a_{ij} b_{pq} \end{pmatrix} \tag{4.39}$$

Using the partition notation, the matrix C can be expressed as follows:

$$C = A \otimes B = \begin{pmatrix} a_{11} B & a_{12} B & \cdots & a_{1n} B \\ a_{21} B & & & \\ \vdots & & & \\ a_{m1} B & a_{m2} B & \cdots & a_{mn} B \end{pmatrix} \tag{4.40}$$

4.3.2 Properties

From the above definition, we can see that, in general,

$$A \otimes B \neq B \otimes A \tag{4.41}$$

It must also be noted that each element of the matrix C is the product of two elements, a_{ij} and b_{kl}, and that the matrix C can be specified by $mn + pq$ elements, instead of $mnpq$ as in the general case. An illustration of this definition in the particular case $n = m = p = q = 2$ is shown in figure 4.1.

$$A = \begin{pmatrix} a_{11} & a_{12} \\ a_{21} & a_{22} \end{pmatrix} \qquad B = \begin{pmatrix} b_{11} & b_{12} \\ b_{21} & b_{22} \end{pmatrix}$$

$$C = A \otimes B = \begin{pmatrix} a_{11}\,b_{11} & a_{11}\,b_{12} & a_{12}\,b_{11} & a_{12}\,b_{12} \\ a_{11}\,b_{21} & a_{11}\,b_{22} & a_{12}\,b_{21} & a_{12}\,b_{22} \\ a_{21}\,b_{11} & a_{21}\,b_{12} & a_{22}\,b_{11} & a_{22}\,b_{12} \\ a_{21}\,b_{21} & a_{21}\,b_{22} & a_{22}\,b_{21} & a_{22}\,b_{22} \end{pmatrix}$$

Fig. 4.1

4.3.3 Remarks

For the rest of this chapter, we will consider only square $n \times n$ matrices, unless otherwise specified. To simplify the notation, the order of a matrix will be given by only one index.

4.3.4 Matrix generated by the Kronecker product

Let us consider a set of n matrices B_r with $r = 1, \ldots, n$, of order ρ and of element $b_{r,i,j}$. The index r denotes a particular matrix of the set. The indices i and j denote the rows and columns, respectively. The matrices B_r are called *basis matrices*.

Using the Kronecker product and the set of matrices B_r, a new set of square matrices A_n can be constructed as follows:

$$A_n = \bigotimes_{i=0}^{n-1} B_i = B_{n-1} \otimes B_{n-2} \otimes \ldots \otimes B_0 \tag{4.42}$$

Due to relation (4.41), the Kronecker product on the right-hand side must be taken from right to left. The matrices A_n can be generated recursively:

$$A_n = B_{n-1} \otimes A_{n-1} \qquad \text{with } A_1 = B_0 \tag{4.43}$$

The matrix A_n generated is of order ρ^n and all its elements may be specified with $n\rho^2$ elements. We say that the matrix A_n has $n\rho^2$ nonredundant elements.

4.3.5 Comments

It is clear that the use of such a matrix as a transform matrix in a processing system is greatly simplified if the algorithm for the generation is memorized instead of the matrix itself. This advantage becomes more important if the order of the transformation matrix is relatively large (on the order of 1000 or more). A transformation processed with a matrix generated in this manner is called a transformation of basis ρ. The dimension of the corresponding vector space, which is at the same time the dimension of the transformation, is $N = \rho^n$.

4.3.6 Elements of the synthesized matrix

An equation expressing any element $a_{n,\alpha,\beta}$ of the matrix A_n, with the elements $b_{r,i,j}$ of the basis matrix B_r, must be established. To represent the index of integer elements, it is more convenient to use the system basis ρ instead of the usual decimal system. The index α (or β) of any element of A_n can be represented in the basis ρ by n digits $\alpha(k)$ or $\beta(k)$, with $k = 0, ..., n - 1$. Each of these digits can take ρ different values. Thus, we may write α and β as follows:

$$\alpha = \rho^{n-1}\alpha(n-1) + \rho^{n-2}\alpha(n-2) + ... + \rho\alpha(1) + \alpha(0)$$

with

$$\alpha(k) = 0, ..., \rho - 1 \tag{4.44}$$

and

$$k = 0, ..., n - 1$$

Similarly, β can be written as

$$\beta = \rho^{n-1}\beta(n-1) + \rho^{n-2}\beta(n-2) + ... + \rho\beta(1) + \beta(0)$$

with

$$\beta(k) = 0, ..., \rho - 1 \tag{4.45}$$

and

$$k = 0, ..., n - 1$$

Equations (4.45) and (4.46) are representations in the basis ρ of an integer in the interval $[0, \rho^n - 1]$. Thus, we have a one-to-one correspondence between α and β and their representations:

$$\alpha \longleftrightarrow \alpha(n-1)\,\alpha(n-2)\ldots\alpha(1)\,\alpha(0)$$
$$\beta \longleftrightarrow \beta(n-1)\,\beta(n-2)\ldots\beta(1)\,\beta(0) \tag{4.46}$$

and $\alpha\,(i)$, $\beta(i) = 0, \ldots, \rho - 1$.

With n digits α_k or β_k, ρ^n integers can be represented, but, in fact, these integers are in the interval $[0, \rho^n - 1]$ and not in the interval $[1, \rho^n]$. Therefore, it is necessary to shift the indices, $0(1)$ representing the first row (or column) and $\rho^n - 1$ the last row (or column).

Using representations (4.44) and (4.45), the general element $a_{1,\alpha,\beta}$ of the matrix A_1 can be written as follows:

$$a_{1,\alpha,\beta} = \prod_{i=0}^{\rho-1} \prod_{j=0}^{\rho-1} b_{0,i,j}^{\delta[\alpha(0),i]\,\delta[\beta(0),j]} \tag{4.47}$$

with

$$\alpha \longleftrightarrow \alpha(0)$$
$$\beta \longleftrightarrow \beta(0)$$

and $\alpha(0)$, $\beta(0) = 0, \ldots, \rho - 1$.

Equation (4.47) indicates that $a_{1,\alpha,\beta}$ is the product of all the elements of the matrix B_0 raised to the *zero*th power, except $b_{0,\alpha,\beta}$, which is the only term that contributes to the double product.

For the general element $a_{2,\alpha,\beta}$ of A_2 we have

$$a_{2,\alpha,\beta} = \prod_{i=0}^{\rho-1} \prod_{j=0}^{\rho-1} b_{1,i,j}^{\delta[\alpha(1),i]\,\delta[\beta(1),j]} \cdot$$
$$\prod_{k=0}^{\rho-1} \prod_{l=0}^{\rho-1} b_{0,k,l}^{\delta[\alpha(0),k]\,\delta[\beta(0),l]} \tag{4.48}$$

with

$$\alpha \longleftrightarrow \alpha(1)\,\alpha(0)$$
$$\beta \longleftrightarrow \beta(1)\,\beta(0) \quad \text{and} \quad \alpha(i), \beta(i) = 0, \ldots, \rho - 1$$

In (4.48), as in (4.47), for a given pair of values α and β, the exponent determines the correct product. More generally, for a_n the following result is obtained:

$$a_{n,\alpha,\beta} = \prod_{k=0}^{n-1} \prod_{i=0}^{\rho-1} \prod_{j=0}^{\rho-1} b_{k,i,j}^{\delta[\alpha(k),i]\,\delta[\beta(k),j]} \tag{4.49}$$

with

$$\alpha \longleftrightarrow \alpha(n-1)\,\alpha(n-2)\ldots\alpha(1)\,\alpha(0)$$
$$\beta \longleftrightarrow \beta(n-1)\,\beta(n-2)\ldots\beta(1)\,\beta(0) \quad \text{and} \quad \alpha(i), \beta(i) = 0, \ldots, \rho - 1$$

Equation (4.49) enables us to generate any of the ρ^{2n} elements of A_n by memorizing only the $n\rho^2$ elements of the basis matrix B_r.

After computing the products this equation can be written as follows:

$$a_{n,\,\alpha,\beta} = b_{n-1,\,\alpha\,(n-1),\,\beta\,(n-1)} \cdot b_{n-2,\,\alpha\,(n-2),\,\beta\,(n-2)} \cdots$$
$$b_{1,\,\alpha\,(1),\,\beta\,(1)} \cdot b_{0,\,\alpha\,(0),\,\beta\,(0)}$$

with

$$\alpha \longleftrightarrow \alpha\,(n-1)\,\alpha\,(n-2)\dots\alpha\,(1)\,\alpha\,(0) \tag{4.50}$$
$$\beta \longleftrightarrow \beta\,(n-1)\,\beta\,(n-2)\dots\beta\,(1)\,\beta\,(0)$$

4.3.7 Particular case

An interesting form of the general equation (4.49) is obtained in the particular case where all the basis matrices are identical. With only one basis matrix B of elements b_{ij}, the general element $a'_{n,\alpha,\beta}$ of a matrix A'_n, generated according to (4.42), is given by

$$a'_{n,\alpha,\beta} = \prod_{i=0}^{\rho-1} \prod_{j=0}^{\rho-1} b_{ij}^{\sum_{k=0}^{n-1} \delta\,[\alpha\,(k),i]\,\delta\,[\beta\,(k),j]} \tag{4.51}$$

with

$$\alpha \longleftrightarrow \alpha\,(n-1)\,\alpha\,(n-2)\dots\alpha\,(1)\,\alpha\,(0)$$
$$\beta \longleftrightarrow \beta\,(n-1)\,\beta\,(n-2)\dots\beta\,(1)\,\beta\,(0) \quad \text{and} \quad \alpha\,(i),\beta\,(i) = 0,\dots,\rho-1$$

In this case, the matrix A'_n has only ρ^2 nonredundant elements.

4.3.8 Remark

In this section, we deliberately limited ourselves to a set of square basis matrices, all of the same order ρ. Our study can also be generalized with a set of square basis matrices B_r, each one of different order ρ_r. A transformation matrix can be synthesized in a similar way. Such a transformation is called a *mixed radix transformation*. The dimension of the transformation is given by

$$N = \prod_{i=0}^{n-1} \rho_i \tag{4.52}$$

However, the expressions become complicated without providing any particular advantages. In almost all practical applications, the order of the basis matrices will be the same.

4.4 PROPERTIES OF MATRICES WITH REDUNDANT ELEMENTS

4.4.1 Definition: Hermitian matrix

By definition, a matrix A is called *hermitian* if its hermitian transpose is identical to itself. In the case where all elements of A are real numbers, a hermitian matrix is a symmetric matrix.

4.4.2 Theorem

If all the basis matrices B_r are hermitian, the matrices A_n generated by the Kronecker product are also hermitian.

4.4.3 Proof

If the matrices B_r are hermitian, the following equation is verified by definition:

$$b_{r,i,j} = b^*_{r,j,i} \tag{4.53}$$

Substituting (4.53) into (4.49) or (4.50), the following result is obtained:

$$a_{n,\alpha,\beta} = a^*_{n,\beta,\alpha} \tag{4.54}$$

4.4.4 Theorem

If all the basis matrices B_r are unitary matrices, the matrices A_n generated by a Kronecker product are also unitary matrices.

4.4.5 Proof

Using the definition of a unitary matrix, we have

$$B^H_r B_r = I = B_r B^H_r \tag{4.55}$$

This leads to [see (4.25)]:

$$\sum_{j=0}^{\rho-1} b_{r,i,j} b^*_{r,k,j} = \delta_{ik} \tag{4.56}$$

The corresponding expression for the matrix A_n is given by

$$\sum_{\beta=1}^{\rho^{n-1}} a_{n,\alpha,\beta}\, a^*_{n,\gamma,\beta} = \sum_{\beta(n-1)=0}^{\rho-1} \sum_{\beta(n-2)=0}^{\rho-1} \cdots \sum_{\beta(0)=0}^{\rho-1} b_{n-1,\alpha(n-1),\beta(n-1)}$$

$$b^*_{n-1,\gamma(n-1),\beta(n-1)} \cdots b_{0,\alpha(0),\beta(0)}\, b^*_{0,\gamma(0),\beta(0)}$$

$$= \sum_{\beta(n-1)=0}^{\rho-1} b_{n-1,\alpha(n-1),\beta(n-1)}\, b^*_{n-1,\gamma(n-1)\beta(n-1)} \cdots$$

$$\sum_{\beta(0)=0}^{\rho-1} b_{0,\alpha(0),\beta(0)}\, b^*_{0,\gamma(0),\beta(0)} \qquad (4.57)$$

However, using hypothesis (4.56), each of these sums is $\delta[\alpha(k),\,\gamma(k)]$. Thus,

$$\sum_{\beta=0}^{\rho^{n}} a_{n,\alpha,\beta}\, a^*_{n,\gamma,\beta} = \delta_{\alpha,\gamma} \qquad (4.58)$$

The same proof can be given for the columns of matrix A_n:

$$A_n^H A_n = A_n A_n^H = I \qquad (4.59)$$

4.4.6 Definition

An *elementary operation* is defined as a multiplication followed by an addition (real or complex, depending on the factors).

4.4.7 Number of operations required for a transformation

One of the most important properties of matrices with redundant elements is the number of operations required when they are used as transformation matrices in a transformation of type (4.15). Such a transformation generally involves a number $N_c = N^2$ of elementary operations. In fact, to compute a coefficient X_i, N multiplications and additions are required. Since the index i varies from one to N, the number of operations is indeed $N_c = N^2$.

This number is significantly diminished if the matrix A_n generated by a Kronecker product is used as a transformation matrix. To simplify the computations, let us consider the case where $N = \rho^2 = \rho \cdot \rho$. We then have the following equation:

$$\alpha \longleftrightarrow \alpha(1)\,\alpha(0)$$
$$\beta \longleftrightarrow \beta(1)\,\beta(0) \qquad \text{and} \quad \alpha(i),\beta(i) = 0,\,...,\,\rho-1 \qquad (4.60)$$

Using the matrix A_2, the general element of which is given by (4.48), relation (4.15) can be written as a series:

$$X_\beta = \sum_{\alpha=0}^{N-1} a_{2,\alpha,\beta} \, x_\alpha \qquad\qquad \text{with } \beta = 0, ..., N-1 \qquad (4.61)$$

After substituting (4.60) and (4.48) into (4.61), the following result is obtained:

$$X_{\beta(1)\beta(0)} = \sum_{\alpha(1)=0}^{\rho-1} \prod_{i=0}^{\rho-1} \prod_{j=0}^{\rho-1} b_{1,i,j}^{\delta[\alpha(1),i]\,\delta[\beta(1),j]}$$

$$\left[\sum_{\alpha(0)=0}^{\rho-1} \prod_{k=0}^{\rho-1} \prod_{l=0}^{\rho-1} b_{0,k,l}^{\delta[\alpha(0),k]\,\delta[\beta(0),l]} \, x_{\alpha(1),\alpha(0)} \right] \qquad (4.62)$$

$$\text{with } \beta(0), \beta(1) = 0, ..., \rho-1$$

The expression in brackets is a transformation of type (4.61) with ρ elements, which can be represented by $z_{\alpha(1)}[\beta(0)]$, with $\beta(0) = 0, ..., \rho - 1$. Thus, we can write

$$z_{\alpha(1)}[\beta(0)] = \sum_{\alpha(0)=0}^{\rho-1} \prod_{k=0}^{\rho-1} \prod_{l=0}^{\rho-1} b_{0,k,l}^{\delta[\alpha(0),k]\,\delta[\beta(0),l]} \, x_{\alpha(1)\alpha(0)} \qquad (4.63)$$

$$\text{with } \beta(0) = 0, ..., \rho-1$$

and

$$\dot{X}_{\beta(1)\beta(0)} = \sum_{\alpha(1)=0}^{\rho-1} \prod_{i=0}^{\rho-1} \prod_{j=0}^{\rho-1} b_{1,i,j}^{\delta[\alpha(1),i]\,\delta[\beta(1),j]} \, z_{\alpha(1)}[\beta(0)] \qquad (4.64)$$

$$\text{with } \beta(0), \beta(1) = 0, ..., \rho-1$$

Expressions (4.63) and (4.64) represent two transformations of type (4.61), each with ρ elements.

The computation of a coefficient $z_{\alpha(1)}$ $[\beta(0)]$ requires ρ elementary operations ($\alpha_0 = 0, ..., \rho - 1$). For ρ coefficients $z_{\alpha(1)}$ $[\beta(0)]$, ρ^2 operations are required. However, there are ρ different sequences $z_{\alpha(1)}$ $[\beta(0)]$ because $\alpha(1) = 0, ..., \rho - 1$. The number of operation needed to compute expression (4.63), is, thus, $\rho \cdot \rho^2 = \rho^3$.

We note that the same result is valid for (4.64). The total number of operations is, then,

$$N_R = \rho^3 + \rho^3 = \rho^2(\rho + \rho) = 2N\rho \qquad (4.65)$$

It is easy to see that if $N = \rho^n$, the total number of operations will be

$$N_R = \underbrace{\rho^n (\rho + \rho + \ldots + \rho)}_{n \text{ terms}} = n N \rho \tag{4.66}$$

Since $N = \rho^n$, we have $n = \log_\rho N$. This leads to the following expression for N_R:

$$N_R = \rho N \log_\rho N \tag{4.67}$$

4.4.8 Computational efficiency

The ratio:

$$\eta = \frac{N_C}{N_R} = \frac{N}{\rho \log_\rho N} \tag{4.68}$$

is a measurement of the gain obtained with a transformation in which the matrix is generated by a Kronecker product. The maximum efficiency for a given value of N is obtained for $\rho = e = 2.718. \ldots$ However, ρ represents a number of elements in a matrix, and thus must be an integer; therefore, the optimal value of ρ is equal to 3. Nevertheless, for programming simplicity or for practical implementation, the value $\rho = 2$ can be used. This results in only a negligible reduction of the efficiency (about 5.6%). It also implies that the number of samples N of the signal $x(k)$ must be a power of 2. In general, this is not a major drawback. With $N = 2^n$, the efficiency becomes

$$\eta = \frac{N_C}{N_R} = \frac{N}{2 \log_2 N} = \frac{2^n}{2n} \tag{4.69}$$

The graph of the efficiency η, expressed as a function of $n = \log_2 N$, is shown in figure 4.2

4.4.9 Theorem

The matrix A_n of order ρ^n, generated from the basis matrices B_r by a Kronecker product, can be decomposed into a product of n matrices $C_r (r = 1, \ldots, n)$ of order ρ^n, where the element $C_{r,\alpha,\beta}$ is given by

$$C_{r,\alpha,\beta} = b_{r,\alpha(n-1),\beta(0)} \, \delta \, [\beta(n-1), \alpha(n-2)] \ldots \delta \, [\beta(2), \alpha(1)] \, \delta \, [\beta(1), \alpha(0)] \tag{4.70}$$

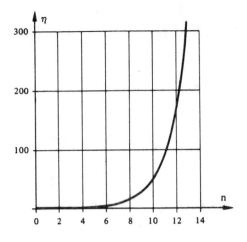

Fig. 4.2

The matrix A_n is, then,

$$A_n = \underset{i=0}{\overset{n-1}{\boxtimes}} B_i = \prod_{i=0}^{n-1} C_i$$

$$= C_{n-1} \cdot C_{n-2} \cdot \ldots \cdot C_1 \cdot C_0 \tag{4.71}$$

For the same reason cited concerning (4.42), the products in (4.71) must be taken from right to left. This theorem is called *Good's theorem* [18].

4.4.10 Proof

For simplicity of writing, only the case $n = 2$ will be considered here. The general element of a matrix M, such that $M = P \cdot Q$, is given by

$$m_{\alpha\beta} = \sum_\gamma p_{\alpha\gamma} q_{\gamma\beta} \tag{4.72}$$

Using equation (4.70), we have

$$p_{\alpha\gamma} = b_{1,\alpha(1),\gamma(0)} \, \delta \, [\gamma(1), \alpha(0)]$$
$$q_{\gamma\beta} = b_{0,\gamma(1),\beta(0)} \, \delta \, [\beta(1), \gamma(0)] \tag{4.73}$$

Replacing these equations in (4.72) leads to

$$m_{\alpha\beta} = \sum_{\gamma(1)=0}^{\rho-1} \sum_{\gamma(0)=0}^{\rho-1} b_{1,\alpha(1),\gamma(0)} \, b_{0,\gamma(1),\beta(0)} \, \delta \, [\gamma(1), \alpha(0)] \, \delta \, [\beta(1), \gamma(0)] \tag{4.74}$$

The sum over $\gamma(0)$ gives only one nonzero term for $\gamma(0) = \beta(1)$, which is $b_{0,\gamma(1),\ \beta(0)}$.

The sum over $\gamma(1)$ also gives only one nonzero term for $\gamma(1) = \alpha(0)$. Thus, the sum is reduced to

$$m_{\alpha\beta} = b_{1\,,\alpha\,(1),\,\beta\,(1)}\, b_{0,\,\alpha\,(0),\,\beta\,(0)} \tag{4.75}$$

This is the same result obtained for $n = 2$ in (4.50)

4.4.11 Comments

By examining (4.70), we can see that in a row of the matrix C_r (α = constant), there can be only ρ nonzero terms. In fact, the $n - 1$ digits $\beta(i)$ used in the Kronecker deltas can form ρ^{n-1} different words. One of these words, corresponding to $\alpha(n-2)\alpha(n-3)\ldots\alpha(1)\alpha(0)$, has a nonzero contribution because it is the only case where all δ_{ij} equal 1. The last digit $\beta(0)$ can take ρ different values while the $n - 1$ others are fixed. This gives ρ nonzero terms per line.

The same relation (4.70) shows that the position of the nonzero elements in the matrix C_r is very structured. Because the matrix C is of order ρ^n, each line can be divided into n adjacent blocks of ρ elements. The nonzero elements of C_r are in the first block of the first row, the second block of the second row, and so on. To illustrate this structure, an example for the case $n = 3$ is shown in figure 4.3. Matrices having such a structure are called *Good matrices.*

$\beta(2)$			0	0	0	0	1	1	1	1
$\beta(1)$			0	0	1	1	0	0	1	1
$\beta(0)$			0	1	0	1	0	1	0	1
$\alpha(2)$	$\alpha(1)$	$\alpha(0)$								
0	0	0	$b_{r,0,0}$	$b_{r,0,1}$	0	0	0	0	0	0
0	0	1	0	0	$b_{r,0,0}$	$b_{r,0,1}$	0	0	0	0
0	1	0	0	0	0	0	$b_{r,0,0}$	$b_{r,0,1}$	0	0
0	1	1	0	0	0	0	0	0	$b_{r,0,0}$	$b_{r,0,1}$
1	0	0	$b_{r,1,0}$	$b_{r,1,1}$	0	0	0	0	0	0
1	0	1	0	0	$b_{r,1,0}$	$b_{r,1,1}$	0	0	0	0
1	1	0	0	0	0	0	$b_{r,1,0}$	$b_{r,1,1}$	0	0
1	1	1	0	0	0	0	0	0	$b_{r,1,0}$	$b_{r,1,1}$

Fig. 4.3

4.4.12 Transformations using Good matrices

Good's theorem is very useful for establishing fast-computing algorithms. These algorithms are generally deduced from *flow charts,* which present an outline of which operations are to be performed. To illustrate the use of Good's theorem for establishing algorithms, let us consider the case of two basis matrices B_0 and B_1 of order 2. Good's theorem allows us to write

$$A_2 = \begin{pmatrix} b_{1,0,0} & b_{1,0,1} \\ b_{1,1,0} & b_{1,1,1} \end{pmatrix} \otimes \begin{pmatrix} b_{0,0,0} & b_{0,0,1} \\ b_{0,1,1} & b_{0,1,1} \end{pmatrix}$$

$$= \begin{pmatrix} b_{1,0,0} & b_{1,0,1} & 0 & 0 \\ 0 & 0 & b_{1,0,0} & b_{1,0,1} \\ b_{1,1,0} & b_{1,1,1} & 0 & 0 \\ 0 & 0 & b_{1,1,0} & b_{1,1,1} \end{pmatrix} \cdot \begin{pmatrix} b_{0,0,0} & b_{0,0,1} & 0 & 0 \\ 0 & 0 & b_{0,0,0} & b_{0,0,1} \\ b_{0,1,0} & b_{0,1,1} & 0 & 0 \\ 0 & 0 & b_{0,1,0} & b_{0,1,1} \end{pmatrix} \tag{4.76}$$

A transformation of type (4.15), using A_2 as a transformation matrix, can be expressed by

$$\begin{pmatrix} X(0) \\ X(1) \\ X(2) \\ X(3) \end{pmatrix} = \begin{pmatrix} b_{1,0,0} & b_{1,0,1} & 0 & 0 \\ 0 & 0 & b_{1,0,0} & b_{1,0,1} \\ b_{1,1,0} & b_{1,1,1} & 0 & 0 \\ 0 & 0 & b_{1,1,0} & b_{1,1,1} \end{pmatrix} \cdot \begin{pmatrix} b_{0,0,0} & b_{0,0,1} & 0 & 0 \\ 0 & 0 & b_{0,0,0} & b_{0,0,1} \\ b_{0,1,0} & b_{0,1,1} & 0 & 0 \\ 0 & 0 & b_{0,1,0} & b_{0,1,1} \end{pmatrix} \begin{pmatrix} x(0) \\ x(1) \\ x(2) \\ x(3) \end{pmatrix}$$

$$\tag{4.77}$$

In such a transformation, there are $n = 2$ matrix vector products denoted by $i = 1, ..., n$. These products are taken according to the rules for the matrix product, from right to left. The result of the matrix vector product for the ith step will be denoted by X'_{ij}, with $j = 0, ..., N - 1$. With this notation, we have

$$\begin{aligned} X'_{0j} &= x(j) \\ X'_{nj} &= X(j) \end{aligned} \qquad \text{with} \qquad j = 0, ..., N-1 \tag{4.78}$$

4.4.13 Flow chart

The flow chart, generally represented from left to right, is obtained by outlining the computations of the different steps in (4.77). Figure 4.4 shows the flow chart corresponding to transformation (4.77).
In this graph, the first column of *nodes* on the left represents the sample of the signal. The second column represents the result of the first matrix vector product:

$$\begin{pmatrix} X'_{10} \\ X'_{11} \\ X'_{12} \\ X'_{13} \end{pmatrix} = \begin{pmatrix} b_{0,0,0} & b_{0,0,1} & 0 & 0 \\ 0 & 0 & b_{0,0,0} & b_{0,0,1} \\ b_{0,1,0} & b_{0,1,1} & 0 & 0 \\ 0 & 0 & b_{0,1,0} & b_{0,1,1} \end{pmatrix} \begin{pmatrix} X'_{00} \\ X'_{01} \\ X'_{02} \\ X'_{03} \end{pmatrix} \tag{4.79}$$

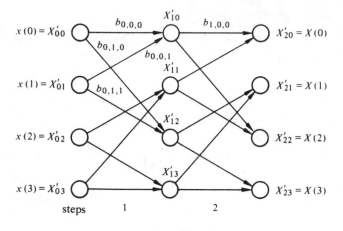

Fig. 4.4

The branches joining the nodes of step 1 are inscribed with the multiplicative coefficient. The values associated with the last column are obtained by the second matrix product:

$$
\begin{pmatrix} X'_{20} \\ X'_{21} \\ X'_{22} \\ X'_{23} \end{pmatrix} = \begin{pmatrix} b_{1,0,0} & b_{1,0,1} & 0 & 0 \\ 0 & 0 & b_{1,0,0} & b_{1,0,1} \\ b_{1,1,0} & b_{1,1,1} & 0 & 0 \\ 0 & 0 & b_{1,1,0} & b_{1,1,1} \end{pmatrix} \begin{pmatrix} X'_{10} \\ X'_{11} \\ X'_{12} \\ X'_{13} \end{pmatrix}
\tag{4.80}
$$

4.4.14 Comments

The flow chart represented in figure 4.4 is deduced directly by applying Good's theorem. We can see that the geometry is identical for each step. This characteristic may be important for some practical implementations. The same computation on a substructure can be repeated, step by step, by changing only the multiplicative coefficients. Moreover, the structure of the branches shows that the access to the data and the memorization of the results in each step can be processed step by step. During the computation indicated by this chart, two rows of memory register of length N can be used, one to store the results, the other to store the data used in the computation of the step. At the end of each step, the roles of these two rows must be reversed. The results of one step become the data of the next step. In this case, the use of a sequential index memory register is advantageous.

4.4.15 Butterfly operation and in-place computation

The important elements in a flow chart are the branches joining the nodes and the multiplicative coefficients. For any ordering of the nodes, the result of the computation will be the same if the branches and the coefficients between two given nodes are the same. Consequently, the flow chart of figure 4.4 can be modified in order to save memory and to speed up the computations. For example, if the second and third rows of the matrix are permuted in (4.79), the intermediate results are obtained in the order X'_{10}, X'_{12}, X'_{11}, and X'_{13}. To account for this change in the order, the second and third columns of the matrix must be permuted in (4.80). The flow chart so modified is shown in figure 4.5.

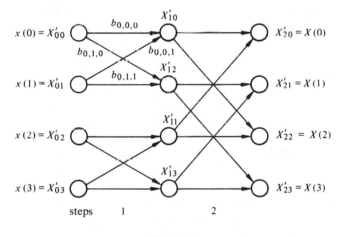

Fig. 4.5

With regard to the speed of the algorithm and the efficient use of memory, the consequences of such a change are important. In the algorithm corresponding to the modified flow chart, the results of one step can be computed by pairs. In this chart, there is a basic structure which is shown in figure 4.6. This structure, due to its shape, is known as a *butterfly operation*. The computations involved in the butterfly operation either can be done simultaneously, if two arithmetic units are available, or in two steps by storing two data in two auxiliary memory registers. The advantage of the butterfly operation is in its efficient use of memory. The data of one step taken by pairs are used only once for one butterfly operation. Consequently, the two corresponding results can be stored in the memory registers used for the data. A computation algorithm having such a characteristic is called *in-place computation*.

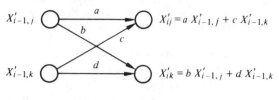

$$X'_{i-1,j} \quad \bigcirc \xrightarrow{\quad a \quad} \bigcirc \quad X'_{ij} = a\, X'_{i-1,j} + c\, X'_{i-1,k}$$

$$X'_{i-1,k} \quad \bigcirc \xrightarrow{\quad d \quad} \bigcirc \quad X'_{ik} = b\, X'_{i-1,j} + d\, X'_{i-1,k}$$

Fig. 4.6

Thus, the total number of registers necessary is reduced to N. However, the advantage of sequential access and storage is lost. Even if a simple algorithm can be written to modify the indexing of a sequential index memory register, a random access memory (RAM) may be convenient. In the case of the flow chart of figure 4.5, the indices of two data or two results of a butterfly operation for step i differ by 2^{i-1}, which is relatively simple.

4.4.16 Remarks

In general, flow charts must be modified to reduce the required memory (for example, in-place computation) and to group the values associated with the nodes — starting samples as well as intermediate or final results — to facilitate and to speed up the addressing and to provide for fast computation. If such modifications alter the starting samples or the final result, a permutation of the values must be carried out to maintain the order in which the data are taken or the results are presented. More specific modifications depend on the particular transformation used and will be treated later.

Good's theorem enables us to verify the number of operations required by a transformation (sect. 4.4.7). The number of elementary operations required to multiply a vector x of $N = \rho^n$ components by a matrix C_r is ρN. For n products of this type:

$$N_R = \rho N n = \rho N \log_\rho N \qquad (4.81)$$

operations are required. This equation is identical to (4.60), established differently in section 4.4.7.

4.4.17 Generalization

The nonzero elements of a matrix C_r, which intervene in Good's theorem, come from only one basis matrix B_r. Another set of matrices D_r of order ρ^n, having the same structure as the matrix C_r, but with nonzero elements coming from ρ^{n-1} different basis matrices can be constructed. For example, for $\rho = 2$ and $n = 3$:

$$
D_r = \begin{pmatrix}
b_{1,0,0} & b_{1,0,1} & 0 & 0 & 0 & 0 & 0 & 0 \\
0 & 0 & b_{2,0,0} & b_{2,0,1} & 0 & 0 & 0 & 0 \\
0 & 0 & 0 & 0 & b_{3,0,0} & b_{3,0,1} & 0 & 0 \\
0 & 0 & 0 & 0 & 0 & 0 & b_{4,0,0} & b_{4,0,1} \\
b_{1,1,0} & b_{1,1,1} & 0 & 0 & 0 & 0 & 0 & 0 \\
0 & 0 & b_{2,1,0} & b_{2,1,1} & 0 & 0 & 0 & 0 \\
0 & 0 & 0 & 0 & b_{3,1,0} & b_{3,1,1} & 0 & 0 \\
0 & 0 & 0 & 0 & 0 & 0 & b_{4,1,0} & b_{4,1,1}
\end{pmatrix}
\tag{4.82}
$$

A matrix A_n can be constructed by successive products of the D_r:

$$
A_n = \prod_{i=1}^{n} D_i = D_n D_{n-1} \dots D_2 D_1
\tag{4.83}
$$

where the matrices D_i are given by (4.82). If a matrix A_n generated in this manner is used as a transformation matrix, the number of elementary operations required is always the same. If the basis matrices are unitary, we can show that A_n is also unitary. However, if the basis matrices are hermitian, the matrix A_n is not hermitian. The usefulness of the transformation matrices generated by (4.83) has still not been shown [19].

4.5 PARTICULAR TRANSFORMATIONS

4.5.1 Introduction

In this section, we will study some particular cases of the general synthesis presented earlier. The emphasis is placed on the case $\rho = 2$, which leads to simple programs and practical implementations. However, the case $\rho \geqslant 3$ is also covered to illustrate the generalized Walsh transform. The most important fast transformation is undoubtedly the Fourier transform. Because of its importance, the Fourier transform will be examined in detail in section 4.6. In particular, we will show how the general synthesis can include the fast Fourier transform algorithm.

4.5.2 Hadamard transform

The matrix of the *Hadamard transform* is a square matrix, in which the elements are $+1$ or -1 and where the rows (or columns) are mutually orthogonal. From this definition, it is clear that interchanging two rows or columns does not alter such a matrix.

The problem of the existence of the Hadamard matrix for any order N has not yet been solved. However, two theorems may be cited for constructing such a matrix for some values of N.

4.5.3 Theorem

If $N = 0 \bmod 4$, then a Hadamard matrix of order N exists.

4.5.4 Proof

Let α_i, β_i, and γ_i be the elements of three different rows of a Hadamard matrix of order $N(i = 1, \ldots, N)$. Let us compute the sum:

$$\sum_{i=1}^{N} (\alpha_i + \beta_i)(\alpha_i + \gamma_i) = \sum_{i=1}^{N} (\alpha_i^2 + \alpha_i \gamma_i + \alpha_i \beta_i + \beta_i \gamma_i)$$

$$= \sum_{i=1}^{N} \alpha_i^2 + \sum_{i=1}^{N} \alpha_i \gamma_i + \sum_{i=1}^{N} \alpha_i \beta_i + \sum_{i=1}^{N} \beta_i \gamma_i$$

The sums over the mixed products are each zero, because of the orthogonality, and the first sum is N. Thus,

$$\sum_{i=1}^{N} (\alpha_i + \beta_i)(\alpha_i + \gamma_i) = N \qquad (4.84)$$

However, this sum is a multiple of 4, because the elements α_i, β_i, and γ_i are either $+1$ or -1. This gives

$$N = 0 \bmod 4 \qquad (4.85)$$

4.5.5 Theorem

If H_a and H_b are two Hadamard matrices, then $H = H_a \otimes H_b$, obtained by the Kronecker product, is also a Hadamard matrix.

4.5.6 Proof

The proof follows from theorems 4.4.2 and 4.4.4.

4.5.7 Normalization

The Hadamard matrices defined in section 4.5.2 are orthogonal, but not orthonormal. They can be made into orthonormal matrices by dividing each element by the norm (the length) of the row or column vectors. Since all elements are either $+1$ or -1, the normalization factor for a Hadamard matrix of order N is $\sqrt{N}$.

4.5.8 General element of a Hadamard matrix

Using theorem 4.5.5, we can see that if the basis matrix B is considered as given by

$$B - \begin{pmatrix} \dfrac{1}{\sqrt{2}} & \dfrac{1}{\sqrt{2}} \\[2mm] \dfrac{1}{\sqrt{2}} & -\dfrac{1}{\sqrt{2}} \end{pmatrix} = \dfrac{1}{\sqrt{2}} \begin{pmatrix} 1 & 1 \\ 1 & -1 \end{pmatrix} \tag{4.86}$$

then, all elements of the Hadamard matrix of order $N = 2^n$ can be generated by using (4.51).

The general element $h_{n,\alpha,\beta}$ of a Hadamard matrix of order $N = 2^n$ is thus given by

$$h_{n,\alpha,\beta} = \prod_{i=0}^{1} \prod_{j=0}^{1} \left(-\frac{1}{\sqrt{2}} \right)^{ij \sum\limits_{k=0}^{n-1} \delta[\alpha(k),i]\,\delta[\beta(k),j]} \tag{4.87}$$

with

$$\alpha \longleftrightarrow \alpha(n-1)\ldots\alpha(1)\,\alpha(0)$$
$$\beta \longleftrightarrow \beta(n-1)\ldots\beta(1)\,\beta(0) \quad \text{and} \quad \alpha(i), \beta(i) = 0, 1$$

Computing the products on i and j, and using the fact that

$$\delta[\alpha(k), 0] = \overline{\alpha(k)} \quad \text{et} \quad \delta[\alpha(k), 1] = \alpha(k) \tag{4.88}$$

where $\overline{\alpha(k)}$ is the complement of $\alpha(k)$ (see D. Mange, vol. V, sect. 1.2.2), then the following result is obtained:

$$h_{n,\alpha,\beta} = \left(\frac{1}{\sqrt{2}} \right)^{\sum\limits_{k=0}^{n-1} [\overline{\alpha(k)}\,\overline{\beta(k)}] + \alpha(k)\beta(k)] + \sum\limits_{k=0}^{n-1} [\overline{\alpha(k)}\,\beta(k) + \alpha(k)\overline{\beta(k)}]}$$
$$(-1)^{\sum\limits_{k=0}^{n-1} \alpha(k)\,\beta(k)} \tag{4.89}$$

The expression $\overline{\alpha(k)}\beta(k) + \alpha(k)\overline{\beta(k)}$ is the logic operation exclusive-or (XOR) of the logic variables $\alpha(k)$ and $\beta(k)$ (vol. V, sect. 1.6.1). It is denoted by $\alpha(k) \oplus \beta(k)$. The expression $\overline{\alpha}(k)\overline{\beta}(k) + \alpha(k)\beta(k)$ is its complement $\overline{\alpha(k) \oplus \beta(k)}$. The sum:

$$\sum_{k=0}^{n-1} (\alpha_k \oplus \beta_k) \tag{4.90}$$

determines the number of different bits in two binary words of n bits by comparing them bit by bit. The complementary sum determines the number of identical bits in the same words. The length n of the logic words is obtained by adding these two numbers.

Expression (4.89) then becomes

$$h_{n,\alpha,\beta} = \left(\frac{1}{\sqrt{2}}\right)^n (-1)^{\sum_{k=0}^{n-1} \alpha(k)\beta(k)} \tag{4.91}$$

with

$$\alpha \longleftrightarrow \alpha(n-1)\ldots\alpha(1)\,\alpha(0)$$
$$\beta \longleftrightarrow \beta(n-1)\ldots\beta(1)\,\beta(0) \quad \text{and} \quad \alpha(k), \beta(k) = 0, 1$$

The orthornormal Hadamard transform can thus be written as

$$X(\beta) = \left(\frac{1}{2}\right)^{n/2} \sum_{\alpha=0}^{N-1} (-1)^{\sum_{k=0}^{n-1} \alpha(k)\beta(k)} x(\alpha) \tag{4.92}$$

with

$$\alpha \longleftrightarrow \alpha(n-1)\ldots\alpha(1)\,\alpha(0)$$
$$\beta \longleftrightarrow \beta(n-1)\ldots\beta(1)\,\beta(0) \quad \text{and} \quad \alpha(k), \beta(k) = 0, 1$$

4.5.9 Decomposition of the Hadamard matrix

Considering that $(-1)^k$, where k is an integer, is ± 1, there are no multiplications involved in the Hadamard transform. The flow chart of the fast computation algorithm can be obtained by using Good's theorem. In the particular case of the basis matrix (4.86), this theorem is expressed as

$$\overset{n}{\underset{i=1}{\boxtimes}}\begin{pmatrix} 1 & 1 \\ 1 & -1 \end{pmatrix} = \begin{pmatrix} 1 & 1 & 0 & 0 & 0 & 0 & \dots & 0 & 0 \\ 0 & 0 & 1 & 1 & 0 & 0 & \dots & 0 & 0 \\ & \cdot & & & & & & & \\ & \cdot & & & & & & & \\ 0 & 0 & 0 & 0 & 0 & 0 & \dots & 1 & 1 \\ 1 & -1 & 0 & 0 & 0 & -0 & \dots & 0 & 0 \\ 0 & 0 & 1 & -1 & 0 & 0 & \dots & 0 & 0 \\ & \cdot & & & & & & & \\ & \cdot & & & & & & & \\ 0 & 0 & 0 & 0 & 0 & 0 & \dots & 1 & -1 \end{pmatrix}^n \qquad (4.93)$$

To simplify the representation of the flow chart, only the case $n = 3$ will be considered here, without any loss of generality. We will assume that, before the transformation, the signal is multiplied by the factor $(1/\sqrt{2})^3$ to guarantee orthogonality. Equation (4.93) then becomes

$$H_8 = \overset{3}{\underset{i=1}{\boxtimes}}\begin{pmatrix} 1 & 1 \\ 1 & -1 \end{pmatrix} = \begin{pmatrix} 1 & 1 & 0 & 0 & 0 & 0 & 0 & 0 \\ 0 & 0 & 1 & 1 & 0 & 0 & 0 & 0 \\ 0 & 0 & 0 & 0 & 1 & 1 & 0 & 0 \\ 0 & 0 & 0 & 0 & 0 & 0 & 1 & 1 \\ 1 & -1 & 0 & 0 & 0 & 0 & 0 & 0 \\ 0 & 0 & 1 & -1 & 0 & 0 & 0 & 0 \\ 0 & 0 & 0 & 0 & 1 & -1 & 0 & 0 \\ 0 & 0 & 0 & 0 & 0 & 0 & 1 & -1 \end{pmatrix}^3 \qquad (4.94)$$

4.5.10 Flow chart with constant geometry

The corresponding flow chart can easily be established. It is shown in figure 4.7. Since no multiplication is involved, no coefficients are given in this figure. The dotted lines indicate a subtraction and the continuous lines signify an addition. We may note immediately that the steps have the same geometrical structure, and access to the data and storage of the results are sequential. However, the corresponding algorithm cannot be processed in place. The number memory cells required is thus $2N = 16$.

4.5.11 Flow chart with butterfly structure

To decrease the number of memory cells required, a butterfly structure must be used in each step. This can be accomplished by changing the order of the rows in the first Good matrix from $0, 1, 2, \dots, 7$ to $0, 4, 1, 5, 2, 6, 3, 7$.

The results of the first step, therefore, will be in the order

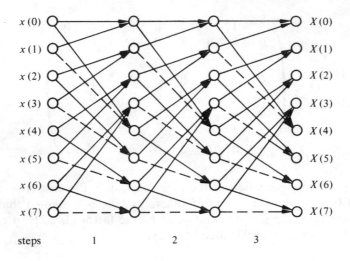

steps   1   2   3

Fig. 4.7

X'_{10}, X'_{14}, X'_{11}, X'_{15}, X'_{12}, X'_{16}, X'_{13}, and X'_{17}. To take this change into account, the columns of the second matrix must be modified in a similar way. The second matrix becomes

$$
\begin{pmatrix}
1 & 0 & 1 & 0 & 0 & 0 & 0 & 0 \\
0 & 0 & 0 & 0 & 1 & 0 & 1 & 0 \\
0 & 1 & 0 & 1 & 0 & 0 & 0 & 0 \\
0 & 0 & 0 & 0 & 0 & 1 & 0 & 1 \\
1 & 0 & -1 & 0 & 0 & 0 & 0 & 0 \\
0 & 0 & 0 & 0 & 1 & 0 & -1 & 0 \\
0 & 1 & 0 & -1 & 0 & 0 & 0 & 0 \\
0 & 0 & 0 & 0 & 0 & 1 & 0 & -1
\end{pmatrix}
\begin{matrix}
\text{rows} \\
0 \\
1 \\
2 \\
3 \\
4 \\
5 \\
6 \\
7
\end{matrix}
\qquad (4.95)
$$

To use as butterfly structure in the second step, the order of the rows of matrix (4.95) must be changed to 0, 2, 4, 6, 1, 3, 5, 7. The results of the second step are then in the order X'_{20}, X'_{22}, X'_{24}, X'_{26}, X'_{21}, X'_{23}, X'_{25}, and X'_{27}. This change must also be performed for the columns of the third matrix, which becomes

$$
\begin{pmatrix}
1 & 0 & 0 & 0 & 1 & 0 & 0 & 0 \\
0 & 1 & 0 & 0 & 0 & 1 & 0 & 0 \\
0 & 0 & 1 & 0 & 0 & 0 & 1 & 0 \\
0 & 0 & 0 & 1 & 0 & 0 & 0 & 1 \\
1 & 0 & 0 & 0 & -1 & 0 & 0 & 0 \\
0 & 1 & 0 & 0 & 0 & -1 & 0 & 0 \\
0 & 0 & 1 & 0 & 0 & 0 & -1 & 0 \\
0 & 0 & 0 & 1 & 0 & 0 & 0 & -1
\end{pmatrix}
\qquad (4.96)
$$

The modified flow chart is shown in figure 4.8. In this chart, the value assigned to a node of step j is obtained by computing the sum or the difference of two results from the previous step $j - 1$, whose addresses differ by 2^{j-1}. This is the repetition of the butterfly structure.

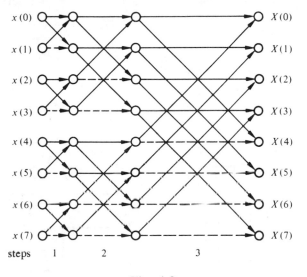

Fig. 4.8

4.5.12 Order of the coefficients

The Hadamard matrix H_8 of order $N = 8$ is derived by computing the products in (4.93). The following result is obtained:

$$
H_8 = \begin{pmatrix}
1 & 1 & 1 & 1 & 1 & 1 & 1 & 1 \\
1 & -1 & 1 & -1 & 1 & -1 & 1 & -1 \\
1 & 1 & -1 & -1 & 1 & 1 & -1 & -1 \\
1 & -1 & -1 & 1 & 1 & -1 & -1 & 1 \\
1 & 1 & 1 & 1 & -1 & -1 & -1 & -1 \\
1 & -1 & 1 & -1 & -1 & 1 & -1 & 1 \\
1 & 1 & -1 & -1 & -1 & -1 & 1 & 1 \\
1 & -1 & -1 & 1 & -1 & 1 & 1 & -1
\end{pmatrix}
\quad
\begin{array}{cc}
\text{row index} & \text{sign changes} \\
0 & 0 \\
1 & 7 \\
2 & 3 \\
3 & 4 \\
4 & 1 \\
5 & 6 \\
6 & 2 \\
7 & 5
\end{array}
\qquad (4.97)
$$

The rows of this matrix, which are obtained by the Kronecker product, are said to be in *natural order*. The coefficients of the transform are also in natural order. (4.97), the number of sign changes for each row is also indicated.

In the literature, the name of Walsh is often associated with the Hadamard transform, since the Walsh functions are the analog versions of the rows of the Hadamard matrix. The Walsh functions can be defined in several ways (vol. V, chap. 3). The first eight Walsh functions are shown in figure 4.9. One possible definition, frequently used because of its practicality, is that derived from the products of Radamacher functions (Fig. 4.9).

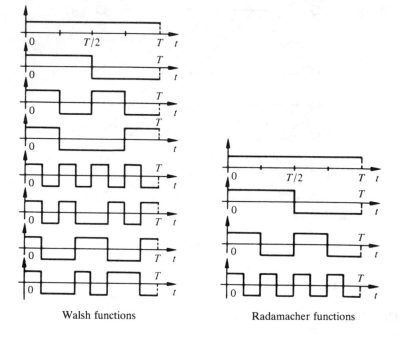

Walsh functions   Radamacher functions

Fig. 4.9

Moreover, by analogy with the frequency of the sine and cosine functions, the number of sign changes in the defined interval of the Walsh functions (or of the rows of the Hadamard matrix) is called the *sequence*. Consequently, it is better to put the Walsh functions, or the rows of the Hadamard matrix, in ascending sequential order. The Hadamard transform, using a matrix modified for this purpose, produces the coefficients of the transform in ascending sequential order. Moreover, it must be noted that, in a transform of the form (4.15), the permutation of the rows of the transformation matrix is equivalent to the identical permutation of the transform coefficients, and the permutation of the columns is equivalent to the identical permutation of the samples of the signal. Consequently, to maintain the advantages of an efficient computation algorithm such as the one shown in figure 4.8, the order of the coefficients or samples may be modified.

4.5.13 Ordering of the coefficients

When the Walsh functions are defined by using a product of Radamacher functions, the first eight indices for the rows of the corresponding Hadamard matrix are in the order:

$$0, 4, 2, 6, 1, 5, 3, 7 \tag{4.98}$$

rather than natural order (4.97). Consequently, the coefficients of the transform are in this order as well. This order is called the *bit-reversed order*. This terminology can be explained as follows: the index β, used to number the coefficients of the transform, can be represented in binary code ($\rho = 2$) by three bits $\beta_2\beta_1\beta_0$ [see (4.45)]. The sequence (4.98) and the natural sequence, as well as their binary representations, are given in table 4.10.

Table 4.10 Natural order and bit-reversed order.

Natural order				Bit-reversed order			
β	β_2	β_1	β_0	β	β_2	β_1	β_0
0	0	0	0	0	0	0	0
1	0	0	1	4	1	0	0
2	0	1	0	2	0	1	0
3	0	1	1	6	1	1	0
4	1	0	0	1	0	0	1
5	1	0	1	5	1	0	1
6	1	1	0	3	0	1	1
7	1	1	1	7	1	1	1

We can see from this table that the coefficient of index $\beta_2\beta_1\beta_0$ in natural order is located at the position of index $\beta_0\beta_1\beta_2$ in bit-reversed order. Thus, to move from natural order to order (4.98), we must reverse the order of the bits in the binary representation of index β. The bit-reversed order also plays an important role in the fast Fourier transform, as we shall see in the next section. The passage from natural order to bit-reversed order is shown in figure 4.11 for the particular case of $N = 8$. The generalization for other values of $N = 2^n$ is straightforward.

4.5.14 Remarks

From figure 4.11, we note that the permutation can be carried out in place. This is performed by using two indices i and j; the first, i, for the natural order, and the second, j, for the bit-reversed order. According to the flow chart, there

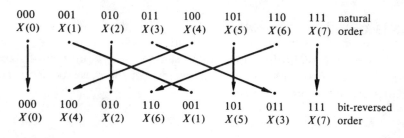

Fig. 4.11

is no change for $i = j$, but for $i \neq j$, $X(i)$ and $X(j)$ must be interchanged. However, this permutation must be carried out only once. This can be guaranteed by comparing i and j, and performing the permutation only when $j > i$. The instructions of a Fortran program for computing these permutations are shown in figure 4.12.

```
C         POUR N=2**M
          J=1
          N1=N-1
          N2=N/2
          DO 1 I=1,N1
          IF(J.LE.I) GO TO 2
          Z=X(J)
          X(J)=X(I)
          X(I)=Z
2         K=N2
3         IF(J.LE.K) GO TO 1
          J=J-K
          K=K/2
1         J=J+K
```

Fig. 4.12

The passage from bit-reversed order to ascending sequential order can be processed by using the Gray code. The Walsh functions and the Hadamard matrix are elements of the *dyadic group*. This is the group of binary vectors governed by addition modulo 2 bit by bit. The space of these vectors, called *dyadic space*, has many dimensions, but only two points on each axis. These points correspond to the binary values 0 and 1. The Gray code, which is a particular case of reflected codes, has the interesting property of only one bit change between one code word and its closest neighbors plus or minus one unit. Integers can be ordered by their arithmetic values on the real axis, such that the length of the path joining them is minimized. Therefore, the Gray code orders the points in the dyadic space according to minimum-length paths [20]. In permutation using the Gray code, the coefficient placed at a given address in the binary code is exchanged with the coefficient at the address having the same code word in the Gray code.

Moreover, using the symmetry of the Hadamard matrix, these permutations can be carried out on the transformation coefficients, on the samples, or one on the coefficients and another on the samples. The complete flow chart of the fast Hadamard transform producing the coefficients in ascending sequential order is shown in figure 4.13. This figure indicates that the number of additions and substractions in the Hadamard transform is $N \log_2 N$, where $N = 2^n$ is the order of the transformation, because no multiplicative coefficients are involved in the steps.

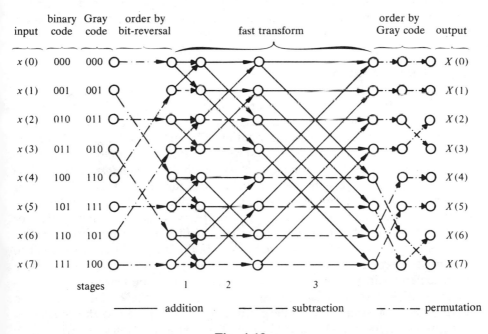

Fig. **4.13**

The two-dimensional Hadamard transform is also used, particularly in image processing and pattern recognition. It can easily be computed by using the fast one-dimensional algorithm described above and the formalism given by (4.36), (4.37) and (4.38).

One last remark is necessary about the correspondence between the Walsh and Hadamard transform. Since a fast computation algorithm is available for the Hadamard transform, and there is a correspondence between the Walsh functions and the rows of the Hadamard matrix, one might be tempted to compute the integral Walsh transform using the Hadamard transform. This is dangerous and can lead to results with large errors if the restrictions of the correspondence between these two transforms are not taken into account. The digital computation of an integral can be processed approximately by decomposing the sought surface into neighboring rectangles (or trapezoids) of width as small as possible to obtain a precise result. The number of rectangles in the integration interval gives the order of the Hadamard transform to be used. To use the Nth-order Hadamard transform to calculate the Nth Walsh coefficient will lead to a very coarse approximation. The Nth coefficient of the Hadamard transformation is meaningful. It is the Nth Hadamard coefficient in a discrete transformation. It must not be identified with the corresponding coefficient of the Walsh transform. In general, if a Hadamard transform of order 10 to 20 times greater than the order of the desired Walsh coefficient is used, then a result can be obtained with acceptable precision.

4.5.15 R transformation

The R transformation [21] is very closely tied to the Hadamard transformation. In a Hadamard transformation, if we take the absolute value after each subtraction, we obtain the R transformation. This indicates the great difference between these two transformations despite their close relationship. The R transformation is essentially nonlinear and *has no inverse transformation*. It is a transformation that permits the extraction of parameters. A very interesting property of this transformation is that the coefficients of the transform are invariant to the circular shift of data. This property is even more interesting for the two-dimensional R transformation. The coefficients are invariant with respect to a translation of the analyzed object in the image plane. If we want to determine whether an object with a certain shape exists in an image where displacements can be made only by translations, the coefficients of the two-dimensional R transform can be used to characterize the desired object. To achieve this, the transform of the given image is compared to the previously computed transform of the desired form. This transformation is invariant only for circular translations and not for rotations or linear translations. Nonetheless, it is a highly useful transformation because it can be computed quickly and the operations involved are simple. The algorithm for the fast computation of the R transformation will not be repeated here, since there is a simple relationship between this and the Hadamard transformation. It is sufficient to consider the algorithm for the computation of the Hadamard transformation and to modify it by introducing the absolute values after each subtraction.

4.5.16 Generalized Walsh transformation

The Walsh functions, which form a complete set of orthogonal functions, are real functions that take only two possible values. They have been generalized on the set of the complex numbers to form a larger set of functions [22]. In the discrete case, the basis matrix B_W of order ρ, which generates, by the Kronecker product, the matrix of the generalized Walsh transformation as follows:

$$B_W = \begin{pmatrix} W^0 & W^0 & W^0 & \dots & W^0 \\ W^0 & W^1 & W^2 & \dots & W^{\rho-1} \\ W^0 & W^2 & & & \\ \vdots & & & & \\ W^0 & W^{\rho-1} & & \dots & W^{(\rho-1)^2} \end{pmatrix} \tag{4.99}$$

where

$$W = \exp\left(j\frac{2\pi}{\rho}\right) \quad \text{and} \quad j = \sqrt{-1}$$

Some redundancies in this matrix can be reduced by noting that

$$W^{nk} = W^{nk \bmod \rho} \tag{4.100}$$

where n and k are integers.

To establish the flow chart of the fast computation algorithm, Good's theorem may again be applied. Let us consider, for simplicity, the case $\rho = 3$ and $N = \rho^2 = 3^2 = 9$. Using (4.99), the corresponding basis matrix is

$$B_W = \begin{pmatrix} W^0 & W^0 & W^0 \\ W^0 & W^1 & W^2 \\ W^0 & W^2 & W^1 \end{pmatrix} \tag{4.101}$$

We can easily verify that this matrix is hermitian and unitary. The transformation matrix, by virtue of Good's theorem, is the product of two identical matrices of order 9 with three nonzero elements in each row. As in the general case (4.77), the flow chart deduced directly from Good's theorem does not allow for computation in place. The rows of the first Good matrix can be permuted in this case to group the data and the results three by three. The same permutation must also be performed on the columns of the second Good matrix. The flow chart obtained is shown in figure 4.14. To simplify the representation, the multiplicative coefficients are not indicated.

It can be noted that in the case where $\rho = 2$, the generalized Walsh transformation is identical to the Hadamard transformation. In fact, for $\rho = 2$, the matrix B_W becomes

$$B_W = \begin{pmatrix} W^0 & W^0 \\ W^0 & W^1 \end{pmatrix} \tag{4.102}$$

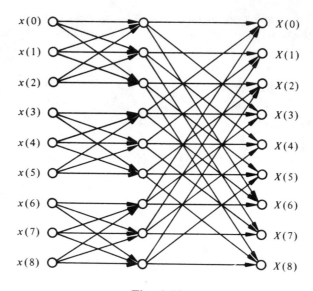

Fig. 4.14

However, $W^0 = 1$ and $W^1 = -1$, which gives us

$$B_W = \begin{pmatrix} 1 & 1 \\ 1 & -1 \end{pmatrix} \tag{4.103}$$

which is a Hadamard matrix.

Moreover, in the case where ρ is greater than 2, the basis matrix B_W given by (4.99) is also the matrix for the discrete Fourier transform of order ρ. However, the matrices generated by the Kronecker product from B_W are no longer the matrices of the discrete Fourier transform. This is examined in section 4.6.

4.5.17 Discrete Haar transformation

The matrix for the discrete Haar transformation can be obtained by a periodic sampling of the Haar functions. However, for the same reasons as those given at the end of section 4.5.14, the discrete Haar transformation must not be identified with the integral Haar transformation. Figure 4.15 shows the first eight Haar functions.

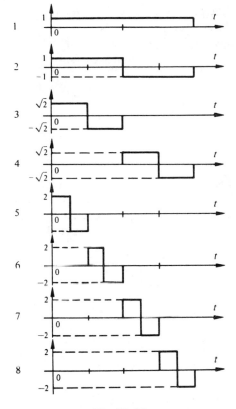

Fig. 4.15

The matrix for the transformation of order $N = 8$ is given by

$$H_a = \begin{pmatrix} 1 & 1 & 1 & 1 & 1 & 1 & 1 & 1 \\ 1 & 1 & 1 & 1 & -1 & -1 & -1 & -1 \\ \sqrt{2} & \sqrt{2} & -\sqrt{2} & -\sqrt{2} & 0 & 0 & 0 & 0 \\ 0 & 0 & 0 & 0 & \sqrt{2} & \sqrt{2} & -\sqrt{2} & -\sqrt{2} \\ 2 & -2 & 0 & 0 & 0 & 0 & 0 & 0 \\ 0 & 0 & 2 & -2 & 0 & 0 & 0 & 0 \\ 0 & 0 & 0 & 0 & 2 & -2 & 0 & 0 \\ 0 & 0 & 0 & 0 & 0 & 0 & 2 & -2 \end{pmatrix} \qquad (4.104)$$

Such a matrix cannot be synthesized by Kronecker products, nor can it be decomposed into a product of Good matrices. However, it can be decomposed into a product of other matrices, which permits a fast computation algorithm that is very similar to the general case (4.77), even if these matrices do not have the structure of Good matrices.

It has been shown [23] that a Haar matrix H_a of order $N = 2^n$ can be decomposed into a product of n matrices D_i of order N, with $i = 0, \ldots, n - 1$. Thus,

$$H_a = \prod_{i=0}^{n-1} D_i$$

$$= D_{n-1} D_{n-2} \cdots D_1 D_0 \qquad (4.105)$$

Again, the order of the decomposition must be preserved. The first matrix D_0 is identical to a Good matrix that may be used for the Hadamard transformation. This matrix will be denoted by $C(N)$, where N represents the order. Using partition notation, the matrix D_1 can be represented by

$$D_1 = \begin{pmatrix} C(N/2) & 0 \\ 0 & \sqrt{2}\,I(N/2) \end{pmatrix} \qquad (4.106)$$

where $I(N/2)$ is the unit matrix of order $N/2$. The matrix D_2 is given by

$$D_2 = \begin{pmatrix} C(N/4) & 0 \\ 0 & \sqrt{2}\,I(N/2 + N/4) \end{pmatrix} \qquad (4.107)$$

In general, for the matrix D_i, we have

$$D_i = \begin{pmatrix} C(N/2^i) & 0 \\ 0 & \sqrt{2}\,I\left(\sum_{j=1}^{i} \dfrac{N}{2^i}\right) \end{pmatrix} \qquad (4.108)$$

The flow chart corresponding to (4.105) is shown in figure 4.16 for the particular case $N = 8$.

The flow chart can also be modified to reflect the butterfly structure. The advantage obtained by this structure however, is lost in the ordering of the coefficients. It must be noted that the number of additions and subtractions in this transformation is

$$N + N/2 + N/4 + \ldots = 2N - 1 \qquad (4.109)$$

The inverse transformation can be obtained by using the same flow chart. It simply must be read in the opposite direction, that is, from right to left.

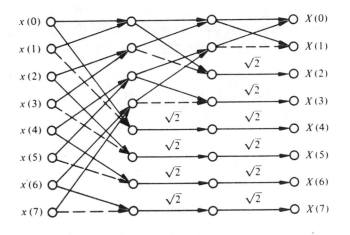

Fig. 4.16

4.6 FAST FOURIER TRANSFORM

4.6.1 Introduction

The fast Fourier transform (FFT) is obviously the most important of the transformations studied in this chapter. Its importance lies not only in its role played in digital signal processing, but also in its possibility for use with analog signals, or, more generally, for continuous functions in several scientific fields. The terminology used in the name FFT can be confusing in some cases. Even though it is called a transformation, it is not different from the discrete Fourier transformation (DFT). It is simply an efficient, economic, and elegant algorithm for computing the DFT. The computation algorithm called FFT and its significance with respect to the general theory previously developed are presented in this section.

4.6.2 Review

The DFT was studied in detail in chapter 3. Its mathematical expression is given by

$$X(n) = \sum_{k=0}^{N-1} x(k) \exp\left(-j\, 2\pi \frac{nk}{N}\right) \qquad (4.110)$$

$$\text{with } n = 0, ..., N-1$$

To simplify the development, the frequency interval $[0, 1]$, corresponding to the domain $[0, N - 1]$ for n, is considered here. The inverse DFT is given by

$$x(k) = \frac{1}{N} \sum_{n=0}^{N-1} X(n) \exp\left(j \, 2\pi \, \frac{nk}{N}\right)$$

with $k = 0, ..., N-1$

(4.111)

4.6.3 Properties

The properties of the DFT are listed below:

- The coefficients $X(n)$ of the DFT are periodic of period N:

  $$X(n) = X(n + iN) \qquad \text{for any integer } i$$

  (4.112)

  This is a consequence of the digital form of the signal $x(k)$

- The signal $x(k)$ is periodic of period N:

  $$x(k) = x(k + iN) \qquad \text{for any integer } i$$

  (4.113)

  This is a consequence of the discrete form of the series $X(n)$.

- The first value of the signal $x(k)$ is taken at $k = 0$ on the digital time scale. Thus, if the signal to be analyzed by the FFT starts at a time other than $k = 0$, this shift must be taken into account in the results obtained by (4.110) using the shift theorem.

- The coefficients of the DFT are computed starting from the discrete frequency $n = 0$. The coefficient $X(0)$ is the continuous component. If a two-sided representation is required on the frequency axis, then the series $X(n)$ must be considered over the period $[-N/2, N/2 - 1]$. It is given by

$$X\left(n - \frac{N}{2}\right) = X\left(n + \frac{N}{2}\right) \quad \text{with } n = 0, \ldots, N-1$$

(4.114)

This equation is a direct consequence of the periodicity of the series $X(n)$.

These properties involved in (4.110) and (4.111), and consequently in the FFT, are of prime importance for the correct interpretation of the results given by the FFT.

4.6.4 Comments

It is clear that (4.110) and (4.111) are transformations of type (4.15) with a unitary matrix of order N (see sect. 4.2.4). The basic idea of the FFT is to decompose the DFT of order N into m DFTs of order N_i, with

$$N = \prod_{i=1}^{m} N_i \qquad (4.115)$$

It is obvious that if N is a prime number, such a decomposition cannot be written. In this case, it is possible, however, to develop the FFT by decomposing the number $N - 1$ and using the primitive roots of prime numbers [24].

The FFT exists in two principal forms, called *decimation in time* and *decimation in frequency*. In the first form, the set of N samples of the signal is divided into several sets, each of N_i samples, while in the second form the same operation is carried out over the coefficients of the DFT. The principle is the same in both cases. Here, the FFT based on decimation in time is presented, while the DFT based on decimation in frequency is left as an exercise for the reader. Moreover, it will be assumed, as before, that the order N of the DFT is a power of 2 and only the case $N = 2^3$ will be considered so as to simplify the expressions and the flow charts.

Using notation (3.10), the matrix of the DFT of order N can be represented by

$$F_N = \begin{pmatrix} W_N^0 & W_N^0 & W_N^0 & \cdots & W_N^0 \\ W_N^0 & W_N^{-1} & W_N^{-2} & & W_N^{-(N-1)} \\ \vdots & & & & \\ W_N^0 & W_N^{-(N-1)} & \cdots & & W_N^{-(N-1)^2} \end{pmatrix} \qquad (4.116)$$

4.6.5 FFT decimated in time

Since N is an even integer, the sequence $x(k)$ can be divided into two sequences of $N/2$ values; the first formed by the values with even indices and the second by those with odd indices. Mathematically, this division can be represented by substituting $k = 2i$ for the even indices and $k = 2i + 1$ for the odd indices. Equation (4.110) becomes

$$X(n) = \sum_{i=0}^{N/2-1} x(2i) W_N^{-2in} + \sum_{i=0}^{N/2-1} x(2i+1) W_N^{-(2i+1)n} \qquad (4.117)$$

with $n = 0, \dots, N-1$

Using the property $W_N^2 = W_{N/2}$ (sect. 3.2.7), relation (4.117) can be written as follows:

$$X(n) = \sum_{i=0}^{N/2-1} x(2i) W_{N/2}^{-in} + W_N^{-n} \sum_{i=0}^{N/2-1} x(2i+1) W_{N/2}^{-in} \qquad (4.118)$$

Each sum of this expression represents the DFT of order $N/2$. The first sum is the DFT of the values with even indices, while the second sum is the DFT of the values with odd indices of the original signal $x(k)$. Thus, we can write

$$X(n) = X_1(n) + W_N^{-n} X_2(n) \qquad (4.119)$$

By virtue of property (4.112), the sequences $X_1(n)$ and $X_2(n)$ are periodic of period $N/2$. The flow chart corresponding to (4.119) is shown in figure 4.17.

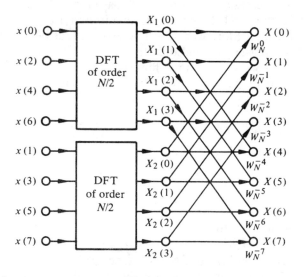

Fig. 4.17

This chart indicates precisely how the periodicity of the sequences $X_1(n)$ and $X_2(n)$ is used.

Because $N/2$ is an even integer, each DFT of order $N/2$ can again be divided into two DFTs of order $N/4$ by grouping the values with even indices and the values with odd indices in each DFT of order $N/2$. The flow chart corresponding to the first DFT of order $N/4$ is shown in figure 4.18.

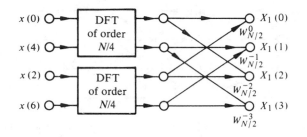

Fig. 4.18

For the DFT of order $N = 8$, which is presented here, the computation is reduced to several DFTs of order 2. The flow chart corresponding to one of these DFTs is represented in figure 4.19.

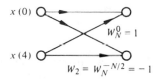

Fig. 4.19

In this figure, we can see a particular form of the butterfly operation studied in the general case of section 4.4.15. For the chosen example of order $N = 8$, no further decomposition is possible. Consequently, the computation indicated by figure 4.19 can be substituted into the computation of figure 4.18, which can then be substituted into the computation of figure 4.17. Finally, we obtain the complete flow chart for the FFT of order $N = 8$ decomposed in time. This chart is shown in figure 4.20.

4.6.6 Remarks

From this figure, we may note that the computation of the FFT can be done in place. By comparing the order of the indices of the input values of this chart to the sequence (4.98), we can see that for the FFT developed in this way the samples must be placed in bit-reversed order. The permutation algorithm presented in section 4.5.13 can be used for this purpose.

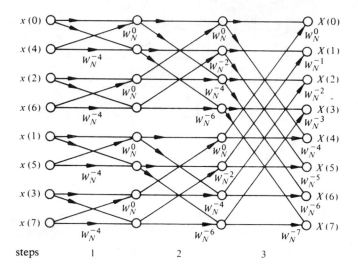

Fig. 4.20

4.6.7 Improvement

The computation involved in a butterfly operation between steps i and $i + 1$ can be written as follows:

$$X_{i+1,k} = X_{ik} + W_N^{-r} X_{ik'}$$
$$X_{i+1,k'} = X_{ik} - W_N^{-r} X_{ik'} \qquad (4.120)$$

The product $W_N^r X_{ik'}$ is computed twice. This can be avoided by computing $Z_{ik'} = W_{\bar{N}}^r X_{ik'}$, once and using the following equations:

$$X_{i+1,k} = X_{ik} + Z_{ik'}$$
$$X_{i+1,k'} = X_{ik} - Z_{ik'} \qquad (4.121)$$

This modification allows us to reduce the number of complex multiplications in half. The flow chart corresponding to this reduction is represented in figure 4.21. In this chart, the number of steps is doubled to show the modification indicated by (4.121). As in the general case, each step is considered to be a matrix vector product. The matrices corresponding to the steps of direct multiplication by powers of W_N are diagonal matrices. The matrices corresponding to the steps containing butterfly operations are the different forms of the Good matrices generated with the basis matrix:

$$B_F = \begin{pmatrix} W_N^0 & W_N^0 \\ W_N^0 & W_N^{N/2} \end{pmatrix} = \begin{pmatrix} 1 & 1 \\ 1 & -1 \end{pmatrix} \qquad (4.122)$$

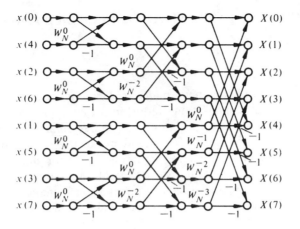

Fig. 4.21

4.6.8 Connection with the general theory

Except for the Good matrix used in the last step, the other Good matrices do not correspond exactly to the general form shown in figure 4.3. This is a result of the different row or column permutations that give the butterfly structure and which order the data or the results. Consequently, the transformation matrix (4.116) of the DFT of order $N = 2^m$ can be decomposed into a product of $2m$ matrices of order N as follows:

$$F_N = \prod_{i=1}^{m} C_i D_i = C_m D_m \ \dots \ C_2 D_2 C_1 D_1 \qquad (4.123)$$

where the D_i are diagonal and the C_i are Good matrices.

We can easily verify by using figure 4.21 and Good's theorem that in the FFT of order $N = 2^m$ developed above, there are $N \log_2 N$ complex additions and $N/2 \log_2 N$ complex multiplications.

To develop the FFT decimated in frequency, a similar development must be followed by grouping the $N/2$ first and then the $N/2$ final values of the signal. This development leads to a similar flow chart, which will be omitted from this text.

As shown for the general case, it is clear that the flow chart of the FFT can be modified to present either a sequential structure or a constant geometry at each step, according to the need of the user.

198

4.7 APPLICATION OF FAST UNITARY TRANSFORMATIONS

4.7.1 Introduction

Unitary transformations have many applications in signal processing. The development of a fast computation algorithm for computing some transformations in an economical, fast, and elegant way has enlarged and continues to enlarge the field of application. More and more, these algorithms are being used, even for the analysis of analog signals after a prior analog-to-digital conversion. In this section, we shall give a general discussion on the applications of the DFT. Then, the applications of unitary transformations to the problem of decorrelation is studied. Finally, more particular applications are briefly mentioned.

4.7.2 Application of the fast Fourier transformation

Obviously, the DFT is the most widely used transformation. It is used in digital as well as analog signal processing. The principal applications of the DFT in these two fields consist of the evaluation of the following functional relations:

- frequency spectra,
- convolution products,
- correlation functions.

The evaluation of the frequency spectra for deterministic signals (digital as well as analog) has been discussed in the previous chapter. Every time the DFT must be computed, the FFT can be used efficiently. However, in the case of analog signals, the rules given for the choice of the parameters must be followed carefully to obtain estimations of sufficient quality. The evaluation of the frequency spectra for random signals, or *power spectral density,* is a complex and important subject. Therefore, chapter 6 is entirely devoted to the detailed study of this problem.

4.7.3 Review

As previously shown, the Fourier transform allows us to express simply the equations corresponding to the convolution products or correlation functions in the frequency domain. The respective expressions for the convolution and the correlation are recalled:

$$y(k) = \sum_{l=0}^{M-1} g(l)x(k-l) \qquad (4.124)$$

$$\overset{\circ}{\varphi}_{xy}(k) = \sum_{l=0}^{M-1} x(l)y(l+k) \qquad (4.125)$$

$$\text{with } k = 0, ..., N-1$$

where $M = N$ if the length of the signal $x(k)$ is greater than that of $g(k)$ or of $y(k)$, which are assumed to be of length N, and $M = N - |k|$ if all these signals have the same length N.

In the frequency domain, the corresponding equations are, respectively,

$$Y(n) = G(n)\,X(n) \qquad (4.126)$$

$$\overset{\circ}{\Phi}_{xy}(n) = X^*(n)\,Y(n) \qquad (4.127)$$

4.7.4 Direct evaluation of a convolution product or correlation function

It is possible to evaluate a convolution product or a correlation function by using (4.124) and (4.125). This method is called *the direct method*. However, the number of operations required is relatively high. In the case where $M = N$, $N_A = N^2$ real operations are needed to compute N values of the convolution product or correlation function of two real signals. In the case where $M = N - |k|$, this number reduces to

$$N_B = N + N - 1 + N - 2 + ... + 1 = (N+1)(N/2) \qquad (4.128)$$

real operations.

4.7.5 Indirect evaluation of a convolution product or correlation function

The method called *indirect* can also be used with (4.126) and (4.127). In both cases, three DFTs must be computed. Two direct DFTs are needed to compute $Y(n)$ or $\overset{\circ}{\Phi}_{xy}(n)$, and an inverse DFT is needed to compute $y(k)$ or $\overset{\circ}{\varphi}_{xy}(k)$. However, since the signals are real, the two direct DFTs of two real signals of N samples can be obtained by a DFT and $2N$ additions (sect. 3.3.5). The computation time of N real additions is generally very short in comparison with the computation time of an FFT. Consequently, this number will be considered as negligible for the rest of this chapter.

The indirect computation can be carried out by using the previously described FFT. The number of operations required for an FFT of order $N = 2^m$ is $N \log_2 N$ complex additions and $N/2 \log_2 N$ complex multiplications. For an accurate comparison, the complex operations must be expressed in terms of real operations. It is obvious that a complex addition A_c involves two real additions, and that a complex multiplication M_c involves four real multiplications M_r and two real additions A_r. Thus,

$$A_c = 2 A_r$$
$$M_c = 4 M_r + 2 A_r \tag{4.129}$$

The number of complex multiplications $N/2 \log_2 N = Nm/2$ of an FFT is, therefore, equivalent to

$$Nm (4 M_r + 2 A_r)/2 = 2 Nm M_r + Nm A_r \tag{4.130}$$

that is, to $2Nm = 2N \log_2 N$ real multiplications and $Nm = N \log_2 N$ real additions. The number of complex additions $N \log_2 N$ of an FFT is equivalent to $2N \log_2 N$ real additions. Consequently, an FFT of order $N = 2^m$ requires $2N \log_2 N$ real multiplications and $3N \log_2 N$ real additions. Since an operation was defined as a multiplication followed by an addition, we can consider, as a first approximation, that an FFT requires the equivalent of $3N \log_2 N$ real operations. This approximation is higher than the actual number of operations needed for the FFT. Moreover, since the signals are real, the DFTs $Y(n)$ and $\overset{\circ}{\Phi}_{xy}(n)$ are conjugate symmetric with respect to $n = 0$. We have

$$Y(-n) = Y^*(n) \tag{4.131}$$

and

$$\overset{\circ}{\Phi}_{xy}(-n) = \overset{\circ}{\Phi}^*_{xy}(n)$$

Thus, the product (4.126) or (4.127) requires $N/2$ complex multiplications, which can be considered as $2N$ real operations. In the indirect computation using the FFT, the total number of real operations is thus given by

$$N_C = 6 N \log_2 N + 2 N \tag{4.132}$$

If we are to evaluate an autocorrelation function, the direct DFT need be computed only once. Since a real signal is involved, its DFT can be obtained by using an FFT of $N/2$ samples. In this case, the number of real operations reduces to

$$N_D = 9/2 N \log_2 N + 2 N \tag{4.133}$$

4.7.6 Comparison between direct and indirect methods

The numbers N_A, N_B, N_C, and N_D are shown in figure 4.22 as functions of N. From these curves, we can see that for signals of length greater than about $N = 30$, the use of the indirect method is obviously more advantageous. However, when using the indirect method, it must not be forgotten that the signals are periodic. The result obtained by the inverse DFT is periodic. Consequently, it can be altered by superposition if the original signals are nonperiodic. This problem was studied in the previous chapter.

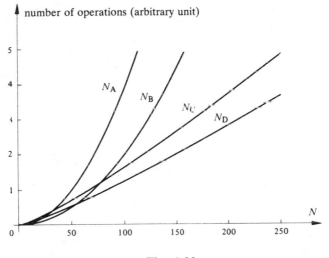

Fig. 4.22

4.7.7 Convolution by recursive filtering

There is another method for the evaluation of a convolution product or correlation function. The method is to use a recursive digital filter. It has some advantages, particularly when considering the number of memory cells used. This method obviously implies that the problem of synthesizing such a filter from one of the signals has already been solved. This method will be studied in more depth in the next chapter (see sections 5.3.10 to 5.3.22).

4.7.8 Interpolation of sampled signals

Let us consider N samples $x(t_k)$ taken periodically, with a period Δt, and according to the sampling theorem (sect. 1.7.2), over a function $x(t)$ on an interval T. We thus have $T = N\Delta t$. The DFT applied to the sequence of the samples gives

$$X(f_n) = \sum_{k=0}^{N-1} x(t_k)\exp(-j\,2\pi f_n t_k)\Delta t \tag{4.134}$$

with

$$t_k = k\,\Delta t, \qquad f_n = \left(n - \frac{N}{2}\right)\Delta f, \qquad \Delta f = \frac{1}{T} \quad \text{and} \quad n = 0,\,...,\,N-1$$

The coefficients $X(f_n)$ are periodically distributed with a period $1/T$ over the interval $[-1/(2\Delta t),\,1/(2\Delta t)]$. A new number of samples $N' > N$ can be artificially considered by adding a series of zeros to the series $X(f_n)$. For practical reasons previously covered, it is useful to have $N' = 2^m$ with $m > 0$. This is equivalent to defining a new sampling period $\Delta t' = T/N' < \Delta t$. Interpolated samples are obtained by computing the inverse DFT of the series $X(f_n)$ over the new interval $[-1/2\Delta t',\,1/2\Delta t']$.

This is equivalent to inserting $2^m - 1$ nonzero samples between the original samples $x(t_k)$ of the signal $x(t)$ and then filtering the signal obtained by an ideal digital low-pass filter, whose cut-off frequency is $1/2\Delta t$ (sect. 1.8.10). If the initial samples are not taken according to the sampling theorem, the interpolation will not be exact. An illustration of this application is shown in figure 4.23.

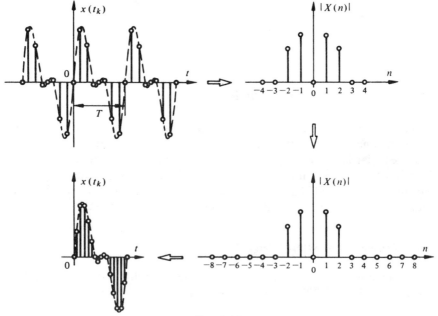

Fig. 4.23

Finally, the important role played by the DFT in a nonlinear processing class called homomorphic processing must be mentioned. This type of processing has interesting applications in audio signal and image processing. It will be studied in detail in chapter 7.

4.7.9 Decorrelation of signals

The crosscorrelation function of two signals $x(k)$ and $y(k)$, or the correlation of two samples $x(k_i)$ and $x(l_i)$, expresses the similarity between these signals or samples. If the value of the correlation is near the maximum possible value, we say that the two signals or samples are correlated. If the correlation is zero, they are said to be uncorrelated.

In the transmission of correlated signals, this correlation is used to reduce the amount of information to be transmitted, and thus use the transmission channel as well as possible. If two signals $x(k)$ and $y(k)$ are correlated, it is possible for example to transmit $x(k)$ and $x'(k) = x(k) - y(k)$ instead of $x(k)$ and $y(k)$ separately. Considering the similarity between $x(k)$ and $y(k)$, their difference generally remains small with respect to the dynamic range of the signal $x(k)$. This is at the heart of the well known technique called differential pulse code modulation (DPCM).

More generally, to transmit correlated signals, we seek a reversible transformation to transform the signals into as few correlated signals as possible. The transformation must be reversible, in order to reconstruct the original signals. This operation is generally called *decorrelation*. The terminology *redundancy reduction* is also sometimes used.

4.7.10 Example

Let us consider two adjacent correlated samples $x(k)$ and $x(k + 1)$ of a signal. Let us assume in addition that the magnitude of the sample is quantized over 10 levels. A particular configuration (digital value) $x(k)x(k + 1)$ is one of the $10 \cdot 10 = 100$ points in the two-dimensional vector space (Fig. 4.24). The correlation between these samples implies that the most probable configurations will be in a region near the line $x(k) = x(k + 1)$. This is the elliptical region indicated. If a 45-degree rotation of the coordinate axis is made, then, in the new axial system $x(k)x(k + 1)$, the same region of the space is now around the axis $y(k)$ (Fig. 4.24). This rotation allows us to change the distribution of the variances with respect to the axes. Due to the unitary nature of the transformation, the following equations are verified by the variances:

$$\sigma^2_{x(k)} + \sigma^2_{x(k+1)} = \sigma^2_{y(k)} + \sigma^2_{y(k+1)} \tag{4.135}$$

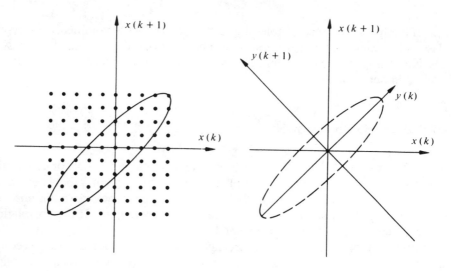

Fig. 4.24

which can be easily verified by a vectorial representation. Before the rotation, the variances were more or less uniformly distributed along $x(k)$ and $x(k + 1)$. After the rotation, we have

$$\sigma^2_{y(k)} > \sigma^2_{y(k+1)} \tag{4.136}$$

This indicates that the dynamic range around $y(k + 1)$ will be shorter than that around $y(k)$. Finally, it must be noted that the inverse rotation enables us to recover the starting axis system. A generalization of the process with N correlated samples is simply obtained by using the transformations of type (4.15) previously studied.

4.7.11 Karhunen-Loeve transformation

If a unitary transformation (4.15) is arbitrarily chosen, then, in general, the transformed coefficients $X(n)$, considered here to be random variables, are correlated. Thus, the problem is to find a particular transformation which leads to uncorrelated coefficients. The uncorrelation of the coefficient is written by the following expected value:

$$E[X(n)X^*(m)] = \lambda_n \delta_{nm} \tag{4.137}$$

where δ_{mn} is the Kronecker delta. The general expression for a transformation of the form (4.15) is

$$X(n) = \sum_{k=0}^{N-1} a_{nk}\, x(k) \tag{4.138}$$

By substituting the general form (4.138) into (4.137), the following result is obtained:

$$E[X(n)X^*(m)] = E\left[\sum_{k=0}^{N-1} a_{nk}\, x(k) \sum_{l=0}^{N-1} a_{ml}^*\, x^*(l)\right]$$

$$= \sum_{l=0}^{N-1} \sum_{k=0}^{N-1} E[x(k)x^*(l)]\, a_{nk}\, a_{ml}^* = \lambda_n\, \delta_{nm}$$

$$(4.139)$$

The expectation $E[x(k)x^*(l)]$ is by definition the general element $\varphi_x(k, l)$ of the correlation matrix of signal $x(k)$ (sect. 6.2.2). Thus,

$$\sum_{l=0}^{N-1} \sum_{k=0}^{N-1} \varphi_x(k,l)\, a_{nk}\, a_{ml}^* = \lambda_n\, \delta_{nm} \qquad (4.140)$$

Comparing this equation to (4.25), we derive the following:

$$\sum_{k=0}^{N-1} \varphi_x(k,l)\, a_{nk} = \lambda_n\, a_{nl} \qquad (4.141)$$

If the vector formed by the a_{nk} with $k = 0, \ldots, N - 1$ is represented by A_n, this equation can be written in the compact matrix form:

$$\varphi_x A_n = \lambda_n A_n \qquad (4.142)$$

In this equation, we recognize the classical problem of eigenvalues and eigenvectors for a matrix φ_x. The rows of the transformation matrix A are, thus, the eigenvectors of the correlation matrix φ_x. The transformation carried out by the matrix A is called the *discrete Karhunen-Loeve transformation*. The correlation matrix φ_x of the coefficients obtained by this transformation is a diagonal matrix (4.137), and the eigenvalues λ_n are the elements of the main diagonal. Unfortunately, the discrete Karhunen-Loeve transformation generally cannot be computed with a fast computation algorithm. Rather, this transformation thus requires N^2 elementary operations. Moreover, it is necessary first to compute the correlation matrix of the signal and then to find its eigenvectors. It is, therefore, not of great practical interest.

However, the Karhunen-Loeve transformation allows us to set the upper limit of the decorrelation that can be achieved. Due to the general property of rotations in an N-dimensional space, and because of the existence of fast computation algorithms for them, the DFT, the Hadamard transformation, and even the discrete Haar transformation have been used as approximations of this theoretical transformation. Two principal fields in which satisfactory results have been achieved are voice transmission with the vocoder and image transmission.

4.7.12 Voice decorrelation

A *vocoder* is an analysis and synthesis system used for the transmission of analog electrical signals representing voice. The well known phase insensitivity of the ear is used by this system, which, in its analysis part (transmitter), determines only the modulus of the *instantaneous spectrum*. The instantaneous spectrum is defined by the Fourier transform for a nonstationary signal observed over a very short interval, in which it can be considered stationary. The observation is then repeated on other segments of the signal to obtain other instantaneous spectra. The instantaneous spectrum is obtained by using a set of bandpass filters, which covers the frequency band of the voice signal in a nonuniform mannner. This band is generally between 0 kHz and 4 kHz. The output signals of these filters are enhanced and filtered by low-pass filters to produce an approximate representation of the spectral envelope (Fig. 4.25). The analog signals obtained are then transformed into digital signals and transmitted after multiplexing. Other complementary information such as the pitch period and the voiced/unvoiced indication (sect. 7.7.6) of the original signals are also transmitted to allow for synthesis at the receiver. The signals representing the spectral envelope exhibit strong correlation between two neighboring bandpass filters. After analog/digital conversion, a linear transformation makes it possible to exploit this correlation. For example, using the Hadamard transformation, the amount of information to be transmitted can be reduced by a factor of about three [25]. At the receiver, the inverse transformation must, of course, be carried out before the synthesis.

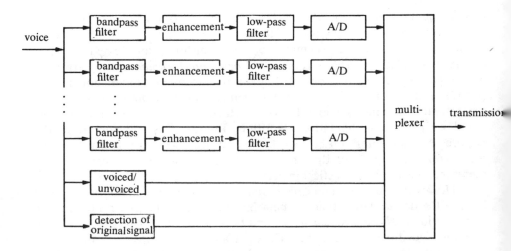

Fig. 4.25

4.7.13 Image decorrelation

The same principle is also used for digital image transmission. A black-and-white image can be represented by a function $B(u, v)$, representing the grey-level value at point (u, v) in the image plane. The discrete version of an image is obtained by two-dimensional periodic sampling (sect. 8.4).

$$x(k, l) = B(k\Delta u, l\Delta v) \tag{4.143}$$

where Δu and Δv are the sampling period along the u and v axies, respectively. The sampling periods must be chosen according to the two-dimensional sampling theorem (sect. 8.5.3). The signal $x(k, l)$ can be transmitted directly, for example, by using pulse code modulation (PCM). However, in a real image, the grey-level value varies little between one sample $x(k, l)$ and its neighbor. This indicates a relatively strong correlation between neighboring samples. Before transmission, a unitary transformation can, therefore, be applied to exploit this correlation. The most frequently used transformations are the DFT and the Hadamard [23]. In general, for practical reasons, the samples $x(k, l)$ are taken on a square domain of dimension $N \times N$. A two-dimensional unitary transformation of order $N \times N$ can be applied, or the image can be divided in two subimages of order $M \times M$, with $M < N$, and where each subimage can be transformed separately. This last method requires less memory for the transformation and provides a reconstructed image of better quality. The reduction of the amount of information to be transmitted depends on many factors, especially the content of the image itself, the particular method chosen to determine M, and the transformation used.

4.7.14 Improvement of the redundancy reduction

In the general problem of decorrelation by unitary transformations (rotations), an additional step can be taken toward reducing the information needed to represent one-dimensional as well as two-dimensional signals. This step consists of neglecting coefficients in the set of the transform coefficients which are insignificant compared to the others. The choice of the criterion for such a selection depends on the particular problems at hand and on the constraints the user has imposed on the final result. The neglected coefficients are simply not transmitted. The inverse transformation at the receiver, therefore, gives an approximation to the original signal. In general, the quality of this approximation is inversely proportional to the number of coefficients dropped.

4.7.15 Application to storage

The applications of unitary transformations to the problem of decorrelation were presented in the context of transmission. Indeed, it is in this context that they are most frequently used. However, thanks to the progress in the technology of electronic memories, the same principles are increasingly being used to store signals in a compact form. These signals are then represented by the transformed coefficients instead of their samples. The representation may be either approximate or exact.

4.7.16 Immunity to perturbations

The inverse transformation reconstructing the original signal can be considered as a weighted sum of the transformed coefficients. Each transformed coefficient is involved in all the samples of the signal. Thus, during transmission, the effect of a perturbation on one coefficient is distributed on all the samples of the reconstructed signal. The effect is thus less significant than that of a perturbation source on a particular sample of the signal. The representation of a signal by its transform has thus a certain immunity to noise in the transmission channel.

4.7.17 Generation of a Gaussian pseudorandom signal

The generation of a Gaussian distributed pseudorandom signal from a uniformly distributed or random signal is a direct application of the well known *central limit theorem* (vol. VI, chap. 15). This theorem indicates that a random variable obtained by the linear combination of independent random variables of any distribution has an approximately Gaussian distribution. This approximation improves as the number of terms in the linear combination increases. Rader has used this theorem with the Hadamard transformation. Starting from N samples of a uniformly distributed pseudorandom signal $u(i)$, which is easily generated numerically, the Nth-order Hadamard transformation produces N samples of a signal $g(i)$. Even if the $u(i)$ are statistically independent, the $g(i)$ are not because of the nature of the linear combination chosen. To overcome this, a random sign change is performed on the $g(i)$. The new signal thusly obtained has a Gaussian distribution. The Hadamard transformation of order $N = 2^m$ requires $N \log_2 N = Nm$ additions and subtractions. Consequently, only N operations are required to generate each $g(i)$. The drawback of this method is the need to store the series $u(i)$ in order to compute the Hadamard transformation.

4.7.18 Simplex code generation

The second application of the Hadamard transformation consists of generating codes which have a special correlation. This application is interesting, especially for long distance communication in space, where the message in a high interference environment is detected by correlation. The codes studied here are binary for obvious reasons. Let us consider an N-dimensional vector space for binary code words of length N. In this space, there are only two possible points along each coordinate axis due to the binary form of the codes.

The correlation of two code words x and y, represented by N component vectors, is by definition:

$$\rho(x,y) = \frac{1}{N} \sum_{i=1}^{N} x_i y_i \tag{4.144}$$

where x_i and y_i are the ith components of the vectors x and y respectively. A code is called simplex if for the entire set of M code words, the correlation of two different words is given by

$$\rho(x,y) = \begin{cases} \dfrac{-1}{M-1} & \text{for } M \text{ even} \\[2mm] \dfrac{1}{M} & \text{for } M \text{ odd} \end{cases} \tag{4.145}$$

If the first column of a Hadamard matrix is suppressed, the remaining row vectors form a simplex code. This is a direct consequence of the orthogonality between rows and columns of the Hadamard matrix and of definition (4.144). If the Hadamard matrix is generated by a Kronecker product from basis matrices of order 2, simplex codes can be generated for any M such that $M = 2^m$. Equation (4.91), therefore, can be used to generate simplex codes quickly and efficiently.

4.8 EXERCISES

4.8.1 Using the property $W_N^{lN} = \exp(j2\pi l) = 1$, express the DFT as a convolution.

4.8.2 Using the result obtained in problem 4.8.1, express the coefficient $X(n)$ using a recursive relation. Deduce the corresponding flow chart.

4.8.3 Consider a digital system, the flow chart of which is given by figure 4.26. What is the output signal obtained after N iterations?

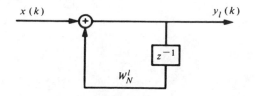

Fig. 4.26

4.8.4 A program was implemented to compute the FFT with N points. The only data given are the samples of a real signal stored in the real part of a complex vector $X(N)$. After computation, this vector contains the coefficients of the DFT. How can the same program be used to compute the inverse FFT?

4.8.5 How can the FFT be used to compute N samples, uniformly distributed on a circle of radius ρ centered at the origin, for the z-transform of a finite-length signal?

4.8.6 Let us consider a finite-length signal $x(k)$ of length K. Its autocorrelation function must be computed up to a maximum shift M. Determine whether computation using the FFT is more efficient than direct computation in the case where $K = 16,384$ and $M = 128$.

4.8.7 Determine whether a program for computing the FFT can be used to compute the Hadamard transformation with a fast algorithm. If so, give the necessary modifications.

4.8.8 Develop basic equations of the fast Fourier transform, decomposed in frequency, for the case $N = 8$. [*Hint:* group the first $N/2$ and last $N/2$ samples of the signals.]

4.8.9 Let $x(k)$ be a signal of finite length N and let $X(z)$ be its z-transform. Can the FFT be used to compute $X(z)$ at the points $z_R = AB_l$ with $R = 0, \ldots, M - 1$ and $A = A_0 \exp(j\alpha_0)$? [*Hint:* use the equation $nk = 1/2(n^2 + k^2 - (k - n)^2)$.]

4.8.10 Show that the interpolation of every second point by the FFT is equivalent to low-pass filtering of the signal obtained by inserting a zero sample between two real samples. What must the cut-off frequency be for the filter used?

Chapter 5

Digital Filtering and Digital Filters

5.1 INTRODUCTION

5.1.1 Preliminary remarks

The previous chapters were devoted to the presentation and development of the principal methods used in digital signal processing. These methods have been illustrated with simple examples, which were occasionally much too theoretical. In this chapter, the principal tools studied previously will be applied to a practical example. Thus, it will be the first chapter in this book devoted to applications. In addition, we shall limit ourselves to a very simple case, which will be that of linear time-invariant digital systems. If such a system is used to modify the frequency distribution of the components of a signal, according to given specifications and using arithmetical operations of limited precision, it is called a *digital filter*. The operation which consists of modifying this distribution by using a digital system is called *digital filtering*. This chapter is devoted to the study of the principal methods of filter design which meet the given specifications, and to the presentation of various ways to implement them.

Due to the size limitations of this volume, problems related to finite precision will not be studied. For this subject, see references [3] and [4].

For historical reasons, digital filters have been studied more than any other branch of digital signal processing. They have been developed and studied to simulate analog filters by computer. This made it possible to verify performance and to optimize the parameters of these filters before their actual implementation. The present progress in the technology of digital integrated circuits has increased the economic interest in digital filters. Therefore, in addition to the well established results for simulation, specific methods for the design of digital filters have also been developed.

5.1.2 Choice of methods

As we may imagine, there is no single filtering method which is the best for all cases and possible constraints. There is a great variety of methods, ranging from the simple use of precomputed analytic results to computer-aided algorithms. We will limit ourselves to the presentation of the principal techniques that are most commonly used and we will omit those which are too specific and too complex because they involve advanced mathematical notions. For a complete study, see chapter 3 and 4 of reference [4].

5.1.3 Organization of the chapter

In this chapter, the principal classifications and structures of digital filters are presented, after a mathematical review. An important distinction is made between finite and infinite impulse-response lengths of a filter. This is justified by the difference between the characteristics of these two types of systems, which require different design methods and implementations. Filters having an impulse response of finite length are presented next. Their important properties and their design using Fourier expansions and frequency sampling are examined as well as the recursive computation for this type of filter. Then, we will study three methods for the design of infinite-length impulse response filters. The methods are the equivalence of the derivation, the equivalence of the integration and the sampling of the impulse response. These methods consist of translating the results computed for the analog case to the digital case. Each method is illustrated with a practical design example. Finally, we study the minimum phase systems that allow us to obtain the frequency phase response for filters which are specified only by the amplitude frequency response. The Hilbert transform (vol. IV, sect. 7.3.33) and the all-pass filter are examined in this context and each are illustrated by an example.

5.2 GENERAL PRINCIPLES

5.2.1 Mathematical review

The frequency (or harmonic) response $G(f)$ of a linear time-invariant system is related to the Fourier transforms $Y(f)$ and $X(f)$ of the input and output signals by the well known equation:

$$Y(f) = G(f) \, X(f) \tag{5.1}$$

This equation indicates that the frequency distributions of the amplitude and phase of the input signal $x(k)$ are modified by the system, according to the particular form of the complex function $G(f)$, to satisfy as well as possible the specific processing requirements. Because it is the particular form of the frequency response $G(f)$ (harmonic response) that determines the attenuation or the amplification of the different frequency components, the corresponding system is called a *filter*. If, for example, the input signal is the sum of a useful signal and an undesired one, with the two signals occupying different frequency bands in $X(f)$, then the useful signal can be isolated by completely attenuating the frequency band of the undesired signal. This is somewhat analogous to the separation of sand and rocks by filtering. However, digital filtering is not limited to this particular processing case.

In the time domain, the equation corresponding to (5.1) is the well known convolution product:

$$y(k) = \sum_{l=-\infty}^{+\infty} g(l) x(k-l) \tag{5.2}$$

where $g(k)$ is the impulse response of the system. The z-transform for the two sides of this equation leads to

$$Y(z) = G(z) X(z) \tag{5.3}$$

where $G(z)$, which denotes the z-transform of the impulse response $g(k)$, is the transfer function of the system.

If we assume that the transfer function can be written as a quotient of two polynomials in z or z^{-1}, the input and output signals are then related by the following difference equation:

$$\sum_{n=0}^{N} a_n y(k-n) = \sum_{m=0}^{M} b_m x(k-m) \tag{5.4}$$

Equations (5.1) to (5.4) are the basic mathematical tools of digital filtering.

5.2.2 Goal

The general problem in digital filtering is to work out a linear shift-invariant system having the required frequency response in such a way that it can be efficiently implemented. Due to this requirement, the ideal frequency response can be obtained only approximately. To be capable of implementation, the system must be simultaneously causal and stable. In addition, the filtering must be performed with a finite number of arithmetical operations of necessarily finite (limited) precision. For example, the form (5.2) of the convolution product for a stable and causal system cannot be used if the impulse response is of infinite length. For such a system, we have

$$y(k) = \sum_{l=0}^{+\infty} g(l) x(k-l) \tag{5.5}$$

5.2.3 Classification of filters

The implementation of (5.5) requires an infinite number of coefficients $g(l)$ characterizing the system. Therefore, a recursive relation of (5.4) must be used to obtain the same result. This shows the importance of the length of the system's impulse response. Therefore, we divide the set of such systems into two large categories, based on the lengths of their impulse responses. These two categories are the following:

- *Infinite impulse response systems* (IIR)—these systems are characterized by infinite-length impulse responses. In other words, the samples $g(k)$ are nonzero on an infinite interval $k_0 \leq k < \infty$.
- *Finite impulse response systems* (FIR)—these systems are characterized by finite-length impulse responses. In this case, the samples $g(k)$ are nonzero only within an interval of finite length L with $k_0 \leq k \leq k_0 + L - 1$.

5.2.4 Classification of the implementations

A second classification permits us to distinguish among the various ways of implementing a given system to perform the filtering operation. Three cases can be distinguished:

- *Tranversal or nonrecursive implementation*—in a system implemented non-recursively, the filtering operation is performed by using the following equation:

$$y(k) = \sum_{l=0}^{L-1} g(l)x(k-l) \tag{5.6}$$

In such systems, as indicated by the above equation, the output signal depends only on the actual and previous values of the input signal.

- *Recursive implementation*—in a system implemented recursively, the filtering is performed by a difference equation of the form (5.4), which, using the hypothesis that $a_0 \neq 0$, can be written as follows:

$$y(k) = \sum_{m=0}^{M} \frac{b_m}{a_0} x(k-m) - \sum_{n=1}^{N} \frac{a_n}{a_0} y(k-n) \tag{5.7}$$

In these systems, the output signal depends not only on the input signal, but also on the previous values of the output signal.

- *Implementation by the DFT*—since a particularly efficient computation algorithm (FFT) for the computation of the DFT exists, it is also possible to compute the DFT of the input signal by using the FFT, to multiply the result by the coefficients of the frequency response of the filter used, and then to compute the inverse DFT by again using the FFT. The operations performed are summarized as follows:

$$X(n) = F[x(k)]$$
$$Y(n) = G(n) X(n) \qquad\qquad (5.8)$$
$$y(k) = F^{-1}[Y(n)]$$

The principles and rules for using the DFT as well as the FFT have been thoroughly covered in the two previous chapters. Consequently, this subject will not be examined again in this chapter, but rather we shall discuss only the recursive and nonrecursive implementations of the IIR and FIR. We must, however, remind the reader that, considering the number of operations, implementation by the DFT using the FFT is more advantageous than a nonrecursive implementation if the length I of the impulse response is greater than about 30 coefficients.

5.2.5 Implementation structures

By examining (5.6) and (5.7), we can observe that recursive as well as the nonrecursive filtering is performed by using three simple elementary operations, which are delay, multiplication, and addition. For the time being, we shall consider subtraction to be addition with a sign change. The transfer function of the system performing a unit delay is $g(z) = z^{-1}$ (see sect. 2.4.3 and fig. 2.3). Therefore, it is common to represent the unit delay in a block diagram of the filter by the notation z^{-1}. Figures 5.1 and 5.2 respectively show the block diagrams of a nonrecursive implementation (5.6) and a recursive implementation

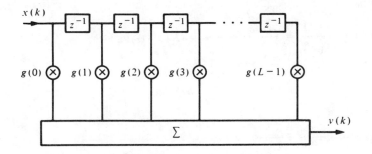

Fig. 5.1

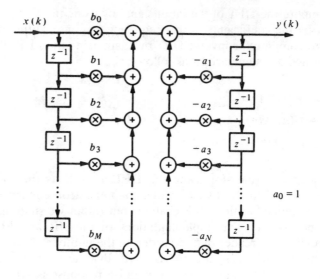

Fig. 5.2

(5.7). In more complicated block diagrams, the multiplication block represented here by $\otimes$ will be omitted. A coefficient written on the branch will indicate a multiplication by this coefficient.

5.2.6 Canonical structures

It's possible to consider the transfer function $g(z)$, corresponding to (5.7), as two cascaded systems. This enables us to obtain another structure for the implementation of the same relation (5.7). The z-transform for both sides of this equation can be written as

$$G(z) = \frac{Y(z)}{X(z)} = \frac{\sum_{m=0}^{M} b_m z^{-m}}{\sum_{n=0}^{N} a_n z^{-n}} = G_1(z)\, G_2(z) \qquad (5.9)$$

with

$$G_1(z) = \frac{1}{\sum_{n=0}^{N} a_n z^{-n}} \qquad (5.10)$$

and

$$G_2(z) = \sum_{m=0}^{M} b_m z^{-m} \tag{5.11}$$

If $w(k)$ represents the signal at the output of the first system, we have

$$G_1(z) = \frac{W(z)}{X(z)} = \frac{1}{\displaystyle\sum_{n=0}^{N} a_n z^{-n}} \tag{5.12}$$

By taking the inverse z-transform of this equation, we obtain

$$\sum_{n=0}^{N} a_n w(k-n) = x(k) \tag{5.13}$$

or

$$w(k) = \frac{1}{a_0} x(k) - \sum_{n=1}^{N} \frac{a_n}{a_0} w(k-n) \tag{5.14}$$

The signal $w(k)$ is also the input signal of the second system. Thus, we have

$$G_2(z) = \frac{Y(z)}{W(z)} = \sum_{m=0}^{M} b_m z^{-m} \tag{5.15}$$

The inverse z-transform of the equation leads to the following convolution product:

$$y(k) = \sum_{m=0}^{M} b_m w(k-m) \tag{5.16}$$

Figure 5.3 shows the global and detailed block diagrams corresponding to (5.9) and to the pairs (5.14) and (5.16), respectively.

Because this structure uses two series of delay elements for the same signal, it can be represented by the form of Fig. 5.4. This is a compact structure for implementing an Nth-order difference equation. It uses the mininum number of delay elements. There are many structures having a minimum number of delay elements and these are called the *canonical forms* of an implementation.

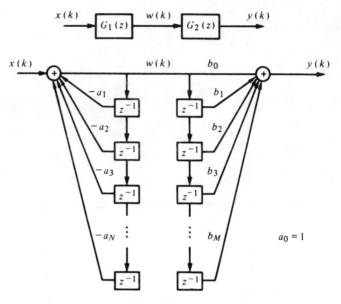

Fig. 5.3

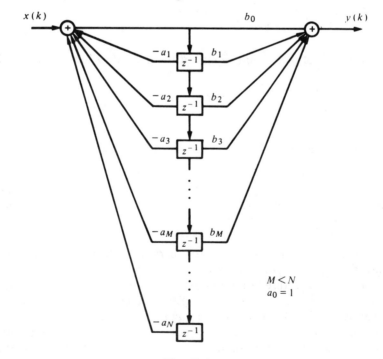

Fig. 5.4

5.2.7 Comments

The recursive implementation structures previously presented are known on the basis of their sensitivity to quantization errors in the finite-precision arithmetical operations because the error on one coefficient affects the positions of all the zeros and poles (see (5.4), (2.105), and (2.106)). This sensitivity increases with the order of the system, that is, with the order of the difference equation. The quantization errors are of secondary importance if the filtering is performed by software on a general-purpose computer that provides a relatively high precision. However, the errors are important if the filtering is performed by a specialized electronic device for which greater precision would be expensive. The way of reducing the sensitivity to quantization errors is to decompose the system into simpler subsystems. Two forms of decomposition may be considered, series and parallel.

5.2.8 Cascaded decomposition

For a cascaded (series) decomposition, the transfer function $G(z)$ of the complete system is expressed as the product of simpler transfer functions (sect. 1.6.6). We have

$$G(z) = C \cdot G_1(z) \cdot G_2(z) \cdot \ ... \ \cdot G_k(z)$$

$$= C \cdot \prod_{i=1}^{k} G_i(z) \tag{5.17}$$

where C is constant.

Figure 5.5 illustrates cascaded decomposition. In most cases, the partial transfer functions are of either first or second order. For a first-order system, we have

$$G_i(z) = \frac{1 + b_{i1} z^{-1}}{1 + a_{i1} z^{-1}} \tag{5.18}$$

To simplify, we consider $a_{i0} = b_{i0} = 1$; the global scaling constant C takes into account the constant terms in $G(z)$. The transfer function of a second-order partial system is the following:

$$G_j(z) = \frac{1 + b_{j1} z^{-1} + b_{j2} z^{-2}}{1 + a_{j1} z^{-1} + a_{j2} z^{-2}} \tag{5.19}$$

A second-order system is more complicated than one of first order. Second-order systems are nonetheless used because they permit us to have complex

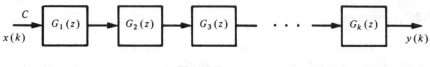

Fig. 5.5

poles and zeros for the global transfer function $G(z)$, with only real coefficients a_{ij} and b_{ij} in the recursive computation of (5.19). Thus, we are not limited to the real axis of the z plane.

5.2.9 Parallel decomposition

A parallel decomposition is obtained by expressing the global transfer function $G(z)$ as the sum of simpler transfer functions. We have

$$G(z) = D + G_1(z) + G_2(z) + \ldots + G_k(z) \tag{5.20}$$

$$= D + \sum_{i=1}^{k} G_i(z)$$

where D is a constant. Figure 5.6 illustrates parallel decomposition. In this case as well, the partial transfer functions are of either first or second order for the same reasons as given above for cascaded decomposition. In this case, however, simpler expressions are obtained due to the presence of the constant D. For a first-order system, we can use the following transfer function:

$$G_j(z) = \frac{b_{j0}}{1 + a_{j1} z^{-1}} \tag{5.21}$$

For a second-order system, we have

$$G_j(z) = \frac{b_{j0} + b_{j1} z^{-1}}{1 + a_{j1} z^{-1} + a_{j2} z^{-2}} \tag{5.22}$$

5.2.10 Remarks

To obtain the cascaded or parallel forms, a complex transfer function must be amenable to decomposition. If its poles and zeros are known, then cascaded decomposition can be obtained by grouping the poles and zeros in pairs, which are complex conjugate for second-order systems, and by grouping the real poles and zeros for first-order systems. For parallel decomposition, a partial fraction decomposition (sect. 2.3.9) must be used, which requires prior knowledge of

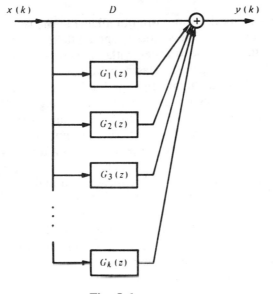

Fig. 5.6

the poles of $G(z)$. Again, the terms can be grouped by pairs of complex conjugate poles for second-order systems, and the terms of real poles can be kept for first-order systems. Because the decomposition into partial fractions of a given transfer function $G(z)$ is unique, the parallel decomposition of a recursive system is also unique. However, several cascaded decompositions may exist, depending on the different grouping of terms.

5.2.11 Specifications

We have previously seen the mathematical expressions for implementation and the principal filter types and structures for performing digital filtering. The main problem is to determine the coefficients in (5.4), or the values $g(k)$ of the impulse response in (5.6), to satisfy the given filtering requirements. The filter specifications are usually given in the frequency domain. Equation (5.1) is therefore of great importance. It allows us to control the filtering completely. For example, in a particular case, the filter may be required to attenuate as much as possible all the components of a signal below a given frequency. Such a filter is called a high-pass filter because it attenuates the low frequencies and does not alter high frequencies.

5.2.12 Classification of ideal filters

It is usual to classify filters that are specified by their frequency response in broad categories, such as low-pass, high-pass, bandpass, and band stop filters.

Moreover, these specifications are often ideal in the sense that all the terms to be attenuated must be done so totally and all the terms to be transferred at the output must undergo the transfer without modification. The ideal frequency responses corresponding to the classes mentioned above are represented in Fig. 5.7. In this figure, only the principal period $[-1/2, 1/2]$ of $G(f)$ is given. If the filtering is performed on a digital signal produced by sampling an analog signal, the sampling period Δt must be used for an exact correspondence of the frequencies. In this case, the principal period becomes $[-1/2\,\Delta t, 1/2\,\Delta t]$. We must also not forget that the impulse response $g(k)$ is a digital signal and its Fourier transform, which is the frequency response $G(f)$, is a periodic function.

The periodic nature of $G(f)$ might seem to make a digital high-pass filter impossible. Indeed, if the entire frequency axis is considered, the frequency response of a digital high-pass filter is similar to that of an analog bandpass filter. In fact, this is due to the digital characteristics of the signal. For analog signals sampled at the frequency $1/\Delta t$, no component of frequency higher than $1/(2\Delta t)$ can exist, according to the sampling theorem. In this case, the digital high-pass filter is effectively high-pass.

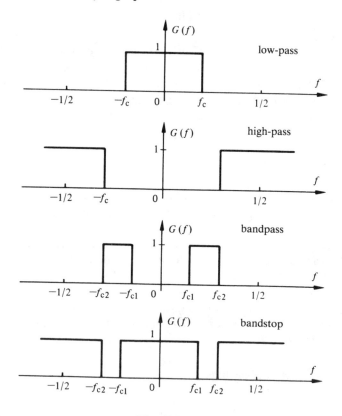

Fig. 5.7

5.2.13 Tolerance band

Due to the idealized nature of these frequency responses, real filters can only approximately satisfy this type of specification. That is why in practical problems the specifications are given with tolerances. This builds a pattern, called the *tolerance band,* into which the frequency response $G(f)$ must fit. For example, in the case of a high-pass filter, one possible pattern is shown in Fig. 5.8. A frequency response satisfying these specifications within the limits of the tolerances is also given. In the part where an ideal total attenuation is required, an approximation error δ_1 can be tolerated. This is the stopband. In the pass-band, where an ideal unit magnitude is required, an error of $\pm \delta_2$ is tolerated. Between those two bands, there is a transition band of finite and nonzero width $(f_2 - f_1)$ in which the frequency response passes from the stopband to the passband.

Most practical filters are given as shown in Fig. 5.8 without any phase specification. This is generally specified by causality and stability constraints, resulting from the restrictions placed on the corresponding transfer function $G(z)$, which must have its poles inside the unit circle.

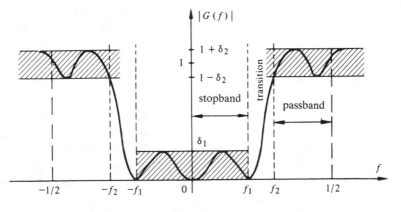

Fig. 5.8

5.2.14 Methodology

After setting the tolerance band, the next problem is to find a linear system having frequency response which satisfies the given conditions. To solve this problem, two possible approaches can be used. The first is to develop typical methods of approximation for digital filters by the use of applied mathematics. The second is to use the previously developed, widely studied and used, and now well known methods of analog filters. In this second case, we can often use pre-established and easy-to-use analytical equations. This allows us simultaneously to simulate analog filters digitally, an operation which has historically been at the heart of digital signal processing.

Because of the different properties of FIR and IIR systems, the methods for designing these filters will hereafter be studied separately.

5.3 FILTERS WITH FINITE-LENGTH IMPULSE RESPONSE (FIR)

5.3.1 General properties

In this section, we present the general properties of FIR filters. Then, we examine the principal design methods. Other more recent methods exist, but these are beyond the scope of this book due to their complexity. A more detailed and complete study can be found in reference [4].

A filter of finite impulse-response length L is always stable, provided that all the values $g(k)$ of its impulse response are finite. In this case, the corresponding transfer function $g(z)$ has its poles at the origin. Moreover, this function is a polynomial of degree $L\text{-}1$ in z or z^{-1}, the coefficients of which are the values of the impulse response $g(k)$. Another advantage of FIR filters is that they are causal if all the values $g(k)$ are zero for $k < 0$. If this is not the case, they can be made causal by using a finite number of unit delays to transfer the origin to the first nonzero value $g(k)$.

5.3.2 Phase-frequency response

It is also possible to design a FIR filter which has a perfectly linear phase-frequency response. Thus, a signal in the passband of the filter is exactly reproduced at the output, with a delay given by the slope of the phase-frequency response. To achieve the necessary conditions to have a linear phase characteristic, let us consider the frequency response of a FIR filter. This response has the following form:

$$G(f) = \sum_{k=k_0}^{k_0+L-1} g(k)\, e^{-j\,2\pi f k} \tag{5.23}$$

Separating the real and imaginary parts, we obtain

$$G(f) = \sum_{k=k_0}^{k_0+L-1} g(k)\cos 2\pi f k - j\left[\sum_{k=k_0}^{k_0+L-1} g(k)\sin 2\pi f k\right]$$

$$= \mathrm{Re}\left[G(f)\right] + j\,\mathrm{Im}\left[G(f)\right] \tag{5.24}$$

To obtain a phase characteristic equal to zero, the imaginary part of $G(f)$ must be zero. This is possible if and only if the impulse response $g(k)$ is an even function. The sine function being an odd function, the terms $g(k)\sin(2\pi fk)$ about the origin $k = 0$ will be identical in absolute value and of opposite sign. Thus, the sum over k will be zero. The condition for which $g(k)$ is even then becomes

$$g(k) = g(-k) \qquad \text{for } k_0 \leqslant k \leqslant k_0 + L - 1 \qquad\qquad (5.25)$$

This symmetry condition also links the parameters k_0 and L. For L odd, we must have $k_0 = -(L - 1)/2$ in order to satisfy the symmetry in (5.25). Thus, we have

$$g(k) = g(-k) \qquad \text{for } |k| \leqslant \frac{L-1}{2} \qquad\qquad (5.26)$$

For L even, the same equation remains valid. In this case, $L - 1/2$ is not an integer. Thus, a filter having an impulse response that satisfies (5.26) has a phase characteristic equal to zero, but it is not causal because half of its impulse response lies on the negative part of the k axis. Such an impulse response can be delayed by $(L - 1)/2$ units toward the positive values of k to transfer the origin to the first nonzero value $g[-(L - 1)/2]$. According to the delay theorem (sect. 1.3.9), this delay corresponds to the product of $G(f)$ in the frequency domain by the term $\exp[-j\pi f(L - 1)/2]$.

Then, the frequency response of the filter after the delay is

$$G'(f) = G(f)\exp[-j\pi f(L-1)] \qquad\qquad (5.27)$$

Because $G(f)$ is a real function, the phase characteristic after delay is given by

$$\theta_g(f) = -\pi(L-1)f \qquad\qquad (5.28)$$

This is a linear equation in f, which remains valid even if the delay is greater than $(L - 1)/2$. In the general case, for any given delay, the symmetry condition (5.26) can be written as follows:

$$g(k) = g(L-1-k) \qquad\qquad (5.29)$$

It must be noted that by filtering with a linear phase filter, half of the multiplications in (5.6) can be avoided if we use the symmetry (5.26). From time and hardware points of view, since multiplication is a more expensive operation than addition, this property of linear phase FIR filters is important.

5.3.3 Design by Fourier series

When a filter is specified in the frequency domain by its frequency response $G(f)$, this response must be considered periodic in order to perform digital filtering. In general, the period corresponds to the frequency interval where the function $G(f)$ is specified. In the principal period, the ideal impulse response is given by

$$g(k) = \int_{-1/2}^{1/2} G(f)\, e^{\,j\,2\pi fk}\, df \tag{5.30}$$

If the frequency response of the filter is one of the ideal responses of Fig. 5.7, or if it is made by line segments, then the analytic evaluation of $g(k)$ by computing integral (5.30) does not involve any problems. However, because the values $g(k)$ are the coefficients of a Fourier series expansion of the periodic function $G(f)$, an infinite number of coefficients is generally required to represent functions $G(f)$ with discontinuities or fast transitions. Consequently, $g(k)$ will be an infinite-length impulse response. To be a suitable response for a FIR filter, its length must be limited in some appropriate manner. This is the classical problem of convergence of a Fourier expansion. This is also the problem of computation of the DFT for an infinite-length signal, which was studied in detail in section 3.7. It is possible to use all the methods described there to multiply the infinite-length impulse response $g(k)$ by a properly chosen window function so as to satisfy the given specification of $G(f)$ within the tolerances for the Fourier transform of the product. Consequently, we will not return to this subject. However, it is useful to discuss some practical problems associated with these methods.

5.3.4 Digital approximation

Design by Fourier series requires an analytical expression for the infinite-length impulse response (5.30). This limits the use of these methods to the case of relatively simple frequency responses for which the integral (5.30) can be evaluated analytically. An approximate way of overcoming this problem is to compute the integral numerically. In this case, the surface of the function $g(f)$ $\exp(-2j\pi fk)$ on the principal period is approximated by the sum of the surfaces of a large number of juxtaposed rectangles. To simplify the computations, it is preferable to choose the same width for all the rectangles. The lengths of the rectangles are the values taken by the integrand in the middle of each interval. If N represents the number of these rectangles, the approximate form of (5.30) is given by

$$\hat{g}(k) = \frac{1}{N} \sum_{n=-N/2}^{N/2-1} G(n)\, \exp\left(j\, 2\pi\, \frac{nk}{N}\right) \tag{5.31}$$

This equation is nothing other than the inverse DFT of $G(n)$. An efficient and powerful algorithm has already been discussed to evaluate the DFT. Moreover, the relationship between $g(k)$ and its approximation $\hat{g}(k)$ is known (sect. 3.2.10). It has the following form:

$$\hat{g}(k) = \sum_{i=-\infty}^{+\infty} g(iN+k) \tag{5.32}$$

For sufficiently large N with respect to the presumed length L of the desired impulse response, we allow the approximation:

$$\hat{g}(k) \cong g(k) \tag{5.33}$$

Thus, the length L of the chosen window function must be small with respect to N if (5.30) is to be computed numerically.

5.3.5 Implementation of filtering

After obtaining the impulse response of the FIR filters, the actual filtering can be done in two ways. Equation (5.6) can be used to compute the convolution product directly, or the DFT can be computed. In the latter case, we must compute the DFT of the signal to be filtered, multiply the result by the DFT of the impulse response, and then calculate the inverse DFT of this last product. The analysis presented in section 4.7.6 shows that for real signals the method using the DFT is faster for signals of length greater than 30. Considering the computation time, (5.6) is only useful for impulse responses of short length. However, from a practical point of view, (5.6) is simpler than the FFT to program and to implement by computer. The price paid for this simplicity is a slow filtering operation. A drawback of design methods using Fourier expansions is their poor flexibility. The product of the signal and the window function in the time domain is equivalently represented by a periodic convolution product, which is often difficult to handle analytically in the frequency domain. This occurs particularly for the exact determination of the cut-off frequencies of the different frequency bands. An example of the use of this method can be found in section 3.7.7.

5.3.6 Frequency sampling

In the design of a FIR filter using frequency sampling, the desired frequency response $G_d(f)$ is represented by its samples $G(n)$. We have established (section 3.2.11) that a finite-length signal is entirely represented by the samples of its Fourier transform. This is also valid for a finite response of length L. It can be totally characterized by L samples of its Fourier transform, that is, of the frequency response. We have

$$g(k) = \sum_{n=-L/2}^{L/2-1} G_d(n) \exp\left(j \frac{2\pi}{L} nk\right)$$ (5.34)

$$\text{with } k_0 \leqslant k \leqslant k_0 + L - 1$$

where $G_d(n)$ represents samples taken periodically on $G_d(f)$ with a period $1/L$ over the principal interval $[-1/2, 1/2]$. We have also showed that it is possible to obtain the z-transform by interpolation between the samples $G_d(n)$ (see (3.56)). The interpolation equation is obtained by substituting (5.34) into the general expression of the z-transform. In this case, using the simplifying hypothesis $k_0 = 0$, the transfer function is given by

$$G(z) = \frac{1 - z^{-L}}{L} \sum_{n=-L/2}^{L/2-1} G_d(n) \frac{1}{1 - \exp(j 2\pi n/L) z^{-1}}$$ (5.35)

By evaluating the transfer function on the unit circle $z = \exp(j2\pi f)$, we obtain the frequency response $G(f)$. Thus,

$$G(f) = \frac{j\pi f(1-L)}{L} \sum_{n=-L/2}^{L/2-1} G_d(n) \frac{\sin \pi f L}{\sin \pi(f - n/L)} e^{-j\pi n/L}$$ (5.36)

This equation represents an interpolation of the form:

$$G(f) = \sum_{n=-L/2}^{L/2-1} G_d(n) \psi\left(f - \frac{n}{L}\right)$$ (5.37)

with

$$\psi\left(f - \frac{n}{L}\right) = \frac{\sin \pi f L}{L \sin \pi(f - n/L)} \exp\{j\pi[(f - n/L) - fL]\}$$ (5.38)

The function ψ is that of interpolation. The interpolation function has the property of being zero at all the sampling frequencies n'/L, except that which is centered and equal to one. In fact, we have

$$\psi\left(\frac{n}{L}\right) = 0 \quad \text{for } n \neq 0$$ (5.39)

due to the terms $\sin(\pi f L)$ in the numerator. We also have

$$\lim_{f \to \frac{n}{L}} \psi \left(f - \frac{n}{L} \right) = 1 \tag{5.40}$$

This can be shown by using the series expension of the function $\sin u$ for small values of the variable u. We can see from the previous development that it is possible to approximate a desired frequency response $G_d(f)$ by a function $G(f)$ obtained through interpolation between the samples $G_d(n)$ taken on $G_d(f)$ at the frequencies $f_n = n/L$. The error of this approximation is zero at the frequency f_n and remains finite for all other frequencies. Thus, it is possible to take L samples of the given frequency response and to obtain the impulse response of the desired filter by computing the inverse DFT using the FFT. Filtering can thus be performed by using the convolution product (5.6), or using the DFT.

5.3.7 Example

Let us consider the design of a high-pass filter. The desired frequency response of such a filter and the $L = 32$ samples taken on the principal period are shown in Fig. 5.9. This figure indicates that the function $G_d(f)$ is an even real function. Consequently, its inverse Fourier transform—that is, the ideal impulse response $g_d(k)$—is a real even function. Such a filter is not causal.

To make this a causal filter, the finite-length impulse response defined over $[-L/2, L/2 - 1]$ must be delayed by $L/2$ steps, which will introduce a perfectly linear phase characteristic, due to the symmetry. The impulse response $g(k)$ of length $L = 32$, obtained by the inverse DFT of the set of samples $G_d(n)$, is also shown in Fig. 5.9.

To make the filter causal, we need only delay this response by 16 steps. To evaluate the quality of the approximation of $G_d(f)$ by $G(f)$, we must compute the Fourier transform of the impulse response $g(k)$. The result of this computation is shown in Fig. 5.9 using a dB scale on the ordinate axis to show more clearly the approximation errors. This figure also shows that, in accordance with properties (5.39) and (5.40), $G(f)$ is identical to $G_d(f)$ at the points where the initial samples were taken.

5.3.8 Remark

If, to compare $G(f)$ and $G_d(f)$, the DFT (whether using the FFT or not) is used with the same number of samples as initially, we will recover initial samples $G_d(f)$ that coincide exactly with the values of $G_d(f)$. This result might lead us to believe that the approximation is perfect, which is totally false. In such a case, before computing the DFT, the length of $g(k)$ must be artificially

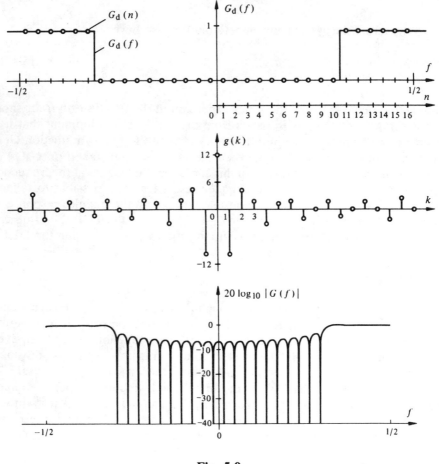

Fig. 5.9

increased with zero samples so as to increase the frequency resolution in the estimation of $G(f)$. Figure 5.9 shows that in the stopband, the approximation $G(f)$ is far from being zero. The oscillations observed are produced by the Gibbs phenomenon. This is caused by the limited length of the impulse response $g_d(k)$, inverse Fourier transform of $G_d(f)$. The length of $g_d(k)$ is necessarily infinite due to the very fast transition in $G_d(f)$.

5.3.9 Improvements

To decrease the importance of the values of $g_d(k)$, in the region of large values of $|k|$, and, consequently, to decrease the magnitude of the Gibbs phenomenon, the transition regions in $G_d(f)$ must be enlarged. This can be performed by introducing samples of intermediate amplitude between the extreme values 0 and 1 of $G_d(f)$ around the transition. In the case of this example, the value of the samples $n = \pm 10$ can be changed from 0 to 0.5. The impulse response $g(k)$ obtained by the inverse DFT of this new set is obviously different from that of Fig. 5.9. The Fourier transform of $g(k)$ modified in this way is shown in Fig. 5.10. The comparison of figs. 5.9 and 5.10 shows the attenuation of the Gibbs phenomenon, which can be achieved by enlarging the transition regions.

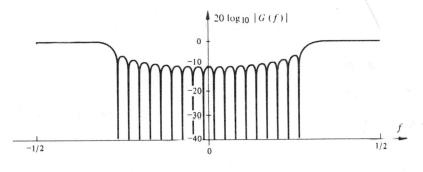

Fig. 5.10

The values 0.5 chosen by common sense for the samples close to the transition region is not an optimal value derived from an optimization process. Equation (5.36) shows that the approximation $G(f)$ is a linear function of the coefficients $G_d(n)$. It is then possible to use linear optimization methods to compute the intermediate value of the samples in the transition regions to obtain the best approximation for the desired frequency response $G_d(f)$. These optimization methods are generally very complicated for an analytic computation. Therefore a computer is often used to perform numerically. Thus, the design method of frequency sampling becomes a computer-aided design method. In this case, the value of the coefficients $G_d(n)$ is fixed in both the passband and stopband, and the value of the coefficient $G_d(n)$ around the transitions is adjusted to minimize the approximation error. This technique is a special case of the more general optimization method known as linear programming. Due to its specialized and advanced nature, this subject is not within the scope of this book, and consequently will be omitted. For more complete study, chapter 3 of reference [4] may be consulted by the reader. Other optimization methods are also presented therein, such as, for example, the method of minimizing the approximation error by distributing it uniformly over the frequency axis.

5.3.10 Recursive implementation of FIR filters

In what we have already seen, filtering using a FIR filter is implemented by either the convolution product (5.6), or the DFT with the help of the FFT. It is also possible to implement FIR filters recursively by using the equation expressing its transfer function $G(z)$ in terms of the samples of its frequency response. In the recursive implementation of a FIR filter, we assume that the finite impulse response $g(k)$ of length L for this filter was obtained by using an appropriate method according to given specifications. Instead of implementing it using a convolution product of the form (5.6) or the DFT, we shall implement it by computing one or several equations of the form (5.4) of order lower than L. The advantages of this technique will be discussed shortly.

5.3.11 Decomposition of the transfer function

Equation (5.35), which expresses the filter's transfer function as one obtained by interpolation between the samples of the frequency response, is recalled below:

$$G(z) = \frac{1 - z^{-L}}{L} \sum_{n=-L/2}^{L/2-1} G(n) \frac{1}{1 - \exp(j 2\pi n/L) z^{-1}} \tag{5.41}$$

where the coefficients $G(n)$ are obtained by computing the DFT of the impulse response $g(k)$. This system of equations can be represented as a series of two systems in the following manner:

$$G(z) = \frac{1}{L} G_a(z) G_b(z) \tag{5.42}$$

with

$$G_a(z) = 1 - z^{-L} \tag{5.43}$$

and

$$G_b(z) = \sum_{n=-L/2}^{L/2-1} G(n) \frac{1}{1 - \exp(j 2\pi n/L) z^{-1}} \tag{5.44}$$

This last equation shows that the partial system represented by the transfer function $G_b(z)$ is formed by L first-order systems in parallel. We have

$$G_b(z) = \sum_{n=-L/2}^{L/2-1} G_n(z) \tag{5.45}$$

with

$$G_n(z) = \frac{G(n)}{1 - \exp(j\, 2\pi n/L)z^{-1}} \qquad (5.46)$$

The decomposition (5.42) and the complete decomposition, which takes (5.45) into account, are shown in Fig. 5.11. To understand how the system works, we must first study the partial systems separately.

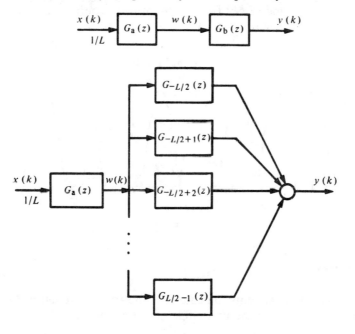

Fig. 5.11

5.3.12 Comb filter

Let us consider a system having a transfer function $G_a(z)$. If $w(k)$ represents the output signal of the system, we have

$$G_a(z) = \frac{W(z)}{X(z)/L} = 1 - z^{-L} \qquad (5.47)$$

The inverse z-transform for the two sides of this equation is simply given by

$$w(k) = \frac{x(k) - x(k-L)}{L} \qquad (5.48)$$

Thus, the output signal of this system is the difference between two samples of the input signal separated by L. The frequency response of this system is given by

$$G_a(f) = 1 - \exp(-j 2\pi f L) \tag{5.49}$$

We can easily verify that $G_a(f)$ is zero for all integer multiples of the frequency $f_0 = 1/L$. The amplitude of this function is given in Fig. 5.12, which shows the periodic shape of $G_a(f)$ over the principal period. Due to its shape, the system is called a *comb filter*.

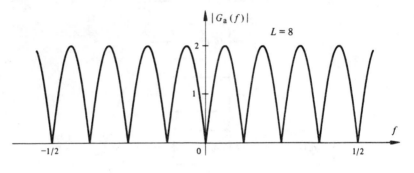

Fig. 5.12

5.3.13 Remarks

The preceding comments indicate that the transfer function $G_a(z)$ has L zeros uniformly spaced on the unit circle at the points:

$$z_i = \exp\left(j \frac{2\pi}{L} i\right) \quad \text{with } i = -L/2, \dots, L/2 - 1 \tag{5.50}$$

Moreover, (5.47) indicates that this function has an Lth-order pole at the origin. Thus, this filter is stable for any value of L. In addition, it is a FIR filter, as can be immediately seen from (5.48). The impulse response of the comb filter is given by

$$g_a(k) = \begin{cases} 1 & \text{for } k = 0 \\ -1 & \text{for } k = 1 \\ 0 & \text{otherwise} \end{cases} \tag{5.51}$$

The length of $g_a(k)$ is $L + 1$. This shows that the first step of the decomposition of section 5.3.1 is very simply performed by a FIR filter using (5.48).

It can be processed with L delays and one subtraction. The recursive part of the decomposition will appear on a system having the transfer function $G_b(z)$.

5.3.14 First-order partial recursive systems

By comparing (5.46) and (2.109), we see that partial systems in parallel forming the system characterized by $G_b(z)$ are first-order recursive systems. Such systems are implemented with a first-order difference equation, such as

$$y(k) - \alpha_n y(k - 1) = \beta_n w(k) \tag{5.52}$$

In this case, the parameters α_n and β_n are different for each partial system characterized by the transfer function $G_n(z)$, where the index n distinguishes the L partial systems. In the decomposition of Fig. 5.11, the output signal $w(k)$ of the first system is now the input signal for all the partial systems. The recursive relation (5.52) allows us to implement each of these partial systems with

$$\alpha_n = \exp\left[j \frac{2\pi}{L} n \right] \quad \text{and } \beta_n = G(n) \tag{5.53}$$

The coefficients α_n are the simple poles of the first-order systems, and are located exactly on the unit circle of the z-plane. Moreover, they are exactly equal to the zeros of the comb filter, so that in (5.42) the zeros of the comb filter are balanced by the poles of the first-order partial system.

5.3.15 Remarks

In general, the coefficients $G(n)$ are complex. However, if the impulse response of the FIR filter that has been implemented recursively is a real signal, then $G(n)$ and $G(-n)$ are complex conjugates of each other (sect. 1.3.8). This is always the case for practical filters. It is clear that α_n and α_{-n} are also complex conjugates. Thus, the L partial systems in the parallel decomposition (5.45) have coefficients that are complex conjugate by pairs. Two cases must be distinguished, depending on whether L is even or odd, to established the precise correspondence of the complex conjugate pairs. If L is even, only the partial systems of indices $-L/2$ and 0 have real coefficients. Since L is even, $L - 2$ is also even. The $L - 2$ other systems have coefficients which are complex conjugate by pairs. For L odd, $L/2$ is not an integer. In this case, the only system having real coefficients is that of index zero. The $L - 1$ others have coefficients which are complex conjugate by pairs.

5.3.16 Second-order partial recursive systems

To avoid the complicated arithmetic in the implementation of (5.52) the complex conjugate terms of (5.45) can be combined. Because $G_n(z)$ and $G_{-n}(z)$ are the transfer functions of two partial systems having complex conjugate coefficients, their sum can be written as follows:

$$G_n(z) + G_{-n}(z) = \frac{G(n)}{1 - \exp\left(j \dfrac{2\pi n}{L}\right) z^{-1}} + \frac{G^*(n)}{1 - \exp\left(-j \dfrac{2\pi n}{L}\right) z^{-1}} \tag{5.54}$$

By giving the two terms a common denominator and expressing the sum of two complex conjugates as twice their real part, we obtain

$$G_n(z) + G_{-n}(z) = 2\,|\,G(n)\,|\, \frac{\cos[\theta(n)] - \cos\left[\theta(n) - \dfrac{2\pi n}{L}\right] z^{-1}}{1 - 2\cos \dfrac{2\pi n}{L} z^{-1} + z^{-2}} \tag{5.55}$$

$$= 2\,|\,G(n)\,|\, H_n(z)$$

where $H_n(z)$ represents the transfer function of a second-order recursive system. Thus, for L even, the transfer function $G_b(z)$ (see (5.45) and (5.46)) can be written as

$$G_b(z) = 2 \sum_{n=1}^{L/2-1} |\,G(n)\,|\, H_n(z) + \frac{G(0)}{1 - z^{-1}} + \frac{G(-L/2)}{1 + z^{-1}} \tag{5.56}$$

This equation shows that $G_b(z)$ is obtained by combining $L/2 - 2$ second-order recursive systems in parallel with two first-order recursive systems, all with real coefficients. For L odd, because there is no sample $G(n)$ with a noninteger index $L/2$, the last term in (5.56) disappears. The only hypothesis which enables us to obtain real coefficients in (5.56) is to have a real finite impulse response. The implementation of decomposition (5.42) using first- and second-order recursive systems by taking (5.56) into account is shown in Fig. 5.13 for L even. One of the second-order systems involved in this implementation is seen in Fig. 5.14.

5.3.17 Particular case: Linear phase response

If, in addition, the real impulse response of finite length L satisfies the condition (5.29), which guarantees a perfectly linear phase-frequency response

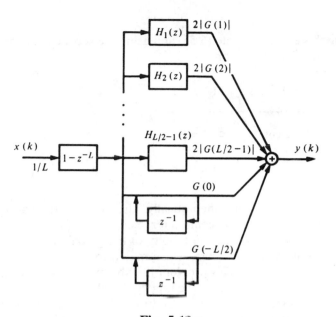

Fig. 5.13

for the global system, then (5.54) can be written in a simpler form. If we assume that the first nonzero value of the impulse response is at the origin ($k = 0$), the phase characteristic is given by (5.28):

$$\theta_g(n) = -2\pi \frac{(L-1)}{2} \frac{n}{L} = -\pi n + \frac{\pi n}{L} \tag{5.57}$$

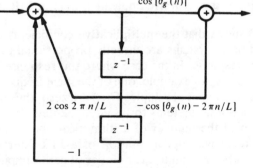

Fig. 5.14

Substituting this expression into (5.55) and using the second trigonometric equation of section 1.2.7, we obtain

$$
H_n(z) = \frac{(-1)^n \cos \dfrac{\pi n}{L} (1 - z^{-1})}{1 - 2 \cos \dfrac{2\pi n}{L} z^{-1} + z^{-2}}
\tag{5.58}
$$

5.3.18 Comment

It has been shown that, theoretically, the poles of the transfer function $G_b(z)$ balance the zeros of the comb filter. Since these poles and zeros are on the unit circle of the z-plane, an instability of the global system may appear in practice due to quantization errors, which may introduce slight differences between the poles and zeros. This problem may be avoided by shifting the poles and the zeros toward the center of the unit circle, along the radii that join them to the origin. This shift is obtained by multiplying the poles and zeros by a real number ρ slightly less than one, or equivalently by replacing z with z/ρ. Thus, the global transfer function (5.41) becomes

$$
G(z) = \frac{1 - \rho^L z^{-L}}{L} \sum_{n=-L/2}^{L/2-1} G(n) \frac{1}{1 - \exp(j\,2\pi n/L)\, \rho\, z^{-1}}
\tag{5.59}
$$

The same substitution may be made directly in (5.55), (5.56), and (5.58). Values of ρ between $1 - 2^{-12}$ and $1 - 2^{-27}$ have been successfully used by Gold and Rader [1] with only minor changes in the behavior of the frequency response.

5.3.19 Advantages of the recursive implementation of FIR filters

Figure 5.13 shows that the multiplicative coefficients at the output of the first- and second-order systems are directly proportional to the samples of the desired frequency response. In the case where this response is that of a selective bandpass filter, such as, for example, one of the ideal frequency responses shown in Fig. 5.7, a certain number of samples $G(n)$ are zero, thus reducing the number of first- and second-order systems to be implemented. It is only in such a case that the reduction of the number of arithmetical operations and the required hardware make recursive implementation of a FIR filter advantageous. Said advantage grows in proportion to the number of zero samples $G(n)$. Due to this dependence, it is not possible in the general case to compare the recursive and nonrecursive implementations of a arbitrary FIR filter. In simple cases, it is a good idea to see if the desired frequency response can be approximated well enough by the one obtained with a first- or second-order recursive system.

5.3.20 Example

Let us consider the rather extreme case of a filter having an impulse response which is the following:

$$g(k) = a^k \, \epsilon(k) \tag{5.60}$$

with $|a| < 1$ to ensure stability. In theory, the length of this response (decreasing exponential) is infinite. In practice, it may be considered finite due to the quantization of the coefficients and the finite precision arithmetic used. Thus, starting from a given value $k_0 = L$, depending on the precision used, $g(k)$ can be considered zero. The nonrecursive implementation of this filter with the convolution product (5.6) requires L multiplications and additions for each output value, and, at the same time, $L - 1$ delays. However, let us not forget that such a filter is described by the following first-order difference equation:

$$y(k) - a\,y(k-1) = x(k) \tag{5.61}$$

This filter requires only a delay element, a multiplication, and an addition for each output value, while producing the same result as the convolution product (5.6). The family of frequency responses which can be obtained with (5.61) is limited to the following particular form:

$$G(f) = \frac{1}{1 - a \exp(-j\,2\,\pi\,f)} = \frac{1 - a \cos 2\,\pi f - j\,a \sin 2\,\pi f}{1 + a^2 - 2\,a \cos 2\,\pi f} \tag{5.62}$$

where a is the only parameter (see Fig. 2.9).

5.3.12 Remarks

Another advantage of the recursive implementation of a FIR filter is that the position of the poles and zeros of the transfer function $G(z)$ given by (5.41) depends only on the length L of the impulse response. Moreover, if the filter has a linear phase, the second-order coefficients $H_n(z)$ also depend only on L. In this case, the implementation of Fig. 5.13 can be used for all linear phase filters of impulse response length L by modifying only the multiplicative coefficients proportional to the samples of a particular frequency response.

Moreover, because the structure of Fig. 5.13 is modular for the second-order partial systems, it is also possible to use the same second-order system in multiplex, which, by sequential modification of its coefficients, can provide the same service. This is valid for linear as well as nonlinear phase filters.

5.3.22 Example

Let us consider the recursive implementation of the bandpass filter having the frequency response shown in Fig. 5.15. The impulse response of this filter, obtained with the same 16 samples, is of finite length $L = 16$. With an appropriate delay of eight steps, the phase characteristics of this filter can be made

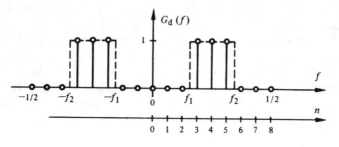

Fig. 5.15

perfectly linear. Moreover, we shall assume that the poles and zeros of the partial systems are on the unit circle, that is, $\rho = 1$. By following decomposition (5.42) in the same order, the filtering must start with the comb filter. If $x(k)$ represents the input signal, the following signal $w(k)$ is obtained at the output of the comb filter

$$w(k) = \frac{1}{16} [x(k) - x(k-16)] \tag{5.63}$$

This output signal is now the common input signal of the partial recursive systems. Because there are only three pairs of nonzero coefficients $G(n)$ and the filter has a linear phase, three second-order recursive systems of the simplified type (5.58) must be used. The transfer functions of the partial systems involved in (5.56) are obtained by substituting the numerical values $L = 16$ and $n = 3, 4$, and 5 in the general expression (5.58). We have in order:

$$H_3(z) = \frac{-0.831\,(1 - z^{-1})}{1 - 0.765\,z^{-1} + z^{-2}}$$

$$H_4(z) = \frac{0.707\,(1 - z^{-1})}{1 + z^{-2}} \tag{5.64}$$

$$H_5(z) = \frac{-0.556\,(1 - z^{-1})}{1 + 0.765\,z^{-1} + z^{-2}}$$

To complete decomposition (5.56), these equations must be multiplied by 2 $|G(n)|$, which is equal to 2 for $n = 3$, 4, or 5. Each of these transfer functions represent a second-order system, which may be implemented by the following second-order recursive equation:

$$y'(k) = a_0 w(k) + a_1 w(k-1) - b_1 y'(k-1) - b_2 y'(k-2) \qquad (5.65)$$

By identifying the coefficients in the functions $H_n(z)$, we obtain the corresponding difference equations:

$$y_1(k) = -1.662 [w(k) - w(k-1)] + 0.765 y_1(k-1) - y_1(k-2)$$
$$y_2(k) = 1.414 [w(k) - w(k-1)] - y_2(k-2) \qquad (5.66)$$
$$y_3(k) = -1.112 [w(k) - w(k-1)] - 0.765 y_3(k-1) - y_3(k-2)$$

Finally, the filtered signal is given by

$$y(k) = y_1(k) + y_2(k) + y_3(k) \qquad (5.67)$$

The nonrecursive implementation of the same filter by using the convolution product (5.6) requires $L = 16$ elementary operations for each value of the output signal. By analyzing the previous equations, we can show that, in the particular case of the example chosen, the recursive implementation of the same filter requires six elementary operations plus six additions for each value of the output signal. This reduction, especially the reduced number of multiplications, is important in practice because the computation time is greater for multiplication than for addition. It must be also noted that the coefficients in the above difference equations were computed for the ideal case $\rho = 1$ (5.59).

An instability of the filter may appear due to the quantization error on the coefficients. As an exercise, the reader may determine the new coefficients with a value of ρ slightly less than one.

5.4 FILTERS WITH INFINITE-LENGTH IMPULSE RESPONSE (IIR)

5.4.1 Introduction

The design method for filters with infinite-length impulse response (IIR) can be grouped into two categories. The first, which is the most traditional and commonly used, includes the techniques transferred from analog filter design. Here, the goal is to benefit from analog filter design methods by establishing a correspondence between the analog and digital domains. In the second category, the error introduced by approximation of the desired filter characteristics using those of a practical filter is minimized, according to some appropriate criterion.

As in the case of FIR filters, the computer-aided optimization techniques are omitted due to their complexity. These methods can be found in chapter 4 of reference [4]. We present in this section the principal methods for applying the techniques of analog filter design to the case of digital filters.

5.4.2 General properties

The IIR filters studied here are linear time-invariant systems governed by Nth-order difference equation of the form:

$$\sum_{n=0}^{N} a_n \, y(k-n) = \sum_{m=0}^{M} b_m \, x(k-m) \tag{5.68}$$

The general problem involved in the design of such filters consists of determining the sets of coefficients $\{a_n\}$ and $\{b_m\}$ such that the frequency response of the filter obtained satisfies the given requirements. Moreover, to be implementable, the filter must be causal and stable. The causality is ensured if the impulse response $g(k)$ satisfies the condition:

$$g(k) = 0 \qquad \text{for all } k < 0 \tag{5.69}$$

The stability condition for IIR filters is more severe than that for FIR filters. It is not only necessary that each value $g(k)$ of the impulse response be finite, but also that

$$\sum_{k=0}^{+\infty} \left| g(k) \right| < \infty \tag{5.70}$$

The transfer function $G(z)$ of an IIR filter, which can be easily obtained from (5.68), is a quotient of two polynomials in z or z^{-1}. We have

$$G(z) = \frac{\displaystyle\sum_{m=0}^{M} b_m \, z^{-m}}{\displaystyle\sum_{n=0}^{N} a_n \, z^{-n}} \tag{5.71}$$

This function generally has M zeros and N poles. In the case of a causal filter, the stability condition (5.70) implies that the N poles must be inside the unit circle in the z-plane (sect. 2.7.5).

In the following, the index a will be used to denote all the functions or variables characterizing analog systems. Moreover, the fundamental notions involved in the study of analog systems, as well as the design methods for analog filters, will be considered as already known. Those readers who are not familiar with these notions may refer to reference [26].

5.4.3 Methodology

In the transformation from an analog filter to a digital one, it is essential to preserve the main properties of the analog filter. In fact, the Laplace transform is for analog systems what the z-transform is for digital systems. A bridge between these two transformations will thus be sought. From a mathematical point of view, such a bridge is a mapping from the s-plane of the Laplace transform to the z-plane. In such a mapping, the imaginary axis of the s-plane must, if possible, be mapped onto the unit circle and the left half-plane of the s-plane inside the unit circle of the z-plane. This guarantees that a stable analog filter is transformed into a stable digital filter. The sampling period Δt of an analog filter is also an important variable which will be introduced in these transformations.

The filtering problem then consists first of finding an analog filter which satisfies the requirements of the given problem. Its transfer function $G_a(s)$, which is the Laplace transform of its impulse response $g_a(t)$ is hence determined. When the mapping from the s-plane to the z-plane is established as a functional relationship $s = f(z)$, this function must be substituted into the expression for $G_a(s)$. The transfer function $G(z)$ of the corresponding digital filter is then obtained:

$$G(z) = G_a(s) \Big|_{s=f(z)} \tag{5.72}$$

Finally, using algebraic manipulations, $G(z)$ must be expressed as the quotient of two polynomials and the coefficients $\{a_n\}$ and $\{b_m\}$ determined by identification. The filtering can then be performed using the difference equation (5.68).

5.4.4 Equivalence of the derivation

A linear time-invariant analog system is characterized by a linear differential equation with constant coefficients. It is, therefore, natural to establish a relationship between derivatives and differences. For example, we have for the first derivative:

$$\frac{d}{dt} x_a(t) \leftrightarrow \frac{x(k) - x(k-1)}{\Delta t} \quad \text{with} \quad x(k) = x_a(k\Delta t) \tag{5.73}$$

where $x_a(t)$ is an analog signal and $x(k)$ is a digital signal. Let us first consider the digital operator for a first-order difference. If $y(k)$ is a signal obtained by the first-order difference of a signal $x(k)$, then

$$y(k) = \frac{x(k) - x(k-1)}{\Delta t} \tag{5.74}$$

The transfer function $G(z)$ of a digital difference operator can be determined by computing the z-transform for both sides of this equation. The following result is obtained:

$$G(z) = \frac{1 - z^{-1}}{\Delta t} \tag{5.75}$$

It is well known that derivation in the time domain corresponds to the product of s and the Laplace transform in the s domain (vol. IV, sect. 8.1.15). Thus, with the identification:

$$s = \frac{1 - z^{-1}}{\Delta t} \tag{5.76}$$

the correspondence (5.73) is established. Equation (5.76) is nothing other than mapping from the s-plane to the z-plane.

5.4.5 Image of the imaginary axis

To obtain the image of the imaginary axis of the s-plane in the z-plane, (5.76) must first be solved for z. We obtain

$$z = \frac{1}{1 - s\,\Delta t} \tag{5.77}$$

The imaginary axis is characterized by $s = j\omega_a$. Thus, for the image:

$$z = \frac{1}{1 - j\omega_a\,\Delta t} = \frac{1}{1 + \omega_a^2\,\Delta t^2} + j\,\frac{\omega_a\,\Delta t}{1 + \omega_a^2\,\Delta t^2} \tag{5.78}$$

To find the locus of z, when s describes the imaginary axis, z can be expressed in polar coordinates with amplitude and phase. We obtain

$$|z| = \sqrt{\text{Re}^2[z] + \text{Im}^2[z]} = \sqrt{\frac{1}{1 + \omega_a^2\,\Delta t^2}} \tag{5.79}$$

and

$$\arg [z] = \arctan \left(\frac{\text{Im } [z]}{\text{Re } [z]} \right) = \arctan (\omega_a \Delta t) \qquad (5.80)$$

$$\frac{\text{Im } [z]}{\text{Re } [z]} = \omega_a \Delta t \qquad (5.81)$$

By substituting this last equation in (5.79), the following result is obtained:

$$(\text{Re}^2 [z] + \text{Im}^2 [z])^2 = \text{Re}^2 [z] \qquad (5.82)$$

The square root of the two sides of this equation can be taken by differentiating the positive and negative roots. However, the real part of z is always positive (5.78). Because the square of the modulus of a complex number is also always positive, only the positive square root need be taken into account. Thus,

$$\text{Re}^2 [z] + \text{Im}^2 [z] = \text{Re} [z]$$

or

$$\text{Re}^2 [z] - \text{Re} [z] + \text{Im}^2 [z] = 0 \qquad (5.83)$$

With the first two terms, a square can be obtained:

$$\left(\text{Re}[z] - \frac{1}{2} \right)^2 + \text{Im}^2 [z] = \frac{1}{4} \qquad (5.84)$$

This is the equation for a circle of radius 1/2 centered at the real point $z = 1/2$. It can easily be verified that the left half of the s-plane is mapped inside this circle in the z-plane. Figure 5.16 gives the correspondence between the s-plane and the z-plane, according to (5.77). This correspondence shows that even if the image of the imaginary axis $s = j\omega_a$ is not the unit circle $|z| = 1$, the mapping (5.77) satisfies the stability condition because it maps the left half of the s-plane inside the unit circle $|z| = 1$.

5.4.6 Remarks

Figure 5.16 also shows a drawback of the equivalence of the derivation. In general, at a particular stage of the mapping from the s-plane to the z-plane, the sampling of an analog function must be performed. This sampling must be done in accordance with the sampling theorem (sect. 1.7.2). During the sampling process, segments of identical length taken on the imaginary axis $s = j\omega_a$ are mapped onto the unit circle. The integral Fourier transform $X_a(g)$ of an analog

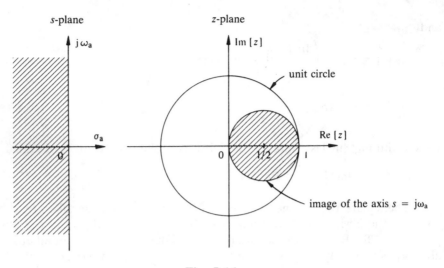

Fig. 5.16

signal is obtained by evaluating its Laplace transform on the imaginary axis $s = j\omega_a$. To avoid the aliasing resulting from the sampling, the length of the segments on this axis must be chosen so as to include all nonzero values of $X_a(f)$. Thus, by sampling, $X_a(f)$ is transposed about the unit circle in the z-plane (sect. 2.5.6).

This result is also valid for the impulse and frequency responses of the system. Increase of the sampling frequency corresponds to increasing the length of the segments of the axis $s = j\omega_a$.

Because such a segment is mapped onto the unit circle regardless of its length, increasing the sampling frequency on the unit circle corresponds to concentrating the Fourier transform in the vicinity of the point $z = 1$. Because (5.77) maps the imaginary axis onto the smaller circle of Fig. 5.16, the approximation of the derivatives with differences is only satisfactory in the vicinity of the point $z = 1$, where the two circles have a common tangent. For a good approximation, this requires a large sampling frequency, or, equivalently, a small period Δt. However, over-sampling is a very inefficient representation of an analog signal, producing a large amount of redundant data. This is the basic limitation to the method of the equivalence of the derivation.

The transfer function (5.75) used for the first difference corresponds to a stable and causal system. It is also possible to use a first difference of the type:

$$\frac{x(k+1) - x(k)}{\Delta t} \tag{5.85}$$

However, this leads to a noncausal system and we may easily show that a stable analog filter can be transformed into an unstable digital filter.

5.4.7 Example

Let us consider an analog filter having a transfer function given by

$$G_a(s) = \frac{1}{s+1} \qquad (5.86)$$

The frequency response of this system is obtained by evaluating $G_a(s)$ on the imaginary axis $s = j2\pi f_a$ of the s-plane. We have

$$G_a(f) = \frac{1}{1+j2\pi f_a} \qquad (5.87)$$

The modulus of this function is shown in Fig. 5.17. To obtain the transfer function $G(z)$ of the corresponding digital filter from $G_a(s)$, we must use substitution (5.76) in the expression for $G_a(s)$. The following result is obtained:

$$G(z) = G_a(s)\Big|_{s=\frac{1-z^{-1}}{\Delta t}} = \frac{\Delta t}{1+\Delta t - z^{-1}} \qquad (5.88)$$

The digital filter can then be implemented by using the following first-order difference equation:

$$y(k) = \frac{1}{1+\Delta t}[\Delta t\, x(k) + y(k-1)] \qquad (5.89)$$

The frequency response of this digital filter is obtained by evaluating its transfer function $G(z)$ on the unit circle of the z-plane. Moreover, because a round trip about the unit circle must correspond to the principal period $[-1/2\Delta t,\ 1/2\Delta t]$ of the frequency axis, $G(z)$ must be evaluated by setting

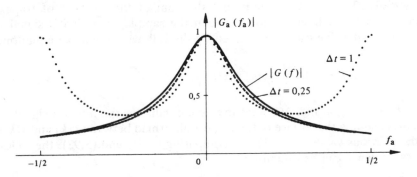

Fig. 5.17

$z = \exp(J2\pi f \Delta t)$ in order to take the sampling period Δt into account. Thus, we obtain

$$G(f) = \frac{\Delta t}{1 + \Delta t - \exp(-j\, 2\pi f \Delta t)}$$

$$= \frac{\Delta t}{1 + \Delta t - \cos 2\pi f \Delta t + j \sin 2\pi f \Delta t} \qquad (5.90)$$

The parameter Δt plays an important role for the function $G(f)$. Δt determines its period, which is $1/\Delta t$, and also the quality of the approximation of $G_a(f)$ by $G(f)$. Theoretical considerations previously presented indicate that Δt must be small for a good approximation. To illustrate this point, the modulus of $G(f)$ is shown in Fig. 5.17 for two values of Δt. For $\Delta t = 1$ s, the period of $1/\Delta t = 1$ Hz of $G(f)$ is relatively small with respect to the length of $G_a(f_a)$. In this case, the approximation is very poor, especially for frequency values higher than about 0.05 Hz. It is only in the vicinity of the origin $f = 0$, corresponding to the point $z = 1$, that the approximation is acceptable over a very small interval. The quality of the approximation is clearly improved by decreasing the value of Δt by a factor of 4. In this case, the interval in which the two functions $G_a(f_a)$ and $G(f)$ are equal is wider.

5.4.8 Equivalence of the integration

If $y_a(t)$, an analog signal, is the integral of a signal $x_a(t)$, we have

$$y_a(t) = \int_\alpha^t x_a(u)\,du \qquad (5.91)$$

The digital equivalent of this equation is the trapezoidal rule for integration. This rule is largely used in the case where the integral cannot be computed analytically. It consists of computing the sum of the surfaces of trapezoids obtained by linear interpolation between the samples of a digital signal $x(k)$. The trapezoidal rule can be expressed by the following recursive equation:

$$y(k) = y(k-1) + \frac{\Delta t}{2}\,[x(k) + x(k-1)] \qquad (5.92)$$

This equation indicates that the actual value of the integral signal is obtained by adding the surface of the trapezoid formed between $x(k)$ and $x(k-1)$ to the previous value $y(k-1)$. By comparing (5.91) and (5.92), the following correspondence can be established:

$$x(k) = x_a(k\,\Delta t) \qquad (5.93)$$

and

$$y(k) \cong y_a(k \, \Delta t)$$

which is illustrated in Fig. 5.18.

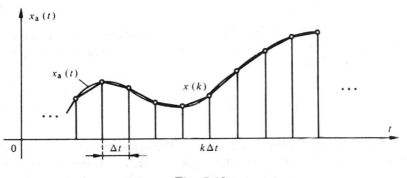

Fig. 5.18

5.4.9 Bilinear transformation

The transfer function of a digital trapezoidal integrator can be deduced from the z-transform for both sides of (5.92). The following result is obtained:

$$G(z) = \frac{\Delta t}{2} \cdot \frac{1 + z^{-1}}{1 - z^{-1}} \tag{5.94}$$

In the s-plane, integration corresponds to the division of the Laplace transform by the variable s. Thus, using the identification:

$$\frac{1}{s} = \frac{\Delta t}{2} \cdot \frac{1 + z^{-1}}{1 - z^{-1}} \tag{5.95}$$

we can thereby establish the correspondence between (5.91) and (5.92). The equivalence relation involved in the integration (5.95) is known as the *bilinear transformation*. This terminology stems from the fact that (5.95) is a linear equation with respect to s as well as z. Indeed, we can write

$$s = \frac{2}{\Delta t} \cdot \frac{1 - z^{-1}}{1 + z^{-1}} \tag{5.96}$$

or, conversely,

$$z = \frac{1 + (\Delta t / 2)s}{1 - (\Delta t / 2)s} \tag{5.97}$$

5.4.10 Image of the imaginary axis

By setting $s = j\omega_a$ in the z-plane, we have from (5.97):

$$z = \frac{1 + j(\Delta t/2)\omega_a}{1 - j(\Delta t/2)\omega_a} \tag{5.98}$$

This expression is nothing other than the quotient of two complex conjugate numbers. Thus, the modulus of z is one and its argument is twice that of $1 + j(\Delta t/2)\omega_a$. Thus,

$$z = \exp\left[j\, 2 \arctan\left(\frac{\Delta t}{2}\, \omega_a \right) \right] \tag{5.99}$$

This equation indicates that if s follows the imaginary axis (or if ω_a follows the real axis), then z follows the unit circle. Moreover, if the real part of s is negative, then (5.97) shows that the modulus of z is less than one. Thus, the left half of the s-plane is mapped inside the unit circle in the z-plane. The mapping defined by (5.96) satisfies the requirement of transforming a stable analog filter into a stable digital filter. This result is illustrated in Fig. 5.19.

5.4.11 Correspondence of the frequencies

The bilinear transformation (5.95) introduces a nonlinear relationship for the correspondence of the frequencies between the s-plane and the z-plane. This

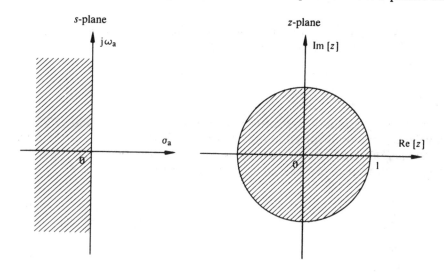

Fig. 5.19

relationship can be established in the following way. The Fourier transform of
a digital signal is obtained by evaluating the z-transform on the unit circle, that
is, by setting $z = \exp(j2\pi f \Delta t)$. In this representation, the continuous variable
f is the frequency in the digital domain. Moreover, the parameter Δt is included
here to indicate that the principal period of the frequency scale in the digital
case is $[-1/2\Delta t, \Delta t]$. By identifying this expression with (5.99), with $\omega_a = 2\pi f_a$,
we obtain

$$ f = \frac{1}{\pi \Delta t} \arctan(\pi \Delta t f_a) \tag{5.100} $$

or, conversely,

$$ f_a = \frac{1}{\pi \Delta t} \tan(\pi f \Delta t) \tag{5.101} $$

5.4.12 Example

The distortions of the frequency axis caused by (5.100) are illustrated in
Fig. 5.20 for the case of an ideal bandstop filter with several bands. This is of

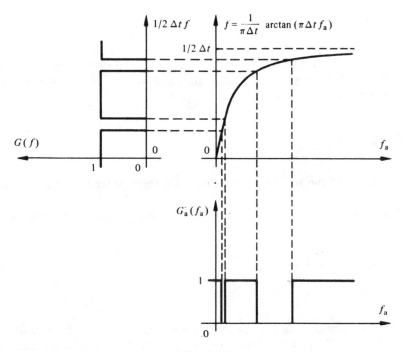

Fig. 5.20

prime importance in the determination of the different cut-off frequencies. In the design of the digital filter, if a desired cut-off frequency is f_c, the cut-off frequency of the analog filter to be transposed is not $f_{ac} = f_c$, but rather $f_{ac} = \tan(\pi f_c \Delta t)/(\pi \Delta t)$.

The transfer function $G_a(s)$ must be determined taking (5.101) into account. As can be seen in fig. 5.20, the frequency intervals in the analog and digital domains are not the same. If we forget this fact, it may lead to severe errors. It is clear that the distortions stemming from (5.100) also affect the phase response.

5.4.13 Chebyshev filters

The analog filters which are the most popular for their frequency selectivity are the Butterworth, Chebyschev, and elliptic filters. The equivalence of the integration method shall be illustrated hereafter by the example of the Chebyschev filter. This is generally characterized by a balanced ripple in the passband and a monotonic decrease in the stopband. The analytical form for the square of the modulus of the frequency response of an analog Chebyschev filter is given by

$$| G_a(f_a) |^2 = \frac{1}{1 + \mu^2 \, T_K^2 \, (f_a)} \tag{5.102}$$

where μ is a real parameter, having a value less than one, which determines the amplitude of the ripples in the passband, and where $T_K(\omega_a)$ is the Kth-order Chebyschev polynomial defined by

$$T_K(f_a) = \begin{cases} \cos(K \arccos 2\pi f_a) & \text{for } 0 \leqslant |f_a| \leqslant 1/2\pi \\ \cosh(K \operatorname{arcosh} 2\pi f_a) & \text{for } |f_a| > 1/2\pi \end{cases} \tag{5.103}$$

The magnitude of the frequency response of a third-order Chebyschev filter is shown in Fig. 5.21.

When the ripples are specified by μ, the analog transfer function can be written as

$$G_a(s) = \frac{\alpha_0}{\displaystyle\sum_{k=0}^{K} \alpha_k \, s^k} \tag{5.104}$$

where the coefficient α_0 in the numerator ensures a unit gain for the dC component ($f_a = 0$). These filters have been so well studied that the coefficients α_k

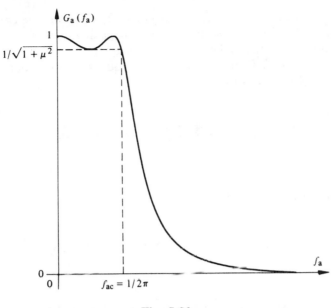

Fig. 5.21

in the denominator of (5.104) can be found in tables [27] for typical values of μ. The order K is generally determined by the attenuation desired in the stopband.

5.4.14 Example

A digital low-pass filter is required, having a maximum ripple in the passband of 0.5 dB, a cut-off frequency of $f_c = 0.1$ Hz, and an attenuation in the stopband of at least 40 dB with respect to unity at a frequency that is three times greater than the cut-off. For simplicity, we shall assume that the sampling period is $\Delta t = 1$ s.

Because the relative amplitudes of the frequency responses are preserved in the equivalence of the integration, the parameter μ can be determined directly:

$$0.5 = 20 \log_{10} \frac{1}{\sqrt{1 + \mu^2}}$$

which gives μ = 0.349.

However, the frequency intervals do not correspond directly (Fig. 5.20). The equivalent of the frequency interval [0.1, 0.3] of the digital filter must be established in the analog domain. By using (5.101), we obtain the corresponding interval [0.103, 0.438]. Because the lower limit of this interval corresponds to the cut-off frequency, the analog filter is required to have an attenuation of 40 dB at 0.438/0.103 = 4.25 times the cut-off frequency. The order of the Chebyschev filter is thus given by

$$0{,}01 = \frac{1}{\sqrt{1 + (0.349)^2 \cosh^2 (K \text{ arcosh } 4.25)}}$$

where

$$\cosh (K \text{ arcosh } 4.25) \cong 286.53$$

from which

$$K = \frac{6.351}{2.126} = 2.99$$

Because K must be an integer, the upper fraction must be rounded to the next highest integer to ensure that the requirements are satisfied. The transfer function of an analog Chebyschev filter of order 3, with a normalized cut-off frequency $f_{ac} = 1/2\pi$ and a ripple of 0.5 dB, is given by [27]:

$$G_a(s) = \frac{0.716}{s^3 + 1.253 s^2 + 1.535 s + 0.716} \tag{5.105}$$

This transfer function is determined by a normalized cut-off frequency $f_{ac} = 1/2\pi$. To obtain the cut-off frequency of 0.1 Hz for the digital filter, the imaginary axis of the s-plane must be contracted so that the cut-off frequency of the analog filter is $f_{ac} = 0.103$ Hz. This is obtained by replacing s with $s/2\pi f_{ac}$ in (5.105). The transfer function of the analog filter having a cut-off frequency of $f_{ac} = 0.103$ Hz is thus given by

$$G_a'(s) = \frac{0.196}{s^3 + 0.814 s^2 + 0.648 s + 0.196} \tag{5.106}$$

To obtain the transfer function of the digital Chebyschev filter of order 3, it is sufficient to replace s by (5.96) in this expression. This leads to

$$G(z) = \frac{0.0154 + 0.0461 z^{-1} + 0.0461 z^{-2} + 0.0154 z^{-3}}{1 - 1.9903 z^{-1} + 1.5717 z^{-2} - 0.458 z^{-3}} \tag{5.107}$$

Digital filtering can thus be accomplished by using the third-order difference equation, which is easily determined from the coefficients of $G(z)$. We then obtain

$$y(k) = 0.0154\,x(k) + 0.0461\,x(k-1) + 0.0461\,x(k-2) + 0.0154\,x(k-3)$$
$$+ 1.9903\,y(k-1) - 1.5717\,y(k-2) + 0.458\,y(k-3)$$

Two multiplications can be avoided by grouping terms having the same coefficient. Finally, the following result is obtained:

$$y(k) = 0.0154\,[x(k) + x(k-3)] + 0.0461\,[x(k-1) + x(k-2)]$$
$$+ 1.9903\,y(k-1) - 1.5717\,y(k-2) + 0.458\,y(k-3)$$

The frequency response of this filter is obtained by evaluating the transfer function (5.107) on the unit circle. The modulus of this response is shown in Fig. 5.22. This figure shows that the specifications are satisfied within the allowable limits.

5.4.15 Sampling of the impulse response

The third and last method for transforming an analog filter into a digital one to be studied in this chapter consists of finding the impulse response of a

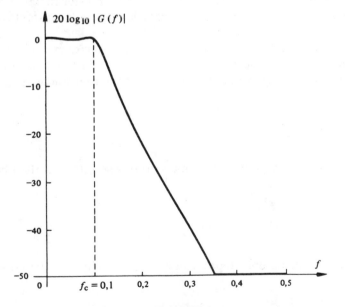

Fig. 5.22

digital filter by periodically sampling the impulse response of an analog filter. This method is a direct application of the sampling theorem (sect. 1.7.2), and uses the relationship between the Laplace transform of a sampled signal and the corresponding function $G(z)$ (sect. 2.5.3).

Although the method consists of sampling the impulse response of an analog filter to obtain the response of the corresponding digital filter, the filtering cannot be accomplished directly with a convolution product. To sample a signal—in this particular case, a continuous impulse response—it is necessary that its Fourier transform be of finite length in the frequency domain. Impulse responses having this property are of infinite length, which, after sampling, leads to digital IIR filters. It is, therefore, necessary to obtain the transfer function $G(z)$ to perform filtering with a difference equation.

Let us consider an analog filter of impulse response $g_a(t)$ and transfer function $G_a(s)$. We need to find the coefficients of the transfer function $G(z)$ of a digital filter, assuming that its impulse response $g(k)$ is given by

$$g(k) = g_a(k \, \Delta t) \tag{5.108}$$

The desired transfer function $G(z)$ can be obtained by computing the z-transform of $g(k)$. In the general case, this computation may be difficult, depending on the form of $g_a(t)$. That is why it is useful to exploit the properties of the Laplace transform and the z-transform. The transfer function of a linear analog system characterized by a differential equation with constant coefficients can be written as the quotient of two polynomials in s.

For simplicity, let us additionally assume that the degree of the denominator is greater than the degree of the numerator, and $G_a(s)$ has only simple poles. In this case, $G_a(s)$ can be decomposed into partial fractions as

$$G_a(s) = \sum_{j=1}^{N} \frac{\alpha_j}{s - s_j} \tag{5.109}$$

where s_j are the simple poles. The impulse response $g_a(t)$ can then be written as follows:

$$g_a(t) = \sum_{j=1}^{N} \alpha_j \exp(s_j t) \epsilon(t) \tag{5.110}$$

where $\epsilon(t)$ is the analog unit-step function. The sampling of $g_a(t)$ gives

$$g(k) = \sum_{j=1}^{N} \alpha_j \exp(s_j k\Delta t) \epsilon(k\Delta t) \tag{5.111}$$

The transfer function of the filter is given by

$$G(z) = \sum_{k=0}^{+\infty} g(k) z^{-k}$$

$$= \sum_{k=0}^{+\infty} \sum_{j=1}^{N} \alpha_j \exp(s_j k\Delta t) z^{-k} \qquad (5.112)$$

By inverting the order of summation, we obtain

$$G(z) = \sum_{j=1}^{N} \alpha_j \sum_{k=0}^{+\infty} \exp(s_j k\Delta t) z^{-k}$$

$$= \sum_{j=1}^{N} \alpha_j \sum_{k=0}^{+\infty} [\exp(s_j \Delta t) z^{-1}]^k \qquad (5.113)$$

The sum with respect to k in this expression is a geometrical series. This leads to the final result:

$$G(z) = \sum_{j=1}^{N} \frac{\alpha_j}{1 - \exp(s_j \Delta t) z^{-1}} \qquad (5.114)$$

5.4.16 Comments

If we compare (5.109) with (5.114), we can see that a pole $s = s_j$ in the s-plane corresponds to a pole $z = \exp(s_j \Delta t)$. We saw earlier that due to the ambiguity in the argument of the complex logarithm (2.81), the equation $z = \exp(s\Delta t)$ maps an infinite number of horizontal bands of width $1/\Delta t$ of the s-plane onto the same z-plane. Although, as a result of this ambiguity, a one-to-one mapping cannot be established between these two planes, the transformation of the poles indicates that a stable analog filter is transformed into a stable digital filter. If the real part of s_j is negative, then the modulus of $z_j = \exp(s_j \Delta t)$ is always less than one.

5.4.17 Aliasing

As a result of the sampling, the frequency response $G(f)$ of the digital filter is obtained by a periodic repetition of the frequency response $G_a(f)$ of the analog filter. From (1.112):

$$G(f) = \frac{1}{\Delta t} \sum_{n=-\infty}^{+\infty} G_a\left(f - \frac{n}{\Delta t}\right) \qquad (5.115)$$

It is only when $G_a(f)$ is limited in the frequency domain within an interval $|f| \leqslant 1/2\Delta t$ that the following equation is verified:

$$G(f) = \frac{1}{\Delta t} G_a(f) \tag{5.116}$$

Because practical analog filters are not limited in the frequency domain, an aliasing error occurs in (5.115). Thus, the response $G(f)$ is not an exact replica of $G_a(f)$, and (5.116) can be used only approximately. Moreover, for higher sampling frequencies (Δt small), the gain of the digital filter may be very large. To avoid the term Δt in (5.116), it is useful to modify (5.116) in the following way:

$$g(k) = \Delta t\, g_a(k\,\Delta t) \tag{5.117}$$

5.4.18 Example

Let us consider the impulse response of a simple analog low-pass filter: the RC filter. We have

$$g_a(t) = \exp(-t/RC)$$

and

$$(5.118)$$

$$G_a(s) = \frac{1}{s + \dfrac{1}{RC}}.$$

The unique pole of the transfer function is given by

$$z_0 = \exp\left(-\frac{\Delta t}{RC}\right) \tag{5.119}$$

from which

$$G(z) = \frac{1}{1 - \exp(-\Delta t/RC)\, z^{-1}} \tag{5.120}$$

Filtering can then be performed by using the following first-order difference equation:

$$y(k) = x(k) + \exp(-\Delta t/RC)\, y(k-1) \tag{5.121}$$

The frequency response of the analog filter is given by

$$G_a(f) = \frac{1}{1 + j\, 2\pi\, RC f_a} \qquad\qquad (5.122)$$

but, for the corresponding digital filter, we have

$$G(f) = \frac{1}{1 - \exp(-\Delta t/RC)\exp(-j\,2\pi f\,\Delta t)} \qquad\qquad (5.123)$$

Figure 5.23 shows the moduli of the functions $G_a(f_a)$ and $G(f)$ for two values of the sampling period Δt to illustrate the aliasing error. This error is significantly decreased if Δt is chosen sufficiently small.

5.5 MINIMUM PHASE SYSTEMS

5.5.1 Introduction

The specifications given for the design of a digital filter very often involve the modulus of the frequency response. The argument of this function, that is,

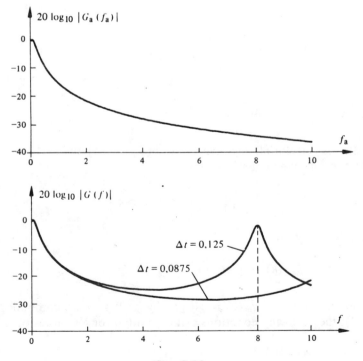

Fig. 5.23

the phase-frequency response, is then determined by stability and causality constraints. Similar situations are encountered in random signal processing, where the characteristics of the desired filter are based only on the square of the modulus of its frequency response. To implement filters specified in this way, it is necessary to have a phase frequency response which cannot be chosen arbitrarly.

In this section, we establish the relationship between the real and imaginary parts of the frequency response of a causal system. This relation is then used to define minimum phase systems, for which the phase-frequency response can be obtained from the logarithm of the amplitude frequency response.

5.5.2 Decomposition of the impulse response of a causal system

Let us consider a stable, causal, and linear time-invariant digital system. The impulse response $g(k)$ of such a system is zero for all the negative values of k. Moreover, a real function can always be expressed as the sum of an even and odd function. By applying the general equations (3.42) to (3.44) to the particular case of the signal $g(k)$, we obtain (vol. IV, sect. 7.2.7):

$$g(k) = g_{ev}(k) + g_{odd}(k) \tag{5.124}$$

with

$$g_{ev}(k) = \frac{1}{2}[g(k) + g(-k)] \tag{5.125}$$

and

$$g_{odd}(k) = \frac{1}{2}[g(k) - g(-k)] \tag{5.126}$$

Because $g(k)$ is zero for all negative values of k, (5.125) and (5.126) imply that for $k > 0$, the following equation gives:

$$g_{ev}(k) = g_{odd}(k) = \frac{1}{2}g(k) \tag{5.127}$$

For k negative, we have

$$g_{ev}(k) = -g_{odd}(k) = \frac{1}{2}g(-k) \tag{5.128}$$

At the origin $k = 0$, $g_{ev}(0) = g(0)$ and $g_{odd}(0) = 0$. By combining (5.127) and (5.128), the following equations can be written for all values of k:

$$g_{odd}(k) = g_{ev}(k)\, h(k) \tag{5.129}$$

$$g_{ev}(k) = g_{odd}(k)\, h(k) + g(0)\, d(k) \tag{5.130}$$

where $d(k)$ is the unit impulse signal, and $h(k)$ is defined by

$$h(k) = \begin{cases} -1 & \text{for } k < 0 \\ 0 & \text{for } k = 0 \\ 1 & \text{for } k > 0 \end{cases} \tag{5.131}$$

5.5.3 Hilbert transform

By computing the Fourier transform for both sides of (5.129) and (5.126), we can see that the real and imaginary parts of the frequency response $G(f)$ are the Fourier transforms of $g_{ev}(k)$ and $g_{odd}(k)$, respectively. Because these two signals are related by the simple products (5.129) and (5.130), the relationship between the real and imaginary parts of $G(f)$ is a periodic convolution. For example, the equivalent of (5.129) in the frequency domain is given by

$$j \, \text{Im}\,[G(f)] = \int_{-1/2}^{1/2} \text{Re}\,[G(u)]\, H(f-u)\, du \tag{5.132}$$

where $H(f)$ is the Fourier transform of the signal $h(k)$. For (5.130), the following expression is obtained in a similar way:

$$\text{Re}\,[G(f)] = j \int_{-1/2}^{1/2} \text{Im}\,[G(u)]\, H(f-u)\, du + g(0) \tag{5.133}$$

Thus, for a causal system, the imaginary part of the frequency response of this system can be obtained from its real part, or, conversely, the real part from the imaginary and the additional constant $g(0)$. Equations (5.132) and (5.133) define the *Hilbert Transform* (vol. IV, sect. 7.3.33).

5.5.4 Fourier transform of the signal $h(k)$

Strictly speaking, it is not possible to compute the Fourier transform of the signal $h(k)$ because it is not absolutely integrable. A direct computation leads to geometrical series having a kernel equal to one. However, this difficulty can be avoided by using the well known idea of considering the signal $h(k)$ as the limit of another signal which has a Fourier transform that is easy to compute. Therefore, let us consider the following exponential signal:

$$h_0(k) = a^k \, \epsilon(k-1) \tag{5.134}$$

where a is a real constant of absolute value less than one. Using this signal and its symmetrical reflection $h_0(-k)$, a new signal $h_1(k)$ can be constructed in the following way:

$$h_1(k) = h_0(k) - h_0(-k) \tag{5.135}$$

It is clear that

$$\lim_{a \to 1} h_1(k) = h(k) \tag{5.136}$$

The Fourier transform of $h_1(k)$ is given by

$$H_1(f) = H_0(f) - H_0^*(f) = 2j \, \text{Im} \, [H_0(f)]$$

where $H_0(f)$ is the Fourier transform of $h_0(k)$. We obtain

$$H_0(f) = \sum_{k=1}^{+\infty} a^k e^{-j2\pi fk} = \sum_{k=1}^{\infty} (a \, e^{-j2\pi f})^k$$

$$= \frac{a \, e^{-j2\pi f}}{1 - a \, e^{-j2\pi f}} \tag{5.137}$$

The limit condition of tending toward one can be taken for any value of f except the origin $f = 0$. We have

$$\lim_{a \to 1} H_0(f) = \frac{\exp(-j2\pi f)}{1 - \exp(-j2\pi f)} = -\frac{j \exp(-j\pi f)}{2 \sin \pi f} \tag{5.138}$$

Two times the imaginary part of this expression is the desired Fourier transform $H(f)$. Thus,

$$H(f) = -j \, \frac{\cos \pi f}{\sin \pi f} = -j \cot \pi f \tag{5.139}$$

The discontinuity at the origin of this function is well known. For greater mathematical precision, (5.132) should be written as follows:

$$\text{Im} \, [G(f)] = -\lim_{\epsilon \to 0} \left\{ \int_{-1/2}^{f-\epsilon} \text{Re} \, [G(u)] \cot [\pi(f-u)] \, du \right.$$

$$\left. + \int_{f+\epsilon}^{1/2} \text{Re} \, [G(u)] \cot [\pi(f-u)] \, du \right\} \tag{5.140}$$

which is also valid for (5.133).

5.5.5 Comments

Because $h(k)$ is an odd signal, its Fourier transform is purely imaginary. Thus, (5.132) and (5.133) are real. To implement these equations, the indirect method of using the Fourier transform with the simple products (5.129) and (5.130) may be simpler due to the discontinuity of $H(f)$ at the origin.

5.5.6 Minimum phase systems

Let us now consider the complex logarithm of the transfer function of the stable, causal, and linear system for which (5.132) and (5.133) were established. From definition (2.81), we have

$$\hat{G}(z) = \ln [G(z)] = \ln |G(z)| + j \arg [G(z)] \tag{5.141}$$

The complex function $\hat{G}(z)$ so defined can be considered as the z-transform of a signal $\hat{g}(k)$, which in turn, can be considered as the impulse response of another linear system. Let us recall that the z-transform is a Laurent expansion defining an analytical function. Thus, to use $\hat{G}(z)$ as a z-transform, that is, for $\hat{G}(z)$ to be an analytical function, the ambiguity of 2π on the argument in (5.141) must be removed. This can be done by defining the argument in (5.141) such that it is a continuous function. If the unit circle is contained within the region of convergence of $\hat{G}(z)$, then it can be evaluated on the unit circle to obtain the frequency response. To apply the previous results (5.132) and (5.133) to this new system, where the transfer function is given by (5.141), the stability and causality conditions must also be imposed. The stability condition implies the following equation:

$$\sum_{k=-\infty}^{+\infty} |\hat{g}(k)| < \infty \tag{5.142}$$

If this condition is satisfied, then $\hat{G}(z)$ converges on the unit circle. This enables us to deduce the existence of the frequency response $\hat{G}(f)$. To be causal as well, the system must have all the poles of $\hat{G}(z)$ inside the unit circle. If these two conditions are satisfied, then (5.132) and (5.133) can be applied to the real and imaginary parts of $\hat{G}(f)$. Because

$$\hat{G}(f) = \ln |G(f)| + j \arg [G(f)] \tag{5.143}$$

we have

$$\arg [G(f)] = \int_{-1/2}^{1/2} \ln |G(u)| \cot [\pi(u-f)] \, du \tag{5.144}$$

and

$$\ln|G(f)| = \int\limits_{-1/2}^{1/2} \arg[G(u)] \cot[\pi(f-u)] \, du + \hat{g}(0) \tag{5.145}$$

Equations (5.144) and (5.145), applied to $\hat{G}(f)$, tie together the logarithm of the amplitude-frequency response and the phase-frequency response of the initial system. The poles of $\hat{G}(z)$ must be inside the unit circle to ensure the validity of these equations. Because $\hat{G}(z)$ and $G(z)$ are related by (5.141), this condition implies that not only the poles, but also the zeros of $G(z)$, must be inside the unit circle. The logarithm diverges for the poles and zeros of $G(z)$. Thus, if the poles and zeros of the transfer function of a linear system are inside the unit circle, then the modulus and the argument are related by (5.144) and (5.145). Such systems are called *minimum phase systems* for reasons which will become apparent later.

5.5.7 Comments

It is clear that the previous development is also valid for a causal system. If the modulus and the argument of the Fourier transform of a causal system are related by the same formulas (5.144) and (5.145), it is called *minimum phase signal*.

If a digital filter to be implemented is specified only by its amplitude-frequency response, then (5.144) can be used to obtain the phase-frequency response. Thus, the poles *and* zeros of the transfer function of this filter are inside the unit circle.

To justify the terminology of minimum phase, it is useful to compare systems which are minimum phase with those which are not. This comparison will yield other properties of minimum phase systems. Since we are trying to obtain the argument of a frequency response from its modulus, a relationship must be established between two systems having the same amplitude-frequency response, with one of them being minimum phase.

5.5.8 All-pass filter

The desired equation is of the form:

$$G(z) = G_{\min}(z) \, G_{\text{ap}}(z) \tag{5.146}$$

where $G_{\min}(z)$ is the transfer function of a minimum phase system such that $|G_{\min}(f)| = |G(f)|$, and where $G_{\text{ap}}(z)$ is the transfer function of a system having an amplitude-frequency response equal to one, that is, $|G_{ap}(f)| = 1$. A system which satisfies this last condition is called an *all-pass filter*. All-pass filters differ only in their phase-frequency response (sect. 1.8.13).

5.5.9 Function of the all-pass filter

Let $G(z)$ be the transfer function of a causal, stable system that is not necessarily minimum phase. If this is not a minimum phase system, it must have some zeros outside the unit circle. To be more precise, let us assume M zeros, Z_n, outside the unit circle. We can then write $G(z)$ as

$$G(z) = G_0(z) \prod_{m=1}^{M} (z - z_m) \tag{5.147}$$

where $G_0(z)$ is the transfer function of a system having no zeros outside the unit circle. Thus, $G_0(z)$ represents a minimum phase system. The decomposition (5.147) is also valid for the transfer function $G(z)$ given by (5.146). Let us note, however, that $|G(f)$ and $|G_0(f)|$ are not identical. Because the systems characterized by $G(z)$ and $G_{min}(z)$ are causal, the only difference is in the positions of the zeros. The function $G_{min}(z)$, by definition, has no zeros outside unit circle, but $G(z)$ has M zeros there. Thus, the transfer function of the all-pass filter must transfer M zeros of $G_{min}(z)$ outside the unit circle without modifying the amplitude-frequency response $|G_{min}(f)|$. A zero may be balanced by a pole. In order to avoid modifying the order of the system and $|G_{min}(f)|$, a new zero must be introduced outside the unit circle. Thus, for only one zero z_i inside the unit circle, the function $G_{ap}(z)$ must have the following form:

$$G_{ap}(z) = \alpha \frac{z - z_i'}{z - z_i} \tag{5.148}$$

where a is a complex scale coefficient.

This system balances the zero z_i by its pole z_i and introduces a new zero z'_i. Indeed, we must have

$$|G_{ap}(f)| = |\alpha| \left| \frac{\exp(j\,2\pi f) - z_i'}{\exp(j\,2\pi f) - z_i} \right| = 1 \tag{5.149}$$

To transfer M zeros outside of the unit circle, it is sufficient to cascade a sequence of M systems of the form (5.149). An all-pass filter thus transfers the zeros outside of the unit circle without modifying the amplitude-frequency response.

5.5.10 Design of the all-pass filter

Assuming a given zero z_i, condition (5.149) enables us to locate the zero z'_i outside the unit circle. By expressing z_i and z'_i from (5.149) in polar coordinates using $r_i \exp(j\theta_i)$ and $z'_i \exp(j\theta'_i)$, the following result is obtained:

$$\frac{1 + r_i'^2 - 2 r_i' \cos(2\pi f - \theta_i')}{1 + r_i^2 - 2 r_i \cos(2\pi f - \theta_i)} = \frac{1}{|\alpha|^2} \tag{5.150}$$

This equation must be satisfied for all values of f within the interval $[-1/2, 1/2]$. By expressing it for two values of f to eliminate the argument θ_i in the denominator and identifying the two terms obtained having the same value $1/|\alpha|^2$, the following equation is obtained:

$$\cos(\theta_i + 2\pi f - \theta_i') = \cos(\theta_i - 2\pi f - \theta_i') \tag{5.151}$$

which implies $\theta_i = \theta'_i$. Thus, the zero z'_i outside the unit circle must have the same argument as z_i. In other words, it must be on the same radial line as z_i. By setting $\theta_i = \theta'_i = 0$ in (5.150) (which corresponds to the rotation of the coordinate axis by θ_i radians) and identifying the two expressions obtained for $f = 0$ and $f = 1/2$, we have

$$\frac{(1 + r_i')^2}{(1 + r_i)^2} = \frac{(1 - r_i')^2}{(1 - r_i)^2} = \frac{1}{|\alpha|^2} \tag{5.152}$$

The square root can be taken with either positive or negative sign. The positive square root leads to the unacceptable solution $r'_i = r_i$, which identifies the two zeros. The negative square root leads to

$$r_i' = \frac{1}{r_i} \tag{5.153}$$

Because z_i is inside the unit circle, r_i is less than one, and consequently r'_i is greater than one. Thus, the zero outside the unit circle is given by

$$z_i' = \frac{1}{z_i^*} \tag{5.154}$$

An immediate calculation of (5.152) shows that $|\alpha|^2 = r_i^2$. Because α is a complex number and the solution is obtained with a negative square root, it is possible to take $\alpha = z_i^*$. Equation (5.148) then becomes

$$G_{\text{iap}}(z) = z_i^* \frac{\dfrac{1}{z_i^*} - z}{z - z_i} = \frac{z^{-1} - z_i^*}{1 - z_i z^{-1}} \tag{5.155}$$

For M zeros $z_i = r_i \exp(j\theta_i)$ inside the unit circle, the transfer function of the all-pass filter is thus given by

$$G_{ap}(z) = \prod_{i=1}^{M} \frac{z^{-1} - z_i^*}{1 - z_i z^{-1}} \qquad (5.156)$$

The all-pass filter of order M so obtained is causal because its M poles are, by definition, inside the unit circle. This result shows that any causal system can be characterized by a sequence of a minimum phase system and an all-pass system.

5.5.11 Interpretation

Let us consider a first-order all-pass filter with a real zero. We have

$$G_{ap}(z) = \frac{z^{-1} - r}{1 - r z^{-1}} = \frac{r\left(\dfrac{1}{r} - z\right)}{(z - r)} \qquad \text{with } r < 1 \qquad (5.157)$$

The phase-frequency response of this filter can be studied by using the geometrical interpretation of section 2.6.3. The vectors representing the numerator and the denominator of (5.157) are illustrated in Fig. 5.24. This figure shows that for values of f within the interval $[0, 1/2]$, corresponding to the upper half

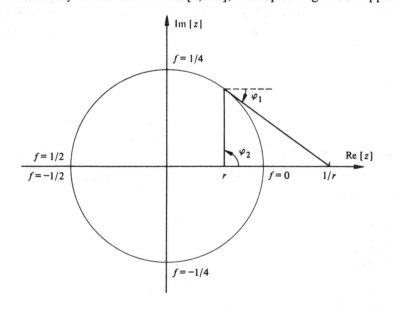

Fig. 5.24

of the unit circle, the phase $\theta(f) = \varphi_1(f) - \varphi_2(f)$ is always negative. Because the phases are added to evaluate the phase of $G(f)$ in the product (5.146), this shows that in said interval the phase of the frequency response $G_{min}(f)$ is always less negative than that of $G(f)$. However $-\theta(f)/f$ is called *phase delay*. Consequently, $G_{min}(f)$ always has a phase delay smaller than that of any system having at least one zero outside the unit circle. Therefore, $G_{min}(z)$ is a minimum phase delay system, commonly abbreviated as *minimum phase system*.

5.5.12 Comments

The preceding results show that it is possible to transform a system of minimum phase into one which is not, without modifying its amplitude-frequency response. It is sufficient to transfer the zeros inside the unit circle to the outside on radial lines at distances given by the inverse distances to these zeros. Conversely, it is possible to transform any system into a minimum phase without modifying its amplitude-frequency response. For a system in which the transfer function $G(z)$ has N zeros, 2^N different phase-frequency responses can be found for the same function $|G(f)|$. Indeed, because each zero can be either inside or outside of the unit circle, the number of different configurations is 2^N.

In contrast to the case of minimum phase, systems in which the transfer function has neither zeros nor poles inside the unit circle are called *maximum phase systems*.

It is clear that the same development as for the all-pass filter can be repeated to move a pole inside the unit circle. This allows us to transform an unstable causal system into a stable one without modifying its amplitude-frequency response.

5.5.13 Example

Let us consider a filter which has a transfer function given by

$$G_0(z) = 1 - 1.386\, z^{-1} + 0.640\, z^{-2} \tag{5.158}$$

This expression clearly shows that it is a FIR filter of length $L = 3$. The impulse response is given by

$$
\begin{aligned}
g_0(k) &= 0 && \text{for } k < 0 \\
g_0(0) &= 1 \\
g_0(1) &= -1.386 \\
g_0(2) &= 0.640 \\
g_0(k) &= 0 && \text{for } k > 2
\end{aligned}
\tag{5.159}
$$

The function $G_0(z)$ has a second-order pole at the origin $z = 0$. Its zeros are

$$z_1 = 0.693 + j\,0.4 = 0.8\,\exp(j\,\pi/6)$$

and

(5.160)

$$z_2 = 0.693 - j\,0.4 = 0.8\,\exp(-j\,\pi/6)$$

Because the poles and zeros are inside the unit circle, $G_0(z)$ represents a minimum phase system. Thus, we have

$$G_0(z) = G_{min}(z) = (1 - z_1\,z^{-1})(1 - z_2\,z^{-1})$$

(5.161)

The frequency response of this filter is given by

$$G_{min}(f) = 1 - 1.386\,\exp(-j\,2\,\pi\,f) + 0.64\,\exp(-j\,4\,\pi\,f)$$

(5.162)

The impulse response, the position of the poles and zeros, and the modulus and argument of $G_{min}(f)$ are shown in Fig. 5.25.

It is clear that because we are dealing with a minimum phase filter, the Hilbert transform of the logarithm of $|G_{min}(f)|$ can be computed to obtain the phase-frequency response. The analytical computation in this case is complicated, but may be carried out by using the FFT.

A new system having the same amplitude-frequency response can be obtained by transferring one or both zeros z_1 and z_2 of $G_{min}(z)$ to the outside of the unit circle. The transfer function of an all-pass filter for the two zeros has the following form:

$$
\begin{aligned}
G_{ap}(z) &= \frac{z^{-1} - z_1^*}{1 - z_1\,z^{-1}} \cdot \frac{z^{-1} - z_2^*}{1 - z_2\,z^{-1}} \\[2mm]
&= \frac{z^{-2} - (z_1^* + z_2^*)\,z^{-1} + z_1^*\,z_2^*}{z_1 z_2\,z^{-2} - (z_1 + z_2)\,z^{-1} + 1} \\[2mm]
&= \frac{z^{-2} - 1.386\,z^{-1} + 0.64}{0.64\,z^{-2} - 1.386\,z^{-1} + 1}
\end{aligned}
$$

(5.163)

It can easily be verified that $|G_{ap}(f)| = 1$ for any frequency. Figure 5.26 shows the positions of the poles and zeros of $G_{ap}(z)$ as well as the phase-frequency response. This figure shows that the poles of $G_{apt}(z)$ coincide with the zeros of $G_{min}(z)$, and its zeros are $1/z_1^*$ and $1/z_2^*$. The phase-frequency response

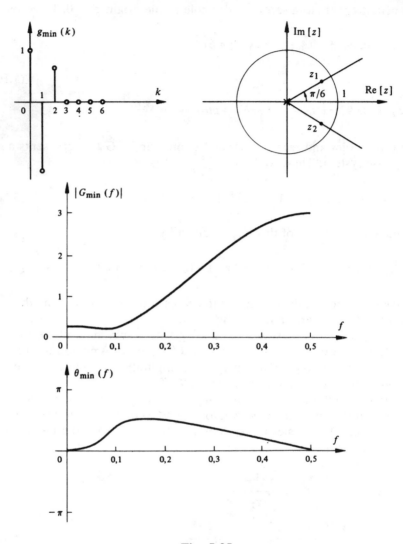

Fig. 5.25

shown in the same figure is negative on the interval $[0, 1/2]$, according to the theory, if the main determination $[-\pi, \pi]$ is not taken into account.

The transfer function for the new system, which is no longer of minimum phase, is given by

$$G(z) = G_{min}(z)\, G_{ap}(z) = 0.64 - 1.386\, z^{-1} + z^{-2} \qquad (5.164)$$

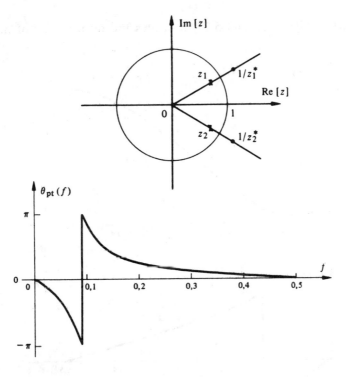

Fig. 5.26

The impulse response of this system flows immediately from its transfer function. We have

$$g(k) = 0 \qquad \text{for } k < 0$$
$$g(0) = 0.64$$
$$g(1) = -1.386 \qquad (5.165)$$
$$g(2) = 1$$
$$g(k) = 0 \qquad \text{for } k > 2$$

The corresponding frequency response is given by

$$G(f) = 0.64 - 1.386 \exp(-j\,2\,\pi f) + \exp(-j\,4\,\pi f) \qquad (5.166)$$

It can easily be verified that $|G(f)| = |G_{min}(f)|$. Figure 5.27 shows the modulus and the argument of $G(f)$ computed directly from (5.166). The comparison between Figs. 5.25 and 5.27 illustrates the terminology of minimum phase.

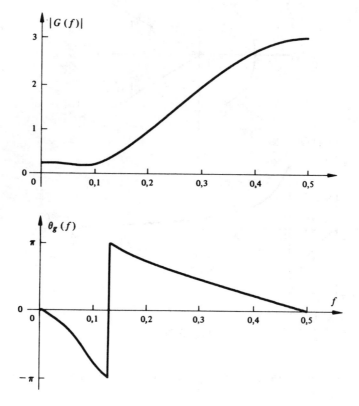

Fig. 5.27

5.6 EXERCISES

5.6.1 Using (5.54), prove (5.55).

5.6.2 What are the amplitude- and phase-frequency responses of the system defined by the equation

$$y(k) = x(k) - x(k - L)$$

5.6.3 Let us consider a system defined by the following difference equation:

$$y(k) = \frac{1}{3} [x(k + m) + x(k + m - 1) + x(k + m - 2)]$$

where m is an integer.

Show that this system is linear and time-invariant. Study the problems of causality and stability with respect to m. Compute the transfer function $G(z)$ for $m = 0$ and $m = 1$, and derive the frequency response. What is the effect of this system?

5.6.4 Find the difference equations for the bandpass filter studied in section 5.3.22 with a coefficient ρ equal to $1 - 1/2^8$.

5.6.5 The transfer function of a second-order analog Chebyshev filter with a normalized cut-off frequency $1/2\pi$ and a ripple of 1 dB is given by

$$G_a(s) = \frac{1.103}{s^2 + 1.098\,s + 1.103}$$

What is the difference equation of the equivalent digital filter obtained by using the equivalence of the derivation. Compare the analog and digital frequency responses obtained for $\Delta t = 1$ and $\Delta t = 1/8$.

5.6.6 With the data from exercise 5.6.5, establish the difference equation of the equivalent digital filter by using the bilinear transformation and a digital cut-off frequency of $f_c = 0.15$. Represent graphically the amplitude-frequency response of the digital filter thus obtained.

5.6.7 Verify that the system having a transfer function:

$$G(z) = \frac{z^{-2} - 1.386\,z^{-1} + 0.64}{0.64\,z^{-2} - 1.386\,z^{-1} + 1}$$

is an all-pass filter. Show that its phase-frequency response is negative in the interval $[0, 1/2]$.

5.6.8 Let us consider a digital filter having a transfer function given by

$$G(z) = \frac{0,0154 + 0,0461\,z^{-1} + 0,0461\,z^{-2} + 0,0154\,z^{-3}}{1 - 1,9903\,z^{-1} + 1.5717\,z^{-2} - 0.458\,z^{-3}}$$

It is a low-pass filter with a cut-off frequency of $f_c = 0.1$ (sect. 5.4.14). What type of filter is obtained if the change of variable

$$z^{-1} = -\frac{w^{-1} + \alpha}{1 + \alpha\,w^{-1}} \qquad \text{with } \alpha = -0.382$$

is done for the given transfer function $G(z)$?

5.6.9 Let us consider two FIR filters having the impulse responses shown in Fig. 5.28. Which of these two responses must be used to process a high-pass filter? Justify your answer.

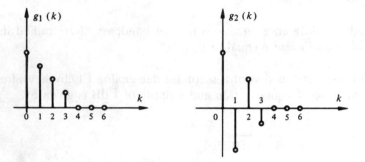

Fig. 5.28

5.6.10 Let $g(k)$ be the impulse response of a causal FIR filter of length L. We additionally assume that the first nonzero sample is $g(0)$. From this response we derive $L - 1$ impulse responses, obtained by cyclic permutation on the interval $(0, L - 1)$. By using the DFT, determine those which correspond to

- a low-pass filter
- an high-pass filter
- a bandpass filter
- a stopband filter

These filters differ by what characteristic?

5.6.11 The transfer function of a filter is given by

$$G(z) = \frac{1}{(z^2 - 2.828\,z + 4)(z^2 + 2.828\,z + 4)} \qquad \text{with } |z| > 2$$

Determine the conditions for which the filter is

- stable
- causal
- stable and causal

In the unstable case, how can the filter be made stable without modifying the amplitude frequency response? Establish the transfer function of the stabilized filter thus obtained.

Chapter 6

Digital Spectrum Analysis

6.1 INTRODUCTION

6.1.1 Preliminary remarks

In chapter 3, the DFT and its use in measuring the amplitude and phase spectra of deterministic signals were studied. The use of the DFT is aided by a fast computation algorithm, the FFT (chapter 4). However, these methods cannot be applied directly to random signals, the frequency representation of which requires a statistical description. Although we will use previously discussed notions, such as the Fourier transform of a finite sequence of samples, the results obtained will be interpreted by using new concepts from estimation theory.

To judge the results quantitatively, theoretical models will be developed by using the linear filtering methods of chapter 5.

6.1.2 Choice of the methods

The choice of the methods in digital spectrum analysis is relatively large. However, there is no single method which is satisfactory in all respects. That is why several methods will be studied, as well as their advantages and disadvantages. The user can choose the method which best meets the requirements of his own problem.

Because of space limitations, the maximum entropy and related methods will not be considered in this chapter.

6.1.3 Organization of the chapter

First of all, the spectral analysis of random signals is an estimation problem. To facilitate the introduction, a short review of estimation theory and its application to spectral analysis shall be given after some fundamental definitions.

Because of the importance of understanding the role played by the parameters of the analysis, each one is illustrated by an example.

When excited by a special random signal, called *white noise,* linear systems allow us to assign a given autocorrelation function or spectral density to the output signal. These methods are studied in detail and used in the examples. They permit us to compare theoretical and experimental results by using the notions of chapter 5.

Estimators for the autocorrelation and crosscorrelation functions are presented. Their properties are studied and illustrated by examples.

Spectral estimators are studied in detail. Their properties are established in simple cases, but some longer proofs of less importance are omitted. All the parameters and their properties discussed theoretically are illustrated by two practical examples of spectral estimation.

The reader who is unfamiliar with the elementary notions of probability theory used in this chapter can refer to *Signal Theory and Processing* by F. de Coulon (Artech House, 1986).

6.2 FREQUENCY DESCRIPTION OF RANDOM SIGNALS

6.2.1 Introduction

A *random signal* is a signal which has evolution in time governed by statistical rules. Such a signal does not have an analytical temporal description. It is considered to be a typical realization caused by a random phenomenon or process. The global behavior of this process is described by statistical rules. A frequency description of random signals can be derived from these rules by using the notion of *power spectrum.*

In this section, we present the functional relationships involved in the spectral analysis of random signals and the models used for the generation of random signals with given frequency distributions. Throughout this chapter, we shall assume that the random signal under consideration is being produced by *stationary* and *ergodic* random processes. Let us recall that these hypotheses allow the system to be invariant with respect to a time delay and also to identify ensemble and time averages.

6.2.2 Definitions

The *autocorrelation function* of a stationary random signal is defined by the following mathematical expectation:

$$\varphi_x(k) = \mathrm{E}\left[x(l)\,x(l+k)\right] \tag{6.1}$$

The ergodicity hypothesis allows us to identify ensemble and time averages. For an ergodic and stationary random signal, we have

$$\varphi_x(k) = \overline{x(l) \, x(l+k)} \tag{6.2}$$

where the bar represents the mean value with respect to the variable l. The *autocovariance function* of an ergodic and stationary random signal is defined by

$$\gamma_x(k) = E\left[(x(l) - m_x)(x(l+k) - m_x)\right]$$
$$= \overline{[x(l) - m_x][x(l+k) - m_x]} \tag{6.3}$$

where m_x is the mean value of the signal $x(k)$. By comparing (6.2) and (6.3), we can show that

$$\varphi_x(k) = \gamma_x(k) + m_x^2 \tag{6.4}$$

The *normalized autocovariance function* is defined by

$$\rho_x(k) = \frac{\gamma_x(k)}{\gamma_x(0)} = \frac{1}{\sigma_x^2}\, \gamma_x(k) \tag{6.5}$$

where σ_x^2 is the variance of the signal $x(k)$. The maximum value of this function is one, which is obtained at the origin. Thus,

$$\rho_x(0) = 1 \tag{6.6}$$

It can easily be shown that $\varphi_x(k)$, $\gamma_x(k)$, and $\rho_x(k)$ are even functions of the variable k.

The *power spectral density* of a random signal is defined by

$$\Phi_x(f) = \sum_{k=-\infty}^{+\infty} \varphi_x(k) \exp(-j\,2\pi fk) \tag{6.7}$$

This is simply the Fourier transform of the autocorrelation function. Because of its frequent use, the term will be shortened to *power spectrum*.

The *crosscorrelation function* of two random signals is defined by

$$\varphi_{xy}(k) = E\left[x(l)\,y(l+k)\right]$$
$$= \overline{x(l)\,y(l+k)} \tag{6.8}$$

The *cross-power spectral density* is defined by

$$\Phi_{xy}(f) = \sum_{k=-\infty}^{+\infty} \varphi_{xy}(k) \exp(-j\,2\pi fk) \tag{6.9}$$

The role of digital spectrum analysis is to estimate functional relationships, such as (6.2), (6.7), (6.8), and (6.9), given a finite number of samples K of the signals $x(k)$ and $y(k)$.

6.2.3 Response of linear systems to random signals

By studying the response of linear systems to random signals, a very useful model can be developed to generate a set of random processes having well-defined characteristics. Let us consider a linear system represented by its impulse response $g(k)$. If such a system is excited by a random signal, the output, which is also a random signal, is given by the following well known equation (see sect. 1.6.5):

$$y(k) = \sum_{l=-\infty}^{+\infty} g(l) x(k-l) \tag{6.10}$$

Because the sequence of samples $x(k)$ is not known, the input signal can be characterized, for example, by its mean value m_x, its autocorrelation function $\varphi_x(k)$, or its power spectrum $\Phi_x(f)$. This is also true for the output signal.

The mean value of the output signal is given by

$$m_y = E[y(k)] = \sum_{l=-\infty}^{+\infty} g(l) E[x(k-l)]$$

$$= m_x \sum_{l=-\infty}^{+\infty} g(l) \tag{6.11}$$

This equation shows that the mean value of the output is constant.

The autocorrelation of the output signal is given by

$$\varphi_y(k) = E[y(l) y(l+k)]$$

$$= E\left[\sum_{u=-\infty}^{+\infty} g(u) x(l-u) \sum_{v=-\infty}^{+\infty} g(v) x(l+k-v) \right]$$

$$= \sum_{u=-\infty}^{+\infty} g(u) \sum_{v=-\infty}^{+\infty} g(v) E[x(l-u) x(l+k-v)] \tag{6.12}$$

Since, by hypothesis, the signal $x(k)$ is stationary, we have

$$E[x(l-u) x(l+k-v)] = E[x(l) x(l+k+u-v)]$$

$$= \varphi_x(k+u-v) \tag{6.13}$$

By substituting this result into (6.12) and using the change of variable $m = v - u$, the following result is obtained:

$$\varphi_y(k) = \sum_{u=-\infty}^{+\infty} g(u) \sum_{m=-\infty}^{+\infty} g(u+m)\, \varphi_x(k-m)$$

$$= \sum_{m=-\infty}^{+\infty} \varphi_x(k-m) \sum_{u=-\infty}^{+\infty} g(u)\, g(u+m)$$

$$= \sum_{m=-\infty}^{+\infty} \varphi_x(k-m)\, \varphi_g(m) \tag{6.14}$$

Thus, the autocorrelation function of the output signal is the convolution product of the autocorrelation functions of the input signal and the impulse response of the system.

By applying the z-transform to both sides of (6.14), we obtain

$$\Phi_y(z) = G(z)\, G\!\left(\frac{1}{z}\right) \Phi_x(z) \tag{6.15}$$

By evaluating this equation on the unit circle, we obtain

$$\Phi_y(f) = G(f)\, G^*(f)\, \Phi_x(f)$$
$$= |G(f)|^2\, \Phi_x(f) \tag{6.16}$$

Equation (6.16) relates the power spectra of the input and output signals to the frequency response of the system.

6.2.4 Difference equation for the correlation

Let us consider a difference equation which links the excitation and the response of an invariant linear system (sect. 1.6.2):

$$\sum_{n=0}^{N} a_n\, y(k-n) = \sum_{m=0}^{M} b_m\, x(k-m) \tag{6.17}$$

This equation is still valid if the input signal $x(k)$ is a random signal. In general, the output signal is also a random signal.

For *zeroth*-order systems ($N = 0$), this equation reduces to

$$y(k) = \sum_{m=0}^{M} b_m\, x(k-m) \qquad \text{with } a_0 = 1 \tag{6.18}$$

This is a particular case of (6.10), where the system has a finite-length impulse response, with $g(l) = b_l$. The length of $g(k)$ is $M + 1$. Therefore.

$$\varphi_g(k) = \sum_{l=0}^{M} b_l\, b_{l+k} \tag{6.19}$$

The length of $\varphi_g(k)$ is $2(M + 1) - 1 = 2M + 1$. The particular form of (6.14) is thus the following:

$$\varphi_y(k) = \sum_{m=-M}^{M} \varphi_g(m)\, \varphi_x(k-m) \tag{6.20}$$

For Nth-order systems, let us consider (6.17) in the particular case of $M = 0$. We have

$$\sum_{n=0}^{N} a_n\, y(k-n) = b_0\, x(k) \tag{6.21}$$

By multiplying both sides by $y(k - l)$ and then taking the expected value, the following result is obtained:

$$E\left[y(k-l) \sum_{n=0}^{N} a_n\, y(k-n) \right] = E[y(k-l)\, b_0\, x(k)]$$

$$\sum_{n=0}^{N} a_n\, E[y(k-n)\, y(k-l)] = b_0\, E[x(k)\, y(k-l)]$$

$$\sum_{n=0}^{N} a_n\, \varphi_y(l-n) = b_0\, \varphi_{xy}(-l) = b_0\, \varphi_{yx}(l) \tag{6.22}$$

Similarly, it can be shown that

$$\sum_{n=0}^{N} a_n\, \gamma_y(l-n) = b_0\, \gamma_{yx}(l) \tag{6.23}$$

and

$$\sigma_y^2 \sum_{n=0}^{N} a_n\, \rho_y(l-n) = b_0\, \sigma_x\, \sigma_y\, \rho_{yx}(l) \tag{6.24}$$

Equations (6.22) to (6.24) are difference equations for the autocorrelation and autocovariance functions of the output signal $y(k)$. The particular case of

$M = 0$ is interesting because it involves only one value of the crosscorrelation, which simplifies the computations.

In the general case $(M \neq 0)$, we have

$$\sum_{n=0}^{N} a_n \varphi_y(l-n) = \sum_{m=0}^{M} b_m \varphi_{yx}(l-m) \qquad (6.25)$$

6.2.5 White noise

By exciting a linear system with a *white noise,* a large set of random signals having a known correlation or power spectral density can be generated. White noise, by definition, is a random signal, the samples of which are not correlated with each other. Thus,

$$\varphi_b(k) = \sigma_b^2 d(k) \qquad (6.26)$$

where the index b represents the white noise and σ_b^2 is its variance. From this equation, we can immediately see that

$$\Phi_b(f) = \sigma_b^2 = \text{const.} \qquad (6.27)$$

Such a signal can be generated by using the method described in chapter 1. It is clear, from definition (6.26), that the mean value of the white noise is zero.

6.2.6 Linear systems excited by white noise

When the input signal of a linear invariant system is white noise, the autocorrelation of the output signal is given by

$$\varphi_y(k) = \sigma_b^2 \varphi_g(k) = \sigma_b^2 \sum_{l=-\infty}^{+\infty} g(l) g(l+k) \qquad (6.28)$$

The power spectrum of the output signal is given by

$$\Phi_y(f) = |G(f)|^2 \sigma_b^2 \qquad (6.29)$$

Equations (6.28) and (6.29) show that by using an appropriate linear system, any desired correlation function or power spectrum can be assigned to the output signal.

If the system is characterized by an impulse response of finite length, we have

$$\varphi_y(k) = \sigma_b^2 \sum_{m=0}^{M} g(m) g(m+k)$$ (6.30)

This results from a *zero*th-order difference equation of the form (6.18).

If the system is characterized by an Nth-order difference equation, with $M = 0$, then (6.22) takes the following particular form:

$$\sum_{n=0}^{N} a_n \varphi_y(l-n) = b_0 \sigma_b^2 g(-l)$$ (6.31)

$$E[y(k-l) x(k)] = E\left[\sum_{m=-\infty}^{+\infty} g(m) x(k-l-m) x(k)\right]$$

$$= \sum_{m=-\infty}^{+\infty} g(m) E[x(k) x(k-l-m)]$$

$$= \sum_{m=-\infty}^{+\infty} g(m) \varphi_x(l+m)$$ (6.32)

Because $\varphi_x(k) = d(k)$, the only nonzero contribution in this sum is for $m = -l$.

If the system is causal, $g(-l)$ is zero for positive values of l. In this case, (6.31) becomes

$$\sum_{n=0}^{N} a_n \varphi_y(l-n) = 0 \qquad \text{for } l \geqslant 1$$

and

$$\sum_{n=0}^{N} a_n \varphi_y(n) = b_0 \sigma_b^2 g(0) \qquad \text{for } l = 0$$ (6.33)

Because $\varphi_y(k)$ is an even function, a system of N equations with N unknowns must be solved to find the initial conditions. Equation (6.33) can be solved using the z-transform. We have

$$\Phi_y(z) = \frac{b_0 \sigma_b^2 g(0)}{\displaystyle\sum_{n=0}^{N} a_n z^{-n}}$$ (6.34)

To find the roots p_i of the polynomial in the denominator, $\Phi_y(z)$ can be decomposed into partial fractions of the following type:

$$\frac{A_i}{1 - p_i z^{-1}} \tag{6.35}$$

The inverse transform of each term is a signal of the type $A_i p_i^k \epsilon(k)$, leading to the general solution:

$$\varphi_y(k) = \sum_{i=1}^{N} A_i p_i^{|k|} \tag{6.36}$$

In this expression, k appears as an absolute value because of the symmetry of the function $\varphi_y(k) = \varphi_y(-k)$.

By using (6.29) and the z-transform of the transfer function, evaluated on the unit circle, the general form of the power spectral density of the output signal can be expressed by

$$\Phi_y(f) = \left| \frac{\displaystyle\sum_{m=0}^{M} b_m \exp(-j\,2\pi f m)}{\displaystyle\sum_{n=0}^{N} a_n \exp(-j\,2\pi f n)} \right|^2 \sigma_b^2 \tag{6.37}$$

6.2.7 Particular case of a first-order system

The difference equation defining a first-order system has the following form:

$$y(k) - a\,y(k-1) = x(k) \tag{6.38}$$

where $x(k)$ is white noise.

The impulse response of this system is given by (sect. 2.7.11):

$$g(k)' = a^k\,\epsilon(k) \tag{6.39}$$

By applying (6.28) or (6.33), we obtained

$$\varphi_y(k) = \sigma_b^2\, a^{|k|} \sum_{l=0}^{+\infty} (a^2)^l \tag{6.40}$$

However,

$$\varphi_y(0) = \sigma_b^2 \sum_{l=0}^{+\infty} (a^2)^l = \frac{\sigma_b^2}{1-a^2} \tag{6.41}$$

thus,

$$\rho_y(k) = a^{|k|} \tag{6.42}$$

The power spectrum is given by

$$\Phi_y(f) = \frac{\sigma_b^2}{|1 - a \exp(-j2\pi f)|^2}$$

$$= \frac{\sigma_b^2}{1 + a^2 - 2a \cos 2\pi f} \tag{6.43}$$

Thus, depending on the value of the parameter a, which, in principle, should be less than one in absolute value to guarantee stability, a family of random processes having correlation functions of the decaying exponential form can be generated.

6.2.8 Particular case of a second-order system

The difference equation of a second-order system has the following form (sect. 2.7.12):

$$y(k) + a_1 y(k-1) + a_2 y(k-2) = x(k) \tag{6.44}$$

where $x(k)$ is white noise.

The corresponding transfer function is given by

$$G(z) = \frac{1}{1 + a_1 z^{-1} + a_2 z^{-2}} \tag{6.45}$$

The impulse response of the stable causal system has the following form:

$$g(k) = \left(\frac{p_1^{k+1}}{p_1 - p_2} + \frac{p_2^{k+1}}{p_2 - p_1} \right) \epsilon(k) \tag{6.46}$$

where p_1 and p_2 are the poles of $G(z)$ with $|p_1| < 1$ and $|p_2| < 2$.

By using (6.33), we obtain

$$\varphi_y(k) + a_1 \varphi_y(k-1) + a_2 \varphi_y(k-2) = 0 \qquad \text{for } k \geqslant 1$$
$$\varphi_y(0) + a_1 \varphi_y(-1) + a_2 \varphi_y(-2) = \sigma_b^2 g(0) \qquad \text{for } k = 0 \tag{6.47}$$

For $k = 1$, this leads to

$$\varphi_y (1) + a_1 \, \varphi_y (0) + a_2 \, \varphi_y (-1) = 0$$
$$\varphi_y (1) (1 + a_2) + a_1 \, \varphi_y (0) = 0$$
$$\varphi_y (1) = \frac{-a_1 \, \varphi_y (0)}{1 + a_2} \qquad (6.48)$$

With $p_1 + p_2 = -a_1$ and $p_1 p_2 = a_2$, the following equation is obtained:

$$\varphi_y (1) = \frac{(p_1 + p_2) \, \varphi_y (0)}{1 + p_1 p_2} \qquad (6.49)$$

Setting $k = 0$ and $k = 1$ in (6.36) gives the following result:

$$\varphi_y (0) = A_1 + A_2$$
$$\varphi_y (1) = A_1 p_1 + A_2 p_2 = \frac{(p_1 + p_2) \, \varphi_y (0)}{1 + p_1 p_2} \qquad (6.50)$$

Solving this system of equations for A_1 and A_2 leads to

$$A_1 = \frac{p_1 (1 - p_2^2) \, \varphi_y (0)}{(1 + p_1 p_2) (p_1 - p_2)}$$

and

$$A_2 = \frac{-p_2 (1 - p_1^2) \, \varphi_y (0)}{(1 + p_1 p_2) (p_1 - p_2)} \qquad (6.51)$$

Thus,

$$\varphi_y (k) = A_1 p_1^{|k|} + A_2 p_2^{|k|} \qquad (6.52)$$

The value at the origin $\varphi_y(0)$ can be obtained by solving (6.47) for $k = 0$. The power spectrum is obtained using (6.37). The result is

$$\Phi_y (f) = \frac{\sigma_b^2}{|1 + a_1 \exp(-j2\pi f) + a_2 \exp(-j4\pi f)|^2}$$

$$= \frac{\sigma_b^2}{1 + a_1^2 + a_2^2 + 2a_1 (1 + a_2) \cos 2\pi f + 2a_2 \cos 4\pi f} \qquad (6.53)$$

Equations (6.52) and (6.53) show that, for a second-order system, a larger family of random processes with given correlation functions and spectral densities can be generated.

Some particular examples of these equations will be used later as an application of the theory.

6.3 ELEMENTS OF ESTIMATION THEORY

6.3.1 Generalities

The digital spectrum analysis of a random signal is essentially an *estimation* problem. In estimation theory, the available data is used to estimate the value of a characteristic parameter and to establish the precision of this estimation. For example, from K samples of a random signal $x(k)$, the mean value can be estimated by

$$\hat{m}_x = \frac{1}{K} \sum_{k=0}^{K-1} x(k) \tag{6.54}$$

and it is possible to say that the true mean value m_x lies in the interval $[\hat{m}_x - a, \hat{m}_x + a]$, with a probability b. Statistically, the precision of the estimation is determined by the values a and b. In this section, we present some of the elementary notions of estimation theory involved in the digital spectrum analysis of a random signal. These notions will be used in later sections to study spectral estimators and their properties.

6.3.2 Definitions

Let $x(k)$ be a stationary and ergodic signal. To implement an arbitrary processing of this signal, a finite set of K samples must be considered. Because $x(k)$ is a random signal, each of these samples can be represented by a random variable. The *estimation* $\hat{\alpha}$ of a parameter α of the random process represented by K samples $x(k)$ is a function of random variables.

$$\hat{\alpha} = S(x(0), x(1), x(2), \dots, x(K-1)) \tag{6.55}$$

The function S is called the *estimator* of the corresponding parameter. Equation (6.55) shows that the estimate $\hat{\alpha}$, being a function of random variables, is also a random variable.

The probability density of the estimate $\hat{\alpha}$ will be denoted by $p_{\hat{\alpha}}(\hat{\alpha})$. Its analytical expression depends on the estimator S and the probability density of the random variables $x(k)$.

Intuitively, an estimator is considered to be satisfactory if there is a large probability that the estimate $\hat{\alpha}$ is close to the true value. Figure 6.1 shows two probability densities corresponding to two estimators.

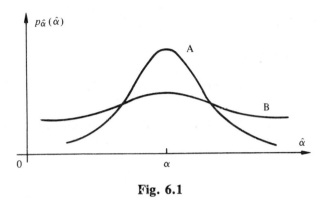

Fig. 6.1

It is clear that the estimator A is better than the estimator B because the density $p_{\hat{\alpha}}(\hat{\alpha})$ is more concentrated in the vicinity of the desired value α.

6.3.3 Bias and variance

For an objective comparison, two parameters are commonly used to characterize an estimator: the *bias* and the *variance*. The bias of an estimator is, by definition, the difference between its expected value and the desired true value; that is

$$B_{\hat{\alpha}} = E[\hat{\alpha}] - \alpha \tag{6.56}$$

If the bias is zero, the probability density of the estimate is centered on the desired value. The corresponding estimator is called an *unbiased estimator*. If the density $p_{\hat{\alpha}}(\hat{\alpha})$ is symmetric, for such an estimator, the desired value is the central value of $\hat{\alpha}$. An estimator whose bias is nonzero is called a *biased estimator*.

The variance of an estimator is a measure of the spread of the probability density $p_{\hat{\alpha}}(\hat{\alpha})$. It is defined by

$$\text{Var}[\hat{\alpha}] = E[(\hat{\alpha} - E[\hat{\alpha}])^2] = \sigma_{\hat{\alpha}}^2 \tag{6.57}$$

A small value of the variance indicates a concentration of the probability density around the mean value $E(\hat{\alpha})$. Thus, an estimator must have a bias and a variance which are as small as possible in order to be satisfactory. In general, a reduction of one induces an increase of the other and *vice versa*. The compromise between bias and variance can be studied by using the *mean square error* of an estimator, which is defined by

$$E_{qm}(\hat{\alpha}) = E[(\hat{\alpha} - \alpha)^2]$$ (6.58)

By expanding this, we obtain

$$E_{qm}(\hat{\alpha}) = E[\hat{\alpha}^2 - 2\alpha\hat{\alpha} + \alpha^2] = E[\hat{\alpha}^2] - 2\alpha E[\hat{\alpha}] + \alpha^2$$ (6.59)

After adding and subtracting $(E[\hat{\alpha}])^2$ and grouping the terms, the following result is obtained:

$$E_{qm}(\hat{\alpha}) = \sigma_{\hat{\alpha}}^2 + B_{\hat{\alpha}}^2$$ (6.60)

If the bias and the variance of an estimator tend toward zero when the number K of observed variable is increased, the estimator is called *consistent*.

6.3.4 Example

To illustrate the previous definitions, let us consider the estimator (6.54) applied to a Gaussian random signal $x(k)$, the samples of which are statistically independent. To compute the bias $B_{\hat{m}_x}$, the following expected value must be computed:

$$E[\hat{m}_x] = E\left[\frac{1}{K}\sum_{k=0}^{K-1} x(k)\right] = \frac{1}{K}\sum_{k=0}^{K-1} E[x(k)]$$

$$= \frac{1}{K}\sum_{k=0}^{K-1} m_x = m_x$$ (6.61)

Thus,

$$B_{\hat{m}_x} = E[\hat{m}_x] - m_x = 0$$ (6.62)

Therefore, the estimator (6.54) is unbiased. Its variance is given by

$$Var[\hat{m}_x] = E[\hat{m}_x^2] - (E[\hat{m}_x])^2$$ (6.63)

However,

$$
\begin{aligned}
E[\hat{m}_x^2] &= E\left[\frac{1}{K}\sum_{k=0}^{K-1}x(k)\cdot\frac{1}{K}\sum_{l=0}^{K-1}x(l)\right]\\
&= \frac{1}{K^2}\left[\sum_{k=0}^{K-1}E[x^2(k)]+\sum_{k=0}^{K-1}\sum_{l=0}^{K-1}E[x(k)]\,E[x(l)]\right]\\
&= \frac{1}{K}E[x^2(k)]+\frac{K-1}{K}m_x^2 \tag{6.64}
\end{aligned}
$$

Thus,

$$
\text{Var}[\hat{m}_x] = \frac{1}{K}(E[x^2(k)]-m_x^2) = \frac{1}{K}\sigma_x^2 \tag{6.65}
$$

This equation shows that the variance of the estimator (6.54) tends toward zero when the number of observations K is increased. Because its bias is zero, the estimator of the mean value is, in this case, a consistent estimator.

6.3.5 Maximum-likelihood estimators

Let $L(x(0), x(1), \ldots , x(K-1), \alpha_1, \alpha_2, \ldots , \alpha_N)$ be the joint probability density of K random variables as a function of N parameters α_n to be estimated. The function L has two interpretations. Before the observation of the K variables $x(k)$, it represents the joint probability density of these variables for given values of the parameters α_n. After the observation, the variables $x(k)$ are known, but the parameters α_n are not. The function of the N parameters obtained by substituting the observed values $x(k)$ into L is called *the likelihood function* of the parameters α_n. The values of the parameters α_n for which the likelihood function is maximum are called *maximum-likelihood estimations*. They are the most preferred values, since they maximize the probability of obtaining the observed values $x(k)$. Maximum-likelihood estimators lead to minimum (or least) mean-square errors for large values of K.

In many cases, the maximum-likelihood estimator can be obtained by derivation. A set of N equations of the form:

$$
\frac{\partial L(\alpha_1, \alpha_2, ..., \alpha_N)}{\partial \alpha_n} = 0 \tag{6.66}
$$

Insert Eq. (6.66) — 5½ pi #

must be solved in such cases. However, it is sometimes more convenient to look for the maximum of the logarithm of the likelihood function. This does not alter the results, since the logarithm is a monotonic function. In this case, the following equations are obtained:

$$L(\alpha_1, \alpha_2, ..., \alpha_N) = \ln L(\alpha_1, ..., \alpha_N) \tag{6.67}$$

$$\frac{\partial L(\alpha_1, \alpha_2, ..., \alpha_N)}{\partial \alpha_n} = \frac{1}{L(\alpha_1, \alpha_2, ..., \alpha_N)} \frac{\partial L(\alpha_1, \alpha_2, ..., \alpha_N)}{\partial \alpha_n} = 0$$

$$\text{with } n = 1, ..., N \tag{6.68}$$

6.3.6 Example

Let us consider, as in the previous example 6.3.4, K statistically independent samples of a Gaussian random signal $x(k)$. The joint probability density of these K random variables is given by

$$L(x(0), ..., x(K-1), m_x, \sigma_x^2) = \prod_{k=0}^{K-1} \frac{1}{\sqrt{2\pi}\,\sigma_x} \exp\left\{ -\frac{[x(k) - m_x]^2}{2\sigma_x^2} \right\} \tag{6.69}$$

In this case, it is more convenient to consider the logarithm of the likelihood function. We have

$$L(m_x, \sigma_x^2) = -\frac{K}{2}\ln 2\pi - K\ln\sigma_x - \frac{1}{2\sigma_x^2}\sum_{k=0}^{K-1}[x(k) - m_x]^2 \tag{6.70}$$

Applying (6.68) gives

$$\frac{1}{\hat{\sigma}_x^2}\sum_{k=0}^{K-1}[x(k) - \hat{m}_x] = 0 \tag{6.71}$$

$$-\frac{K}{\hat{\sigma}_x} + \frac{1}{\hat{\sigma}_x^3}\sum_{k=0}^{K-1}[x(k) - \hat{m}_x]^2 = 0 \tag{6.72}$$

The solution of (6.71) is given by

$$\hat{m}_x = \frac{1}{K}\sum_{k=0}^{K-1} x(k) \tag{6.73}$$

The solution of (6.72) is given by

$$\hat{\sigma}_x^2 = \frac{1}{K} \sum_{k=0}^{K-1} [x(k) - \hat{m}_x]^2 \tag{6.74}$$

This example shows that the estimator (6.54), or (6.73), is a maximum-likelihood estimator for a Gaussian signal $x(k)$, the samples of which are statistically independent. The same is true for the estimator (6.74), which estimates the variance of such a signal.

6.3.7 Confidence intervals

Another way to express the concentration of the probability density of an estimator is to use a probability of the form:

$$P(-a_1 < \hat{\alpha} - \alpha \leqslant a_2) = 1 - \beta \tag{6.75}$$

or, of the equivalent form:

$$P(\hat{\alpha} - a_2 \leqslant \alpha < \hat{\alpha} + u_1) = 1 - \beta \tag{6.76}$$

These probabilities can be determined from the density $p_{\hat{\alpha}}(\hat{\alpha})$, as shown in Fig. 6.2.

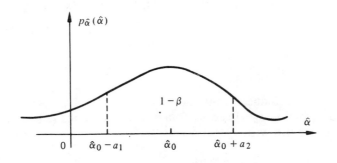

Fig. 6.2

The probability $1 - \beta$ represents the surface under the curve $p_{\hat{\alpha}}(\hat{\alpha})$, bounded by a_1 and a_2. The probability β represents the sum of the two lateral surfaces, which are shown unshaded. Before the estimation, (6.75) and (6.76) represent probabilities based on the function $p_{\hat{\alpha}}(\hat{\alpha})$. After the estimation, the interpretation of these equations must be modified because we have $P[\hat{\alpha}_0] = 1$, where $\hat{\alpha}_0$ is a particular estimate. After the estimation, the following equation is written:

$$C(\hat{\alpha}_0 - a_2 \leqslant \alpha < \hat{\alpha}_0 + a_1) = 1 - \beta \tag{6.77}$$

and is interpreted as the *confidence interval* in which the true value α can be found, with a probability $1 - \beta$, or, better still, with a confidence $1 - \beta$. This interval is based on the estimation of $\hat{\alpha}_0$. It is bounded by $\hat{\alpha}_0 - a_2$ and by $\hat{\alpha}_0 + a_1$. In general, to increase the confidence $1 - \beta$, a_1 and a_2 must be increased.

6.3.8 Example

To illustrate the notion of confidence interval, let us consider again the estimation of the mean value of a random Gaussian signal having samples which are statistically independent. We know, by using the central limit theorem, that the random variable $\hat{m}_x$ given by

$$\hat{m}_x = \frac{1}{K} \sum_{k=0}^{K-1} x(k) \tag{6.78}$$

has a Gaussian probability density. Because this estimator is unbiased, its mean value is that of the signal, i.e., m_x. The variance of this estimator has already been calculated (6.65). Consequently, the distribution $p_{\hat{m}_x}(\hat{m}_x)$ is entirely known. By setting:

$$\mu = \frac{\hat{m}_x - m_x}{\sigma_x/\sqrt{K}} \tag{6.79}$$

we thereby obtain a reduced centered Gaussian variable. The corresponding normalized distribution $p_u(u)$ can be found in many books [28]. For example, with $\beta = 0.05$, we can write

$$P(-U < u < U) = 1 - \beta = 0.95 \tag{6.80}$$

The calculation of the normalized repartition function (expected value of $p_u(u)$) gives $U = 1.96$. After computing an estimate $\hat{m}_x$, we have

$$C\left(-1.96 < \frac{\hat{m}_x - m_x}{\sigma_x/\sqrt{K}} < 1.96\right) = 0.95 \tag{6.81}$$

or

$$C\left(\hat{m}_x - \frac{1.96\,\sigma_x}{\sqrt{K}} < m_x < \hat{m}_x + \frac{1.96\,\sigma_x}{\sqrt{K}}\right) = 0.95 \tag{6.82}$$

Equation (6.82) indicates that we have a 95% confidence of finding the true mean value of m_x in the interval:

$$\left[\hat{m}_x - \frac{1.96\,\sigma_x}{\sqrt{K}} \; ; \; \hat{m}_x + \frac{1.96\,\sigma_x}{\sqrt{K}}\right] \tag{6.83}$$

The confidence level can be increased. For example, with $\beta = 0.01$, $U = 2.58$. This widens the interval (6.83).

6.4 ESTIMATORS FOR THE CORRELATION

6.4.1 Introduction

In this section, we establish the different estimators used for the correlation functions and their properties. To simplify the development, the mean values of the signals considered are assumed to be zero. This hypothesis, which is not restrictive and is valid in practice, allows for an adequate estimate of the covariance and correlation functions.

6.4.2 Unbiased estimator

Equation (6.1), which expresses the autocorrelation function of an ergodic random process, can be written in the following form:

$$\varphi_x(k) = \overline{u_k(l)}$$

with

$$u_k(l) = x(l)\,x(l+k) \tag{6.84}$$

For a given value of k, the problem is to estimate the mean value of a signal $u_k(l)$. Because only K values of $x(k)$ are available, the mean value of $u_k(l)$ can be estimated with only $(K - k)$ values $u_k(l)$. Applying the average value estimator (6.54) to the signal $u_k(l)$ yields

$$c'_x(k) = \frac{1}{K - |k|} \sum_{l=0}^{K-|k|-1} u_k(l) = \frac{1}{K - |k|} \sum_{l=0}^{K-|k|-1} x(l)\,x(l+k) \tag{6.85}$$

The estimator (6.85) must be applied for the $2K - 1$ possible values of k, with $|k| < K$. The signal given by (6.85), for all these values of k, is an estimate of the autocorrelation function $\varphi_x(k)$. If the signal $u_k(l)$ is Gaussian, then (6.85) is a maximum-likelihood estimate.

In general, it is difficult and complicated to apply the formalism of section 6.3.5 to establish the form of a maximum likelihood estimator, even if the distribution of the signal $x(k)$ is known. The estimator (6.85), if it is not optimal, is useful.

6.4.3 Bias and variance of the estimator $c_x'(k)$

The bias of the estimator can easily be computed. We have

$$
E[c_x'(k)] = \frac{1}{K - |k|} \sum_{l=0}^{K - |k| - 1} E[x(l)\, x(l+k)]
$$

$$
= \frac{1}{K - |k|} \sum_{l=0}^{K - |k| - 1} \varphi_x(k) = \varphi_x(k) \tag{6.86}
$$

Thus,

$$
B_{c_x'(k)} = E[c_x'(k)] - \varphi_x(k) = 0 \tag{6.87}
$$

The estimator $c_x'(k)$ is thus unbiased.

The computation of the variance of the estimator $c_x'(k)$ is very long and complicated, and thus has been omitted from this text. This calculation, the details of which can be found in references [29] and [30], leads to the following result:

$$
\text{Var}[c_x'(k)] = \frac{1}{(K - |k|)^2} \sum_{l=-(K-k)}^{K-k} (K - k - |l|)[\varphi_x^2(l) + \varphi_x(l+k)\, \varphi_x(l-k)] \tag{6.88}
$$

For values of K much larger than k, the approximate expression yields

$$
\text{Var}[c_x'(k)] \cong \frac{K}{(K - |k|)^2} \sum_{l=-\infty}^{+\infty} [\varphi_x^2(l) + \varphi_x(l+k)\, \varphi_x(l-k)] \tag{6.89}
$$

Expressions (6.88) and (6.89) show that, in general, the variance of the estimator $c_x'(k)$ is proportional to $1/K$. We have

$$
\lim_{K \to \infty} \text{Var}[c_x'(k)] = 0 \tag{6.90}
$$

Because the bias is also zero (see eq. (6.87)), the estimator $c'_x(k)$ is consistent.

6.4.4 Biased estimator

Another commonly used estimator is given by

$$c_x(k) = \frac{1}{K} \sum_{l=0}^{K-|k|-1} x(l)\, x(l+k) \qquad (6.91)$$

We may observe that it differs from the unbiased estimator simply by a scale factor. Comparison of (6.85) and (6.91) leads to

$$c_x(k) = \frac{K-|k|}{K}\, c'_x(k) \qquad (6.92)$$

6.4.5 Bias and variance of the estimator $c_x(k)$

With (6.86) and (6.92), we obtain

$$B_{c_x(k)} = E[c_x(k)] - \varphi_x(k) = \left(\frac{K-|k|}{K} - 1\right)\varphi_x(k)$$

$$= -\frac{|k|}{K}\, \varphi_x(k) \qquad (6.93)$$

Therefore, the estimator $c_x(k)$ is biased, hence its name.

We may compute its variance as before. The following result is thus obtained:

$$\text{Var}[c_x(k)] = \frac{1}{K^2} \sum_{l=-(K-k)}^{K-k} (K-k-|l|)[\varphi_x^2(l) + \varphi_x(l+k)\,\varphi_x(l-k)] \qquad (6.94)$$

For values of K greater than k, the following approximate expression can be used:

$$\text{Var}[c_x(k)] \cong \frac{1}{K} \sum_{l=-\infty}^{+\infty} [\varphi_x^2(l) + \varphi_x(l+k)\,\varphi_x(l-k)] \qquad (6.95)$$

Notice that this variance tends toward zero as K tends toward infinity. The same is true for the bias, therefore, $c_x(k)$ is a consistent estimator.

6.4.6 Comments

The choice between the estimators $c_x(k)$ and $c_x'(k)$ usually leads to the unbiased consistent estimator, $c_x'(k)$. However, for values of k close to K, the variance of this estimator increases dramatically (see eqs. (6.88) and (6.89)). This results from the fact that the mean value (6.85) is computed with a small number of terms when k is close to K. Thus, in this case, the unbiased estimator is not very useful because its variance is too large. The variance of the biased estimator $c_x(k)$, however, does not depend directly on the delay k (see eq. (6.95)). When k is close to the observation length K, this variance does not increase as is the case for the unbiased estimator. Rather, in this case, the bias increases with the delay k (see eq. (8.39)). This estimator is not very useful for values of k close to K because of its bias.

For a fair comparison of these two estimators, we must compare their mean square errors. Jenkins and Watts [30] have shown that in many cases, the mean square error is smaller for the biased estimator. The choice should then be made for the biased estimator, which is paradoxical. In both cases, the estimations can be improved by increasing the observation length K. The biased estimator is asymptotically unbiased.

6.4.7 Example

Let us consider a random signal $x(k)$, having a normalized autocovariance function, of the form:

$$\rho_x(k) = a^{|k|} \quad \text{with } a = 0.75 \tag{6.96}$$

Such a signal can be generated, starting from white noise, by using (6.38). Figure 6.3 shows the results obtained for different values of the observation length K, with the estimators $c_x(k)$ and $c_x'(k)$. The curves show that the variance of the unbiased estimator decreases for small values of k with increasing K, but it increases for values of k close to K. The same comment can be made for the bias of the estimator $c_x(k)$.

6.4.8 Estimators for the crosscorrelation

The methods developed in sect. 6.4.2 and 6.4.4 can be applied with slight modifications to the estimation of crosscorrelation functions. Let $x(k)$ and $y(k)$

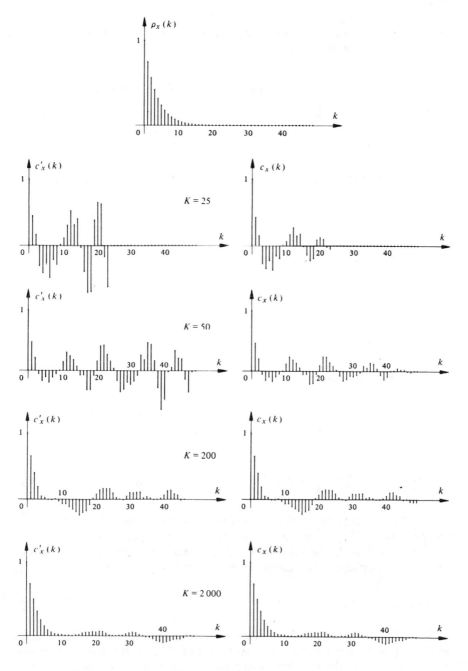

Fig. 6.3

be two stationary, ergodic, random signals with zero mean. The estimator of the crosscorrelation is given by

$$
c_{xy}(k) = \begin{cases} \dfrac{1}{K} \displaystyle\sum_{l=0}^{K-k-1} x(l)\,y(l+k) & \text{for } 0 \leqslant k < K \\[4mm] \dfrac{1}{K} \displaystyle\sum_{l=-k}^{K-1} x(l)\,y(l+k) & \text{for } -K < k \leqslant 0 \end{cases}
\tag{6.97}
$$

The second expression, for negative values of k, can be written as

$$
c_{xy}(k) = \frac{1}{K} \sum_{l'=0}^{K+k-1} x(l'-k)\,y(l') \qquad \text{for } -K < k \leqslant 0
\tag{6.98}
$$

The expectation of this estimator is given by

$$
E[c_{xy}(k)] = \frac{1}{K} \sum_{l=0}^{K-k-1} E[x(l)\,y(l+k)]
$$

$$
= \left(1 - \frac{k}{K}\right) \varphi_{xy}(k) \qquad \text{for } 0 \leqslant k < K
\tag{6.99}
$$

Using (6.98), we obtain

$$
E[c_{xy}(k)] = \left(1 + \frac{k}{K}\right) \varphi_{xy}(k) \qquad \text{for } -K < k \leqslant 0
\tag{6.100}
$$

Thus,

$$
E[c_{xy}(k)] = \left(1 - \frac{|k|}{K}\right) \varphi_{xy}(k) \qquad \text{for } -K < k < K
\tag{6.101}
$$

Equation (6.101) shows that the estimator $c_{xy}(k)$ is biased. As in the case of the estimator $c_x(k)$, its variance is inversely proportional to K.

It is clear that similar equations can be established for an unbiased estimator $c'_x(k)$ by replacing the multiplicative factor $1/K$ in (6.97) by $1/(K - |k|)$.

6.4.9 Example

One of the most common applications of the crosscorrelation is to determine the impulse response of a linear system. Let $g(k)$ be this desired response. If the system is excited with white noise $x(k)$, the response $y(k)$ is given by

$$
y(k) = x(k) * g(k)
$$

with

$$\varphi_x(k) = \sigma_b^2 d(k) \tag{6.102}$$

The crosscorrelation of the input and output signals is given by (sect. 1.6.6):

$$
\begin{aligned}
\varphi_{xy}(k) &= x(-k) * y(k) \\
&= x(-k) * x(k) * g(k) \\
&= \varphi_x(k) * g(k) \\
&= \sigma_x^2 d(k) * g(k) \\
&= \sigma_x^2 g(k)
\end{aligned} \tag{6.103}
$$

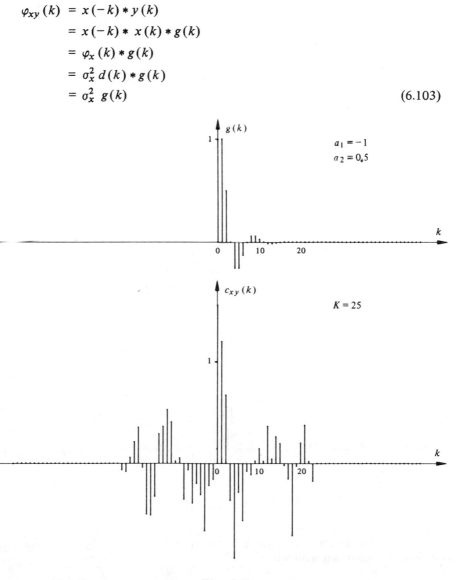

Fig. 6.4·

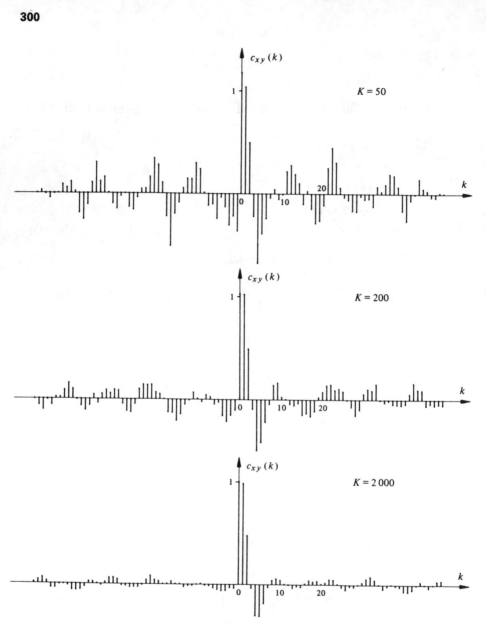

Fig. 6.4 (cont'd)

Let us consider, as an example, a linear system defined by the following second-order difference equation:

$$y(k) + a_1 y(k-1) + a_2 y(k-2) = x(k) \tag{6.104}$$

The impulse response of this system was previously established (sect. 2.7.16). Using white noise for $x(k)$ in (6.104), the output signal $y(k)$ is obtained. The results obtained for the estimator $c_{xy}(k)$, for different values of the observation length K, are shown in Fig. 6.4. We can see behavior similar to the case of the autocorrelation (sect. 6.4.7).

6.5 SPECTRAL ESTIMATORS

6.5.1 Preliminary remarks

In the previous section, two asymptotically consistent estimators were established for the estimation of the autocorrelation function. In light of definition (6.7), it is tempting to think that the Fourier transform of a good estimator for the correlation is also satisfactory for the power spectrum. Unfortunately, such is not the case. In this section, we will study the properties of such an estimator and of other, better estimators, derived from the first.

6.5.2 Simple spectral estimator

The simple spectral estimator is the Fourier transform of the estimator $c_x(k)$. We have

$$S_x(f) = \sum_{k=-(K-1)}^{K-1} c_x(k) \exp(-j 2\pi f k) \qquad (6.105)$$

An old terminology still occasionally used refers to $S_x(f)$ as a periodogram. By using definition (6.91) in (6.105), the following equation is obtained:

$$S_x(f) = \frac{1}{K} |X_K(f)|^2 \qquad (6.106)$$

with

$$X_K(f) = \sum_{k=0}^{K-1} x(k) \, e^{-j 2\pi f k} \qquad (6.107)$$

6.5.3 Bias of the simple estimator

The bias of the simple estimator is the difference between $E[S_x(f)]$ and the desired spectrum $\Phi_x(f)$. The expectation of the simple spectral estimator is given by

$$E[S_x(f)] = \sum_{k=-(K-1)}^{K-1} E[c_x(k)] \exp(-j\,2\pi fk)$$

$$= \sum_{k=-(K-1)}^{K-1} \frac{K-|k|}{K} \varphi_x(k) \exp(-j\,2\pi fk) \qquad (6.108)$$

Thus, because of the term $(K - |k|)/K$, the expected value of $S_x(f)$ is not the Fourier transform of the autocorrelation function. Accordingly, $S_x(f)$ is a biased estimator.

6.5.4 Variance of the simple estimator

It is very difficult to compute the variance of the estimator $S_x(f)$ applied to a signal with an arbitrary distribution. That is why it will be computed in a simple particular case, which is that of Gaussian noise. The result will be generalized by approximation. Let us consider K samples $x(k)$ of a Gaussian white noise. By using (6.106) and (6.107), we obtain, at the harmonic frequencies $f_n = n/K$:

$$S_x(n) = \frac{1}{K}[A_x^2(n) + B_x^2(n)] \qquad (6.109)$$

with

$$A_x(n) = \sum_{k=0}^{K-1} x(k) \cos\frac{2\pi nk}{K}$$

$$B_x(n) = \sum_{k=0}^{K-1} x(k) \sin\frac{2\pi nk}{K} \qquad (6.110)$$

and

$$n = -K/2, ..., K/2 - 1$$

The functions $A_x(n)$ and $B_x(n)$ defined in this way are linear combinations of Gaussian random variables. Consequently, they also have Gaussian distributions. Their mean values are zero for a signal $x(k)$ having a mean value of zero. The variance of $A_x(n)$ is given by

$$\text{Var}\,[A_x(n)] = \text{E}\,[A_x^2(n)]$$

$$= \text{E}\left[\sum_{k=0}^{K-1} x^2(k)\,\cos^2\left(2\pi\,\frac{nk}{K}\right) + 2\sum_{\substack{k=0 \\ k \neq l}}^{K-1}\sum_{l=0}^{K-1} x(k)\,\cos\left(2\pi\,\frac{nk}{K}\right) x(l)\,\cos\left(2\pi\,\frac{nl}{K}\right)\right]$$

$$(6.111)$$

Because $x(k)$ is white noise, we have

$$\text{E}\,[x(k)\,x(l)] = \begin{cases} 0 & \text{for } k \neq l \\ \sigma_x^2\,d(l) & \text{for } k = l \end{cases} \qquad (6.112)$$

Thus,

$$\text{Var}\,[A_x(n)] = \sigma_x^2 \sum_{k=0}^{K-1} \cos^2\left(\frac{2\pi nk}{K}\right)$$

$$= \begin{cases} \sigma_x^2\,\dfrac{K}{2} & \text{for } n = -\dfrac{K}{2}+1, ..., -1, 1, ..., \dfrac{K}{2}-1 \\[3mm] \sigma_x^2\,K & \text{for } n = -\dfrac{K}{2}\,,\,0 \end{cases} \qquad (6.113)$$

Equivalently, the following equation is obtained:

$$\text{Var}\,[B_x(n)] = \begin{cases} \sigma_x^2\,\dfrac{K}{2} & \text{for } n = -\dfrac{K}{2}+1, ..., -1, 1, ..., \dfrac{K}{2}-1 \\[3mm] 0 & \text{for } n = -\dfrac{K}{2}\,,\,0 \end{cases} \qquad (6.114)$$

Moreover, it can easily be shown that the related second-order statistics are zero:

$$\text{Cov}\,[A_x(n)\,A_x(m)] = \text{E}\,[A_x(n)\,A_x(m)] = 0$$
$$\text{Cov}\,[A_x(n)\,B_x(m)] = 0$$
$$\text{Cov}\,[B_x(n)\,B_x(m)] = 0 \qquad (6.115)$$

Therefore, the random variables $A_x(n)$ and $B_x(m)$ are independent. The normalized variables:

$$A'_x(n) = \frac{A_x^2(n)}{\text{Var}[A_x(n)]} \quad \text{and} \quad B'_x(n) = \frac{B_x^2(n)}{\text{Var}[B_x(n)]} \tag{6.116}$$

thus have χ_1^2 distributions with one degree of freedom (Rayleigh distribution). Their sum:

$$D_x(n) = A'_x(n) + B'_x(n) = \frac{2 S_x(n)}{\sigma_x^2} \tag{6.117}$$

has a χ_2^2 distribution with two degrees of freedom. The variance of a χ_ν^2 distribution with ν degrees of freedom is given by

$$\text{Var}[\chi_\nu^2] = 2\nu \tag{6.118}$$

Application of this to the variable $D_x(n)$ finally gives

$$\text{Var}[S_x(n)] = \sigma_x^4 = \Phi_x^2(n) \tag{6.119}$$

The calculation (6.119) can be done for any frequency. However, due to the length of this calculation, it will be omitted here. The result, valid for any frequency, is [30]:

$$\text{Var}[S_x(f)] = \sigma_x^4 \left[1 + \left(\frac{\sin 2\pi f K}{K \sin 2\pi f} \right)^2 \right] = \Phi_x^2(f) \left[1 + \left(\frac{\sin 2\pi f K}{K \sin 2\pi f} \right)^2 \right] \tag{6.120}$$

We can easily see that for $f = f_n = n/K$, this equation reduces to (6.119). This reduction is sufficient because, in practice, the calculations are performed by using the FFT at the harmonic frequencies.

6.5.5 Comments

Equation (6.119) shows that *the variance of the simple spectral estimator does not depend on the observation length K.* For any length, the variance of this estimator remains proportional to the square of the desired spectrum. Consequently, (6.105) is not a consistent estimator. However, this was established with the hypothesis of white Gaussian noise, which is a particular signal. If the analyzed signal is not Gaussian, $A_x(n)$ and $B_x(n)$ are, nevertheless, approximately Gaussian due to the central limit theorem. Thus, (6.12) remains valid.

If the analyzed signal is not white noise, the model (6.29) can be used to represent the desired spectral density. Equation (6.16) can be established approximately by using estimators $S_x(f)$. We have

$$S_y(f) \approx |G(f)|^2 S_x(f) \qquad (6.121)$$

If $x(k)$ is white noise, $S_x(n)$ has a χ_2^2 distribution with two degrees of freedom. Thus, the variable:

$$\frac{2 S_y(n)}{|G(n)|^2 \sigma_x^2} = \frac{2 S_y(n)}{\Phi_y(n)} \qquad (6.122)$$

has approximately a χ_2^2 distribution with two degrees of freedom. This shows that (6.119) remains approximately valid for a signal of any arbitrary distribution.

Because we have established the distribution of the simple spectral estimator, it is possible in principle to define a confidence interval with the help of numerical tables (sect. 6.3.7). This is not very useful, however, because it is not possible to change the variance of the estimator.

In definition (6.105), the biased estimator $c_x(k)$ was used. The same results are valid to within a multiplicative factor if the unbiased estimator $c_x'(k)$ used.

Because the simple spectral estimator is not consistent and its variance does not depend on the observation length K, we must define other estimators. Among all the possibilities, those which consist of modifying the simple spectral estimator are preferred. This results from the fact that the FFT can be used to evaluate $S_x(f)$ quickly by using (6.106).

6.5.6 Averaged spectral estimator

The direct method for reducing the variance is to compute a mean value over several independent estimators. To apply this idea to the simple estimator, the signal observed over a length K must be divided into L sections $x_l(k)$, each of length M. Then,

$$x_l(k) = x(k + (l-1)M) \qquad (6.123)$$

with $k = 0, \ldots, M-1$, $K = ML$ and $l = 1, \ldots, L$

L simple estimators of the following type are hence evaluated:

$$S_{x_l}(f) = \frac{1}{M} \left| \sum_{k=0}^{M-1} x_l(k) e^{-j 2\pi f k} \right|^2 \qquad (6.124)$$

with $l = 1, \ldots, L$

The averaged spectral estimator is thus given by

$$\bar{S}_x(f) = \frac{1}{L} \sum_{l=1}^{L} S_{x_l}(f) \tag{6.125}$$

6.5.7 Bias of the averaged estimator

The bias of the mean estimator is the difference between $E[\bar{S}_x(f)]$ and the desired spectrum $\Phi_x(f)$. The expected value of (6.125) is given by

$$E[\bar{S}_x(f)] = \frac{1}{L} \sum_{l=1}^{L} E[S_{x_l}(f)] = E[S_{x_l}(f)] \tag{6.126}$$

To compare the estimators $S_x(f)$ and $\bar{S}_x(f)$, their bias can be compared. To this end, (6.108) can be interpreted as the Fourier transform of the autocorrelation as viewed across the triangular window (sect. 3.7.9). In the frequency domain, this leads to the following convolution product:

$$E[S_x(f)] = \int_{-1/2}^{1/2} W_T(g)\,\Phi_x(f-g)\,dg \tag{6.127}$$

The same equation remains valid for the expected value (6.126), but with a window $W_T(f)$, the central peak of which is L times wider. Thus, the bias of the mean estimator is larger.

6.5.8 Variance of the averaged estimator

If the L simple averaged estimators in (6.125) are assumed do be independent, then, $\bar{S}_x(f)$ is the mean value of L observations $S_{xl}(f)$. If we apply (6.65), this leads to

$$\text{Var}[\bar{S}_x(f)] = \frac{1}{L} \text{Var}[\bar{S}_{x_l}(f)]$$

$$\cong \frac{1}{L}\,\Phi_x^2(f) = \frac{M}{K}\,\Phi_x^2(f) \tag{6.128}$$

This equation shows that the variance of the averaged estimator is inversely proportional to the observation length K.

6.5.9 Comments

We saw (sect. 6.5.7) that the averaged estimator can be interpreted as the convolution product between the desired spectrum and a spectral window function, which, in general, has a main central peak and secondary sidelobes. The width at the base of the central peak, which is inversely proportional to the observation length, determines the bias. The narrower is this central peak, the smaller is the bias, because the expected value (6.127) tends toward the desired spectrum. Thus, to decrease the bias of $\bar{S}_x(f)$, the length M of the sections $x_l(k)$ must be increased. Conversely, to decrease the variance of this estimator with respect to that of the simple estimator, the number L of sections contributing to the mean value (6.125) must be increased. For a fixed overall observation length $K = ML$, this leads to a compromise.

In an experiment, the choice of K and M is influenced by the information available about the signal. For example, if narrow peaks must be detected in the spectrum, M must be chosen relatively large so that the resolution given by (6.127) is sufficient. The global observation length is then determined by the acceptable variance chosen.

6.5.10 Smoothed spectral estimator

Another way to reduce the variance of the simple spectral estimator is to filter it. Let us consider $S_x(f)$ to be the input signal of a filter having an impulse response of $W(f)$. It is, of course, a filtering performed as if frequency and time were interchanged. A smoothed estimator is then obtained:

$$\tilde{S}_x(f) = \int_{-1/2}^{1/2} S_x(g)\, W(f-g)\, dg \tag{6.129}$$

As before, the function $W(f)$ is called a spectral window (sect. 3.7.4). The smoothed estimator can be interpreted as the Fourier transform of the product of the estimator $c_x(k)$ and a window function $w(k)$ of finite length $2M - 1$. We have

$$\tilde{S}_x(f) = \sum_{k=-(M-1)}^{M-1} w(k)\, c_x(k)\, e^{-j 2\pi f k} \tag{6.130}$$

This equation shows the constraints to be imposed on the window functions. Because the power spectrum, and any estimation of it, must be positive real functions, the function $w(k)$ must be even and its transform $W(f)$ must be positive.

6.5.11 Bias of the smoothed estimator

The expected value of (6.129) leads to

$$E[\tilde{S}_x(f)] = \int_{-1/2}^{1/2} E[S_x(g)]\, W(f-g)\, dg \tag{6.131}$$

Because $E[S_x(f)]$ is a convolution product (see eq. (6.127)), the expected value of the smoothed estimator is a double convolution product. It can be also written as the Fourier transform of the double simple product:

$$E[\tilde{S}_x(f)] = \sum_{k=-(M-1)}^{M-1} w_T(k)\, w(k)\, \varphi_x(k)\, e^{-j 2\pi f k} \tag{6.132}$$

with $w_T(k) = 1 - \dfrac{|k|}{K}$ and $|k| < K$

If M is relatively small with respect to K, it is possible to write approximately by neglecting the term $|k|/K$:

$$E[\tilde{S}_x(f)] \cong \sum_{k=-(M-1)}^{M-1} w(k)\, \varphi_x(k)\, e^{-j 2\pi f k}$$

$$\cong \int_{-1/2}^{1/2} W(g)\, \Phi_x(f-g)\, dg \tag{6.133}$$

The larger is M, the smaller is the bias $E[\tilde{S}_x(f)] - \Phi_x(f)$. Conversely, decreasing M implies a loss in frequency resolution.

6.5.12 Variance of the smoothed estimator

The variance of the smoothed estimator is given by

$$\mathrm{Var}[\bar{S}_x(f)] = E[(\tilde{S}_x(f) - E[\tilde{S}_x(f)])^2] \tag{6.134}$$

By using (6.129) and (6.131), we have

$$\tilde{S}_x(f) - E[\tilde{S}_x(f)] = \int_{-1/2}^{1/2} \{S_x(g) - E[S_x(g)]\}\, W(f-g)\, dg \tag{6.135}$$

Thus,

$$\text{Var}\,[\widetilde{S}_x\,(f)] = \int\limits_{-1/2}^{1/2}\int\limits_{-1/2}^{1/2} W(f-g)\,W(f-u)\,\text{Cov}\,[S_x(g)\,S_x(u)]\,dg\,du$$

$$(6.136)$$

The covariance of the simple spectral estimator can be computed, but the computations are long and relatively complex [30]. Here, we will give only the result:

$$\text{Cov}\,[S_x\,(g)\,S_x\,(u)] \;=\; \Phi_x\,(g)\,\Phi_x\,(u)\left\{\left[\frac{\sin\,(u+g)K/2}{K\sin\,(u+g)/2}\right]^2 + \left[\frac{\sin\,(u-g)K/2}{K\sin\,(u-g)/2}\right]^2\right\}$$

$$(6.137)$$

Substituting (6.137) into (6.136) leads to a very complex expression. It can be written in a simpler, but approximate, form by making several hypotheses. If K is relatively large, the terms in brackets in (6.137) tend toward Dirac impulses. Additionally assuming that the variations of $\Phi_x(f)$ are small with respect to those of the functions in brackets, we obtain:

$$\text{Var}\,[\widetilde{S}_x\,(f)] \cong \frac{1}{K}\int\limits_{-1/2}^{1/2} \Phi_x^2(g)\,W^2(f-g)\,dg$$

$$(6.138)$$

Finally, as a last approximation, it can be assumed that the spectral window $W(f)$ is narrow with respect to the variations of the desired spectrum. Equation (6.138) then becomes

$$\text{Var}\,[\widetilde{S}_x\,(f)] \cong \frac{1}{K}\,\Phi_x^2(f)\int\limits_{-1/2}^{1/2} W^2(g)\,dg$$

$$(6.139)$$

This approximate equation is valid if and only if the length $2M-1$ of the window $w(k)$ is such that its transform $W(f)$ is narrow with respect to the variations of $\Phi_x(f)$, and that, at the same time, $W(f)$ is wide enough with respect to the bracketed term in (6.137).

Once the window used for smoothing has been chosen, its energy can be computed. Thus, expression (6.139) can be compared with the other, previously established variances.

6.5.13 Distribution of the smoothed estimator

If the distribution of the soothed estimator is known, the confidence interval can be established. This distribution can be computed approximately by considering the smoothed estimator at the harmonic frequencies $f_n = n/K$. We have

$$\widetilde{S}_x(n) = \frac{1}{K} \sum_{m=-K/2}^{K/2-1} S_x(m) W(n-m) \tag{6.140}$$

This equation shows that, at the harmonic frequencies, the smoothed estimator is a weighted sum of the random variables $S_x(m)$. These variables have a χ_2^2 distribution with two degrees of freedom. This suggests approximating the distribution of $\widetilde{S}_x(n)$ by an $a\chi_\nu^2$ distribution with ν degrees of freedom, which has the same first- and second-order moments as the variable $\widetilde{S}_x(n)$. The parameter a is a normalization coefficient so that the total probability is equal to one.

The first two moments of an $a\chi_\nu^2$ distribution are given by

$$\begin{aligned}
\mathrm{E}\,[\,a\chi_\nu^2] &= a\nu \\
\mathrm{Var}\,[\,a\chi_\nu^2] &= 2a^2\nu
\end{aligned} \tag{6.141}$$

After solving these equations for ν and a, we obtain

$$\nu = \frac{2\,(\mathrm{E}\,[\,a\chi_\nu^2])^2}{\mathrm{Var}\,[\,a\chi_\nu^2]}$$

$$a = \frac{\mathrm{E}\,[\,a\chi_\nu^2]}{\nu} \tag{6.142}$$

For the random variable $\widetilde{S}_x(n)$:

$$\nu = \frac{2\,(\mathrm{E}\,[\,\widetilde{S}_x(n)])^2}{\mathrm{Var}\,[\,\widetilde{S}_x(n)]}$$

$$a = \frac{\mathrm{E}\,[\,\widetilde{S}_x(n)]}{\nu} \tag{6.143}$$

If in (6.132) we assume that the variations of $\Phi_x(f)$ are small with respect to those of the window $W(f)$:

$$\mathrm{E}\,[\,\widetilde{S}_x(n)] \cong \Phi_x(n) \tag{6.144}$$

Substitution of (6.144) and (6.139) into (6.143) finally leads to

$$
\nu \cong \frac{2K}{\int\limits_{-1/2}^{1/2} W^2(g)\, dg} = \frac{2K}{W_w}
$$

$$
a \cong \frac{\Phi_x(n)}{\nu} \tag{6.145}
$$

6.5.14 Confidence intervals for the smoothed estimator

For the normalized variable $\nu \tilde{S}_x(n)/\Phi_x(n)$, the following probability (sect. 6.3.7) can be written

$$
P\left(a_1 < \frac{\nu \tilde{S}_x(n)}{\Phi_x(n)} < a_2 \right) = F(a_2) - F(a_1) \tag{6.146}
$$

where $F(\chi_\nu^2)$ is the repartition function (vol. VI, chap. 15). If we consider $F(a_1)$ and $F(a_2)$ to be symmetrically placed with respect to the probability $F(\chi_\nu^2) = 0.5$, the following results are obtained:

$$
F(a_1) = \alpha/2, \quad F(a_2) = 1 - \alpha/2, \qquad F(a_2) - F(a_1) = 1 - \alpha \tag{6.147}
$$

The confidence interval is then given by

$$
C\left(a_1 < \frac{\nu \tilde{S}_x(n)}{\Phi_x(n)} < a_2 \right) = 1 - \alpha
$$

or

$$
C\left(\frac{\nu \tilde{S}_x(n)}{a_2} < \Phi_x(n) < \frac{\nu \tilde{S}_x(n)}{a_1} \right) = 1 - \alpha \tag{6.148}
$$

The functions ν/a_1 and ν/a_2 can be computed by using numerical tables [28]. These functions are given in fig. 6.5 for the most common values of α.

It is often useful to give the power spectrum $\Phi_x(f)$ on a logarithmic scale. In this case, the interval equivalent to (6.148) is

$$
C\left(\log \tilde{S}_x(n) + \log \frac{\nu}{a_2} < \log \Phi_x(n) < \log \tilde{S}_x(n) + \log \frac{\nu}{a_1} \right) = 1 - \alpha \tag{6.149}
$$

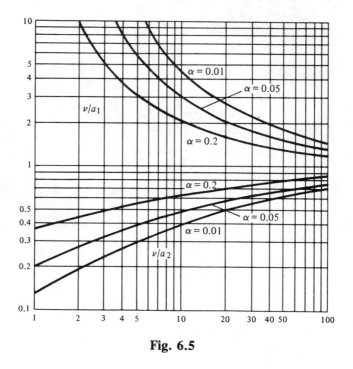

Fig. 6.5

Thus, if $\tilde{S}_x(n)$ is represented, the confidence interval is a constant bandwidth, independent of frequency. This means that among 100 events, there are 100(1 − α) chances that $\Phi_x(f)$ is within this band.

6.5.15 Modified spectral estimator

To decrease the variance of the simple spectral estimator, two methods, leading to the averaged and smoothed estimators, have been considered. These two methods can be combined to develop another spectral estimator, which is · particularly well oriented for use of the FFT. This estimator, introduced by Welch [31], is called the *modified spectral estimator*. The observed signal is again divided into L sections of length $M = K/L$ (6.123). However, in this case, each section is multiplied by a window function $w(k)$ before computing the simple estimator. The equivalent of the simple estimator is then given by

$$R_{xl}(f) = \frac{1}{MP} \left| \sum_{k=0}^{M-1} x_l(k) \, w(k) \, e^{-j2\pi fk} \right|^2$$

with $l = 1, ..., L$

$$(6.150)$$

and

$$P = \frac{1}{M} \sum_{k=0}^{M-1} w^2(k)$$

The normalization factor P is necessary to guarantee an asymptotically unbiased estimator. The estimator is hence given by

$$\bar{R}_x(f) = \frac{1}{L} \sum_{l=1}^{L} R_{x_l}(f) \tag{6.151}$$

6.5.16 Bias and variance of the modified estimator

We can see that the expected value of the modified estimator is given by

$$E[\bar{R}_x(f)] = \int_{-1/2}^{1/2} \Phi_x(g)\ \Phi_w(f-g)\ dg$$

where

$$\tag{6.152}$$

$$\Phi_w(f) = \frac{1}{MP} \left| \sum_{k=0}^{M-1} w(k)\ e^{-j2\pi f k} \right|^2$$

The above expression shows that when M tends toward infinity, $\Phi_w(f)$ tends toward a Dirac impulse. In this case, the expected value is equal to the desired spectrum $\Phi_x(f)$. It can also be shown that the variance of the modified estimator is given by

$$\text{Var}[\bar{R}_x(f)] \cong \frac{1}{L}\ \Phi_x^2(f) \tag{6.153}$$

In contrast with the case of the smoothed estimator, in this case there is no constraint on the chosen window. Because the smoothing is done by a spectral window function proportional to the modulus of a Fourier transform squared, the modified estimator will never lead to negative values.

6.5.17 Estimation of the cross-power spectrum

The methodology used for spectral estimators can be followed to determine cross-spectral estimators. The simple cross-spectral estimator is the Fourier transform of the estimator $c_{xy}(k)$ (see eq. (6.97)):

$$S_{xy}(f) = \sum_{k=-(K-1)}^{K-1} c_{xy}(k)\, e^{-j2\pi f k} \qquad (6.154)$$

By substituting (6.97) in (6.154), we obtain

$$S_{xy}(f) = \frac{1}{K} X^*(f)\, Y(f) \qquad (6.155)$$

Similarly, we can show that the estimator $S_{xy}(f)$ is asymptotically unbiased and its variance does not depend on the observation length K. Therefore, we often define a mean cross-spectral estimator:

$$\bar{S}_{xy}(f) = \frac{1}{L} \sum_{l=0}^{L-1} S_{x_l y_l}(f) \qquad (6.156)$$

or a smoothed cross-spectral estimator:

$$\tilde{S}_{xy}(f) = \int_{-1/2}^{1/2} S_{xy}(g)\, W(f-g)\, \mathrm{d}g$$

$$= \sum_{k=-(K-1)}^{K-1} w(k)\, c_{xy}(k)\, e^{-j2\pi f k} \qquad (6.157)$$

so as to decrease the variance at the price of a bias, which can be controlled by the shape of the chosen window functions.

6.6 APPLICATIONS OF THE FFT IN SPECTRAL ANALYSIS

6.6.1 Estimation of autocorrelation functions

In chapters 3 and 4, we saw that the FFT is well suited to evaluate estimators of the form $c_x(k)$, or $c'_x(k)$. Because these estimators are even functions, the correlation can be evaluated for only positive values of the delay k. However, there is a precaution to take due to the periodicity of the correlations computed by using the FFT (sect. 3.5.7). To avoid aliasing, zero-valued samples must be added to the observed signal. The number of zero samples to add depends on the maximum delay M for which we must evaluate the correlation. It is clear that M cannot be greater than the observation length K. A minimum of $M - 1$ zero samples must be added to the observed signal so that the periodicity generated by the FFT does not give rise to aliasing errors between two neighboring periods. The computation rules can be summarized as follows:

- Set the maximum delay M for which we must evaluate the correlation.
- Increase the length of the singal $x(k)$ observed over a length K to a length $K' = K + M - 1$ by adding $M - 1$ zero samples.
- Compute the FFT with K' points:

$$X(n) = \sum_{k=0}^{K'-1} x(k) \, e^{-j2\pi nk/K'} \tag{6.158}$$

$$\text{with } n = 0, ..., K' - 1$$

- Compute the inverse FFT with K' points:

$$u(k) = \frac{1}{K'} \sum_{n=0}^{K'-1} |X(n)|^2 \, e^{j2\pi nk/K'} \tag{6.159}$$

$$\text{with } k = 0, ..., K' - 1$$

- The correlation is given by either

$$c'_x(k) = \frac{1}{K - |k|} \, u(k) \tag{6.160}$$

or

$$c_x(k) = \frac{1}{K} \, u(k) \tag{6.161}$$

The FFT is more efficient than the direct method when K is greater than about 30 (sect. 4.7.6).

6.6.2 Sectioned correlation

To decrease the variance of the correlation estimators, the observation length K can be increased. For very large values of K, the computation of the DFT can be difficult, if not impossible, due to the amount of memory required. In general, only delays much smaller than K are considered. The methods presented in section 3.6. can then be used. Because the signal is considered to be zero outside the interval $[0, K-1]$, the upper limit of the sum can be modified in (6.91), which becomes

$$c_x(k) = \frac{1}{K} \sum_{l=0}^{K-1} x(l) x(l+k)$$

(6.162)

By considering L sections of length M, it can be written as

$$c_x(k) = \frac{1}{K} \sum_{l=0}^{L-1} \sum_{m=0}^{M-1} x(m+lM) x(m+lM+k)$$

$$= \frac{1}{K} \sum_{l=0}^{L-1} y_l(k)$$

(6.163)

with

$$y_l(k) = \sum_{m=0}^{M-1} x(m+lM) x(m+lM+k)$$

(6.164)

To evaluate the signals $y_l(k)$ by the FFT, the following auxiliary signals must be used:

$$u_l(k) = \begin{cases} x(k+lM) & \text{for } 0 \leqslant k \leqslant M-1 \\ 0 & \text{for } M \leqslant k \leqslant 2M-2 \end{cases}$$

and

$$v_l(k) = x(k+lM) \qquad \text{for } 0 \leqslant k \leqslant 2M-2$$

(6.165)

Let $U_1(n)$ and $V_1(n)$ be the DFTs obtained by the FFT with $2M - 1$ points from these signals. We have

$$Y_l(n) = U_l^*(n) V_l(n)$$

and

$$y_l(k) = \sum_{n=0}^{2M-2} Y_l(n) \exp\left[j \frac{2\pi nk}{2M-1}\right]$$

$$\text{for } k = 0, ..., M-1$$

(6.166)

Equation (6.166) can then be substituted in (6.163). With this method, two direct and one inverse DFT must be computed on each section. The method can be made more efficient by calculating:

$$Y(n) = \frac{1}{K} \sum_{l=0}^{L-1} Y_l(n) \qquad (6.167)$$

with $n = 0, ..., 2M - 2$

Then,

$$c_x(k) = \frac{1}{K} y(k) = \frac{1}{(2M-1)K} \sum_{n=0}^{2M-2} Y(n) \exp\left[-j\frac{2\pi nk}{2M-1}\right] \qquad (6.168)$$

This version involves only one inverse DFT. Moreover, for real signals, two transforms can be computed with only one DFT (sect. 3.3.5).

Usually, zero samples must be added to the sections to form the auxiliary signals of length $2M - 1$ (6.165). Because any length greater than $2M - 1$ is also valid, it is more convenient to add samples until the length $2M$ is reached. In this case, it must be noted that the first half of $u_l(k)$ is identical to the second half of $v_{l-1}(k)$. Thus,

$$v_l(k) = u_l(k) + u_{l+1}(k - M) \qquad (6.169)$$

By applying the DFT with $2M$ points to both sides of this equation, we obtain the following result:

$$V_l(n) = U_l(n) + (-1)^n U_{l+1}(n) \qquad (6.170)$$

with $n = 0, ..., 2M - 1$

Thus, the calculation can be significantly accelerated if the terms $V_l(n)$ are computed by using (6.170) instead of another FFT.

6.6.3 Estimation of the crosscorrelation

The formalism described in the two previous sections can also be applied to the estimation of crosscorrelation functions. If we have two signals $x(k)$ and $y(k)$ of the same length K, then $K - 1$ zero samples are added to both. Hence, two DFTs with $2K - 1$ points are evaluated to obtain $X(n)$ and $Y(n)$. The crosscorrelation is thus given by the inverse DFT:

$$c_{xy}(k) = \frac{1}{K} \frac{1}{2K-1} \sum_{n=0}^{2K-2} X^*(n) Y(n) \qquad (6.171)$$

with $k = 0, ..., 2K - 1$

For negative values of k, $x(k)$ and $y(k)$ must be interchanged, because $c_{yx}(k) = c_{xy}(k)$.

If the length K is such that it becomes necessary to use the sectioned correlation, the method of section 6.6.2 remains valid. However, the auxiliary signals (6.165) must be defined by

$$u_l(k) = \begin{cases} x(k+lM) & \text{for } 0 \leqslant k \leqslant M-1 \\ 0 & \text{for } M \leqslant k \leqslant 2M-2 \end{cases}$$

and

$$v_l(k) = y(k+lM) \qquad 0 \leqslant k \leqslant 2M-2 \tag{6.172}$$

6.6.4 Evaluation of the spectral estimators

The FFT is also well suited for the evaluation of spectral estimators. The simple spectral estimation $S_x(f)$ is obtained directly from (6.106) by evaluating $X_K(f)$ at the harmonic frequencies $f_n = n/2K$ by using the FFT and then applying (6.106).

At the harmonic frequencies, the mean spectral estimator is given by

$$\bar{S}_x(n) = \frac{1}{L} \sum_{l=1}^{L} S_{x_l}(n) \tag{6.173}$$

L simple estimators and their mean values must hence be evaluated. Each of these estimators is of the type:

$$S_{x_l}(n) = \frac{1}{M} |X_{Ml}(n)|^2$$

$$= \frac{1}{M} \left| \sum_{k=0}^{2M-1} x_l(k) e^{-j2\pi nk/2M} \right|^2 \tag{6.174}$$

This type can thus be evaluated, as before, with the FFT. Moreover, if the signal analyzed is real, two transforms $X_{Ml}(n)$ can be computed by using one FFT (sect. 3.3.5). This makes the computation twice as fast. This same process is also valid for the modified estimator. However, the equivalent of (6.174) has the following form:

$$R_{x_l}(n) = \frac{1}{MP} \left| \sum_{k=0}^{2M-1} x_l(k) w(k) e^{-j2\pi nk/2M} \right|^2 \tag{6.175}$$

Before evaluating the DFT, each section $x_l(k)$ of the analyzed signal must be multiplied by the temporal window $w(k)$.

6.6.5 Evaluation of the smoothed spectral estimator

The smoothed spectral estimator can be evaluated by using two different methods. The first consists of using definition (6.129), which can be evaluated numerically, starting from a simple spectral estimator. However, in most cases, the number of samples taken on $S_x(f)$ and $W(f)$ is large. The direct computation of the convolution product is inefficient compared with that via the DFT.

The second possibility, which is more efficient, consists of using the indirect method with the DFT. The equivalent of (6.129) is the Fourier transform of the product $c_x(k)w(k)$ (6.130). This requires the prior evaluation of the estimator $c_x(k)$ with the methods described above (sect. 6.6.1 and 6.6.2).

This estimation must then be multiplied by the appropriate window function $w(k)$. The DFT of this product gives the smoothed estimator. To increase the number of points in the representation of $\tilde{S}_x(n)$, the DFT can be computed on a length longer than $2M - 1$.

$$\tilde{S}_x(n) = \sum_{k=-L}^{L-1} c_x(k)\, w(k)\, \exp\left(-j\,2\pi\,\frac{nk}{2L}\right) \qquad (6.176)$$
$$\text{with } L > M$$

However, it must be noted that the frequency resolution and the quality of the estimator do not depend on L, but rather on the shape and length of the window $w(k)$.

6.7 EXAMPLES OF SPECTRUM ANALYSIS

6.7.1 Preliminary remarks

The use of spectral estimators is illustrated in this section with two examples. To compare the different estimators and to show the role of the different parameters, it is useful to know *a priori* the real spectrum $\Phi_x(g)$. Therefore, in both examples, a pseudorandom signal generated by the methods of section 6.2.6 is used.

In the first example, we consider the bias and the variance. In the second one, we examine the frequency and dynamic resolutions.

6.7.2 Data for example 1

In this example, a pseudorandom Gaussian white noise of variance one, generated with the methods of section 1.2.10 is considered. This signal is then filtered by using a filter having a transfer function given as

$$G(z) = \frac{0.0154 + 0.0461\,z^{-1} + 0.0461\,z^{-2} + 0.0154\,z^{-3}}{1 - 1.9903\,z^{-1} + 1.5717\,z^{-2} - 0.458\,z^{-3}} \qquad (6.177)$$

We have a third-order Chebyshev filter (sect. 5.4.14) with a normalized cut-off frequency $f_c = 0.1$ and a ripple of 0.5 dB in the passband.

According to (6.29), the power spectrum of the output signal is given by

$$\Phi_x(f) = |G(f)|^2 \tag{6.178}$$

This function is shown in Fig. 6.6. It will be used as a basis for the comparison of the different results obtained.

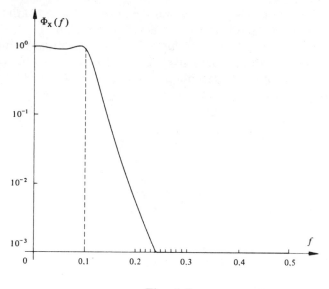

Fig. 6.6

6.7.3 Results of the simple estimator

Figure 6.7 shows the results obtained with the simple estimator $S_x(f)$ for various values of the observation length K. These results show that, for any K, $S_x(f_n)$ fluctuates as a function of frequency about an average value $E[S_x(f)]$ (6.108).

Due to the bias, this mean value is the desired value of the real spectrum $\Phi_x(f)$ only for the limit of K tending toward infinity.

The bias is significant for small values of K ($K = 128$) and decreases slowly as a function of K. It can be considered negligible in the case $K = 2048$.

The fluctuations are caused by the variance of the estimator. We can see that they are independent of K, as previously shown (6.119). Representation on a logarithmic scale shows that the variance is approximately proportional to the square of the spectral power density. As a result of this property, the simple estimator gives results which are difficult to interpret and to use.

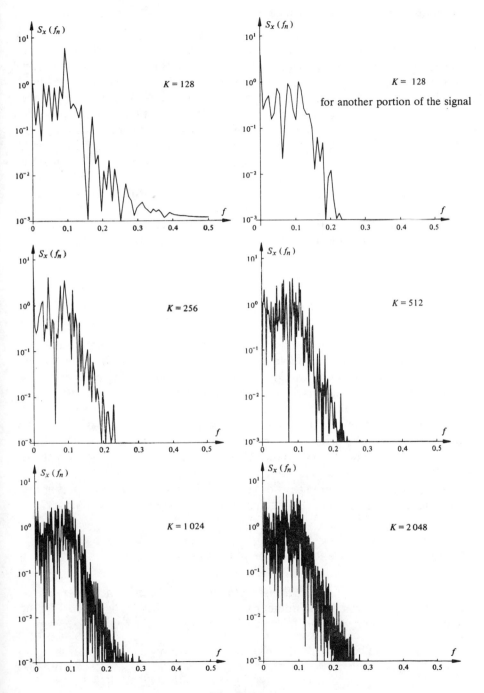

Fig. 6.7

6.7.4 Results of the averaged estimator

The results obtained with the averaged estimator in the analysis of the previously considered signal are given in Fig. 6.8. In this case, two parameters are involved: the length M of the sections and their number L. The observation length is simply the product ML. The bias decreases with M and the variance decreases with L. Thus, for a given observation length K, there is a compromise between bias and variance. This compromise is clearly illustrated in Fig. 6.8. The variance, which is relatively large in the beginning ($L = 2$), decreases with L. It is significantly attenuated for $L = 32$. However, the bias, which is negligible for $M = 1024$, becomes very important for $M = 64$.

The only method to decrease one without increasing the other consists of using a larger observation length K. This is shown in the case $K = 4096$, $M = 256$, and $L = 16$. The results obtained are very satisfactory in comparison with those of Fig. 6.6. The improvement obtained compared to the case of the simple estimator is obvious.

6.7.5 Results of the smoothed estimator

The smoothed estimator involves a window function having parameters and shape which dramatically influence the results. Figure 6.9 shows the results obtained with the Hamming window (3.116). By slowly enlarging the central peak of the function $W_h(f)$ (see eq. (6.129)), the filtering effect is increased. This decreases the variance, but increases the bias, and can be seen in the example with $M = 65$. In this case, the presence of secondary peaks of the function $W_h(f)$ may be observed. For small values of M, the Fourier transform (6.130) depends not on the estimator $c_x(k)$, as it should, but rather on the shape of the function $w_h(k)$.

An acceptable compromise is obtained with $M = 129$. By comparing this result with those of Fig. 6.6, the estimation can be considered satisfactory.

6.7.6 Confidence intervals for the smoothed estimator

The important parameter in the evaluation of the confidence intervals for the smoothed estimator is the number of degrees of freedom v (6.148). In the case of the Hamming window, from (6.145), we have

$$v \cong \frac{2K}{\displaystyle\sum_{k=-M/2}^{M/2} w_h^2(k)} = \frac{2K}{\displaystyle\int_{-1/2}^{1/2} W_h^2(f)\,df} = \frac{2K}{\alpha^2 M + (1-\alpha)^2 M/2} \tag{6.179}$$

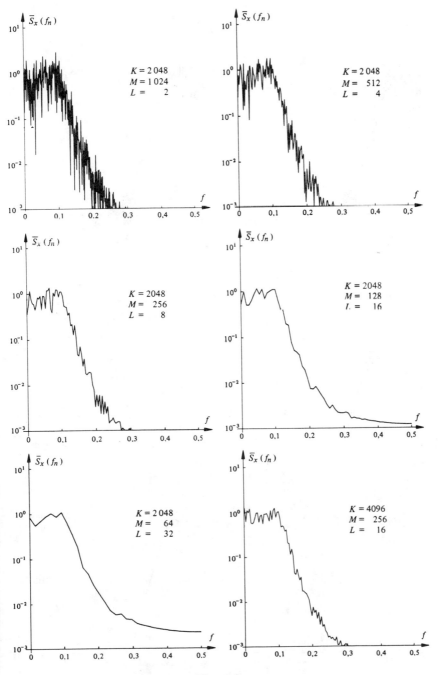

Fig. 6.8

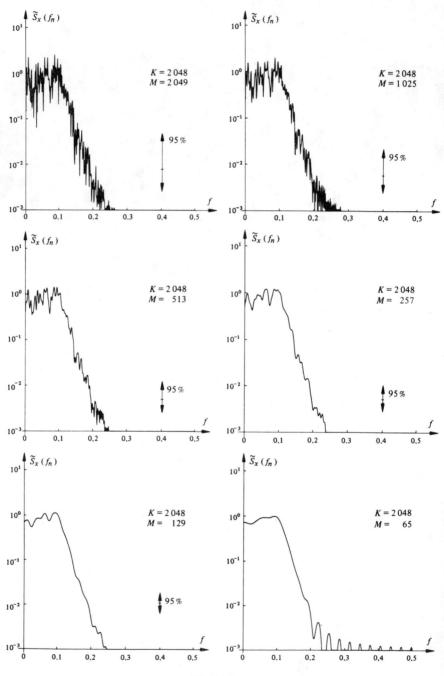

Fig. 6.9

With $\alpha = 0.54$, the following value is obtained:

$$\nu = \frac{2K}{0.4\,M} = 5.05\,\frac{K}{M} \tag{6.180}$$

The functions in Fig. 6.5 enable us to obtain the confidence intervals for each value of K and M. In Fig. 6.9 only the intervals of 95% confidence are mentioned.

In the last case, $M = 65$, the peak of the spectral window is so wide that the hypotheses under which the confidence intervals were established are no longer valid.

6.7.7 Other window functions

We may wonder how the shape of the window function $w(k)$ influences the results of the smoothed estimator. The result obtained with a rectangular window of length $M = 129$ is shown in Fig. 6.10. The bias caused by the secondary peaks of the function $W_R(f)$ is obvious. For the same length, $M = 129$, the Hamming window leads to a much better result (Fig. 6.9). In fact, the bias introduced by $W_h(f)$ exists, but it is smaller than 10^{-3} in this example. For a smaller length ($M = 65$), the result obtained with a triangular window is shown in Fig. 6.11. The previous remarks remain valid in this case.

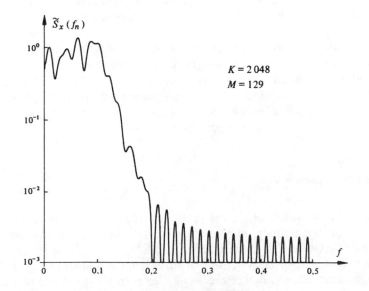

Fig. 6.10

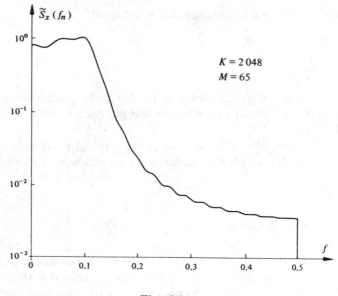

Fig. 6.11

6.7.8 Modified estimator

The modified estimator involves a window function directly on the observed signal. The results obtained for the analysis of the same signal are shown in Fig. 6.12.

The window function used is the Hamming window. These results are obviously better than those of the averaged estimator, which is a particular case of the modified estimator with a rectangular window. When the number L of sections increases, the variance decreases significantly without greatly increasing the bias. The result obtained with $M = 64$ and $L = 32$ is very good. The bias exists, but it does not appear for powers less than 10^{-3}.

6.7.9 Data for example 2

This example will illustrate the problems of dynamic range and frequency resolution. The signal analyzed consists of white noise having a variance 10^{-5}, and two sine waves having amplitudes $\sqrt{2}$ and $\sqrt{2} \cdot 10^{-2}$, respectively.

The theoretical spectral density of this signal is shown in Fig. 6.13. The spectral estimator should detect the weak signal in the presence of noise and the strong signal.

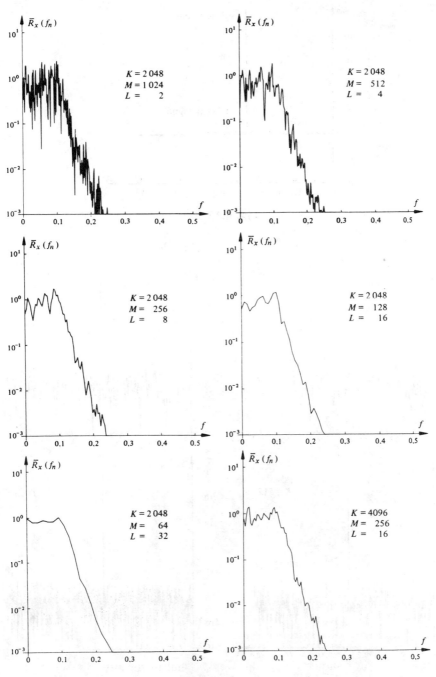

Fig. 6.12

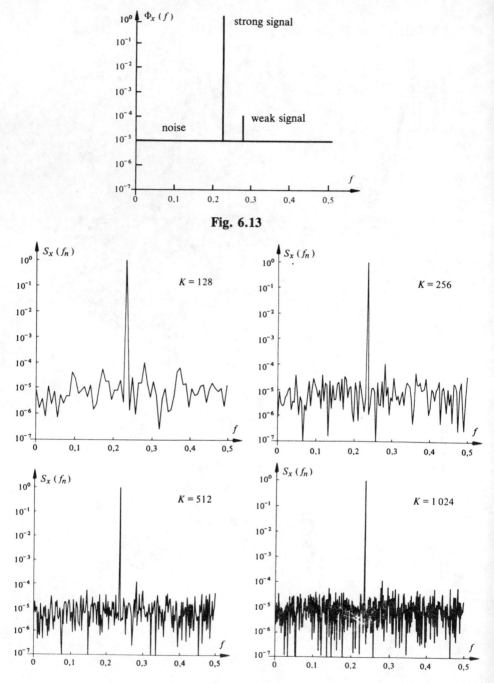

Fig. 6.13

Fig. 6.14

6.7.10 Results of the simple estimator

As a first case, the frequencies of the sine waves have been chosen to be identical to the harmonic frequencies $f_n = n/K$. The results obtained for different values of K are shown in Fig. 6.14. As in the previous example, for any K, the variance of the noise contribution is not attenuated. Therefore, the typical peak of the weak signal does not appear clearly. A peak of power 10^{-4} exists, but it is not interpretable because of the variance of the estimator. For any K, the strong signal is well detected because its frequency is equal to one harmonic frequency. Thus, with a noise of variance 10^{-5}, the simple estimator does not permit the detection of a signal of power 10^{-4}.

In the second case, the frequency of the strong signal was shifted to be in the middle of two consecutive harmonic frequencies. The results obtained for the same values of K are shown in Fig. 6.15.

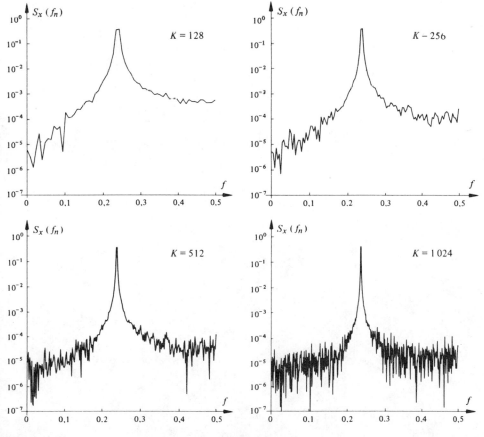

Fig. 6.15

We see that even the strong signal is not correctly detected. The reason for this is the distribution phenomenon of one component over all frequencies because of the limited observation length (sect. 3.7.21 and eq. (3.127)). Only the components at the harmonic frequencies are not subject to this phenomenon.

Increasing the observation length permits us to narrow the peak caused by the function $W_R(f)$ and to improve the estimation of the power of the strong signal. The weak signal is totally drowned by noise. The only possible improvement is an increase of the length K for better detection of the strong signal.

6.7.11 Results of the mean estimator

When the input signal is analyzed by using the averaged estimator, the results seen in Fig. 6.16 are obtained. In this case, the frequencies of the sine

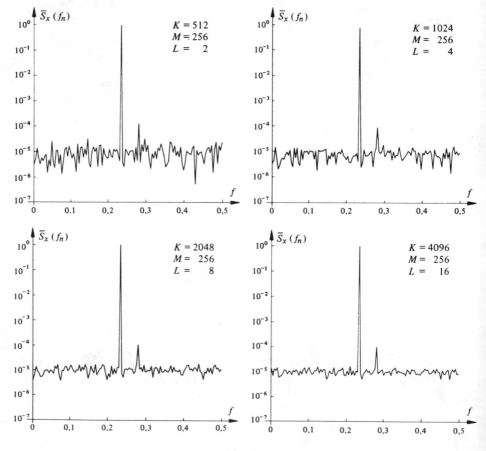

Fig. 6.16

waves are equal to the harmonic frequencies $f_n = n/M$. For a given frequency resolution (M constant), an increase in the number L of sections dramatically attenuates the variance, and therefore enables us to distinguish between the two signals.

Shifting the frequency of the strong signal leads to the results of Fig. 6.17. The same frequency distribution phenomenon exists for each section. The estimate is not improved by taking the mean. Figure 6.18 shows the result of delaying the frequency of the weak signal. The previous remarks are also valid in this case.

The solution is to improve the frequency resolution by increasing the length M of the sections, even at the price of an increase of the variance, if the total length K cannot be increased (Fig. 6.19).

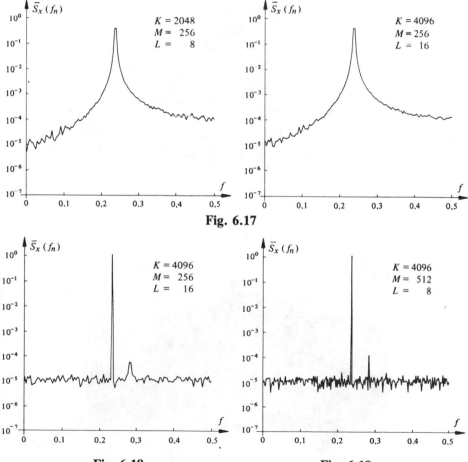

Fig. 6.17

Fig. 6.18   **Fig. 6.19**

6.7.12 Results of the smoothed estimator

Figure 6.20 shows the results obtained with a rectangular window. The same distribution phenomenon of the frequency components appears in this case. However, it is not at the level of a signal and its Fourier transform, but at that of an autocorrelation function and its transform. Because of the con-

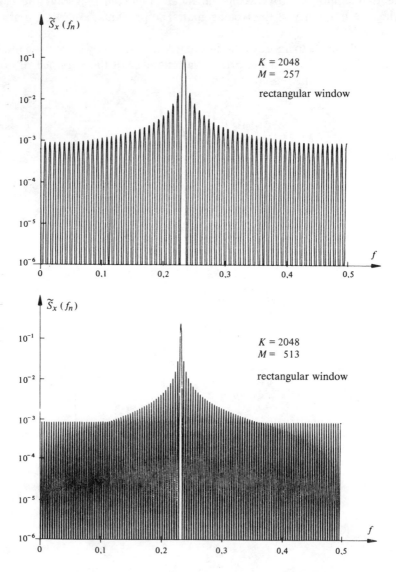

Fig. 6.20

volution product (6.129), the peak of the strong signal is distributed over the entire frequency axis by the function $W_R(f)$. The contributions of the noise and the weak signal are respectively 20 dB and 10 dB smaller than the secondary peaks of $W_R(f)$.

This situation is improved with the Hamming window (Fig. 6.21), but the improvement is not sufficient to bring out the typical peak of the weak signal.

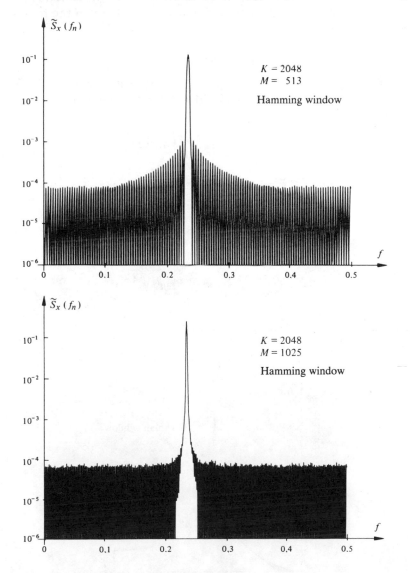

Fig. 6.21

One of the window functions having the smallest secondary peaks is the Blackman window. It is only in this case that the 10^{-5} level of the noise contribution is reached.

However, the detection of the weak signal is not clear (Fig. 6.22). It remains subject to the distribution phenomenon (6.129). To evaluate the effect of noise in these results, the case without noise can be considered (Fig. 6.23). In this case, the weak signal is detected, but the estimation of the power is still subject to the distribution phenomenon.

6.7.13 Results of the modified estimator

As a first case where the frequencies of the sine waves are equal to harmonic frequencies $f_n = n/M$, the results obtained with the Hamming window are shown in Fig. 6.24.

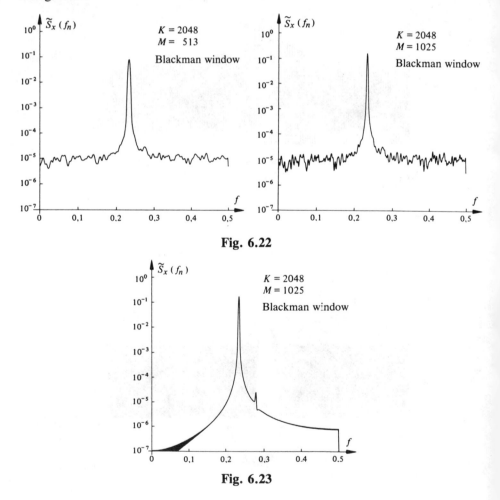

Fig. 6.22

Fig. 6.23

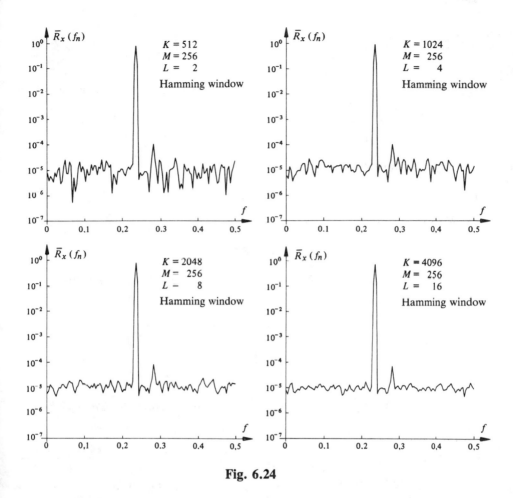

Fig. 6.24

Increasing the number of sections L allows us to decrease the variance of the estimator. For $L = 16$, the estimation is very satisfactory. By comparing these results with those of Fig. 6.16, we can see the enlargement of the peaks due to the main peak of the function $W_h(f)$. This can be a drawback in the estimation of the frequency of the peaks, but the inverse appears when the frequencies of the sine waves are not equal to the harmonic frequencies.

Figure 6.25 shows the results obtained for this case. The comparison with fig. 6.25 shows the advantage of using a window other than a rectangular one. The weak signal, although difficult to distinguish, does appear with the modified estimator. Shifting only the frequency of the weak signal leads to the result of fig. 6.26. By contrast with the averaged estimator, we have for the modified estimator two ways of improving the estimation. The first consists of improving the frequency resolution by increasing M (Fig. 6.27) and the second consists of

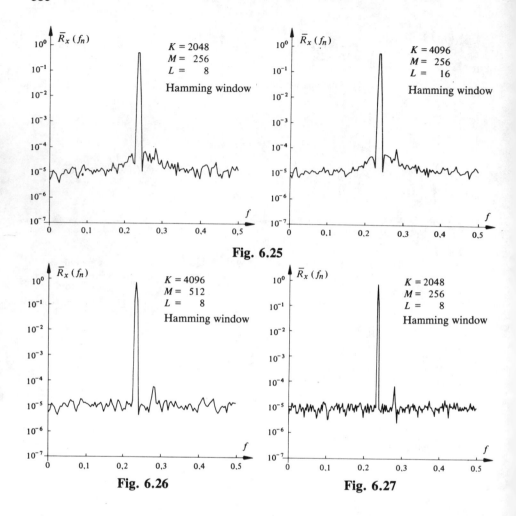

Fig. 6.25

Fig. 6.26

Fig. 6.27

using another, more powerful window function. Figure 6.28 shows the results obtained with the Blackman window in the case where the frequencies of the sine waves are not equal to harmonic frequencies. With this window, the two signals are well detected at the price of an enlargement of the peaks. This results from the main peak of the function $W_B(f)$, which is three times wider than the main peak of $W_R(f)$.

6.7.14 Remarks

The two previous examples illustrate the properties of the different estimators in the analysis of random signals. The simple estimator leads to the worst results due to its uncontrollable variance and the implicit rectangular window. It can only be used to get a very rough idea of the desired spectral density.

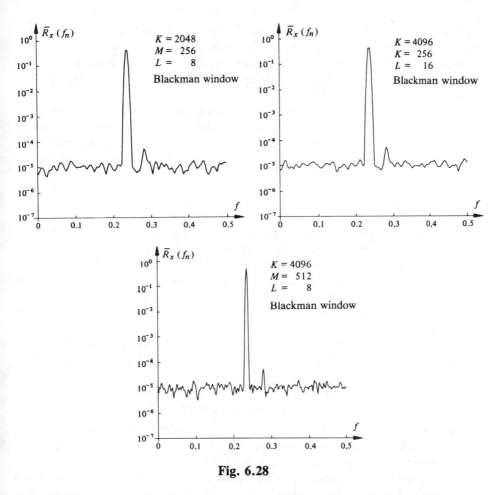

Fig. 6.28

It is clear that the averaged estimator is a particular case of the modified one in which a rectangular window is used. Because a rectangular window has important secondary peaks, the modified estimator will always be better in terms of dynamic range resolution. The price of this superiority is paid in the frequency resolution, which is directly related to the width of the main peak of the function $W_i(f)$ used. Only the rectangular function has a main peak where the width coincides with the harmonic frequencies. All other windows lead to wider main peaks and, consequently, to a smaller frequency resolution. To satisfy the constraints of dynamic range and frequency resolutions simultaneously, the only solution is to increase the length M of the sections and the total observation length.

The smoothed spectral estimator is the least efficient with respect to dynamic range resolution. The reason is that, in the modified estimator, the signal is multiplied by a window function $w_i(f)$. To calculate the power spectral density, the square of the function $W_i(f)$ is used. This square increases the difference between the main and the secondary peaks. With the smoothed estimator, however, it is the estimation of the correlation $c_x(k)$ that is multiplied by the window function $w_i(k)$. In the estimation of the spectral density, $W_i(f)$ is involved instead of $W_i^2(f)$.

Moreover, the distribution phenomenon of the frequency components is more obvious due to the convolution product (6.129). However, this same phenomenon allows for a better reduction of the variance in the estimation. It can be controlled, not only by way of the length M of the window, but also by its shape.

6.7.15 Conclusion

In this chapter, we have presented the application of the discrete Fourier transform to the spectral analysis of random signals. Because this is, by definition, an estimation problem, some elements of estimation theory were reviewed and illustrated by examples. The use of linear systems was studied in detail to assign an autocorrelation function or power spectrum to a given random signal. These methods are very useful, for example, for the simulation of communication systems.

The estimators for the autocorrelation and crosscorrelation functions were examined and illustrated. The importance of the bias and the variance as well as the ways to control both were discussed.

The bulk of this chapter was devoted to spectral estimators and their properties. The following main points can be noted:

- The simple estimator has a variance proportional to the square of the desired spectrum. It cannot be reduced by increasing the observation length. Moreover, the simple estimator is subject to the distribution phenomenon of the frequency components on the total frequency axis, except at the harmonic frequencies. This is caused by the rectangular window implicitly included in the finite-length observation.
- The averaged estimator allows us to reduce the variance by changing the number of sections L. However, it has the same behavior as the simple estimator with respect to the distribution phenomenon. If, for a given observation length K, the variance is reduced too much, the bias increases significantly. The optimal compromise is hence not always easy to determine.
- The smoothed estimator also allows us to decrease the variance significantly by modifying the length and shape of the window function used. In general, it has good frequency resolution, but a poor dynamic range resolution.
- The modified estimator is that which allows for the best dynamic range resolution. Nonetheless, for the same observation length, its frequency resolution is smaller than that of the simple estimator. However, it is less sensitive to the distribution phenomenon.

These points are summarized qualitatively in Fig. 6.29. To obtain satisfactory results with any estimator, the parameters of the analysis must be carefully chosen and their roles and properties well understood.

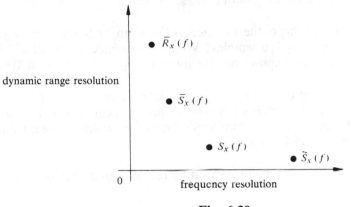

Fig. 6.29

6.8 EXERCISES

6.8.1 Let us consider Gaussian white noise of variance σ_6^2. This signal is filtered by a second-order recursive filter defined by the equation:

$$y(k) - y(k-1) + 0.5\, y(k-2) = x(k)$$

Determine analytically and represent graphically the power spectrum of the output signal.

6.8.2 Let us consider the signal $y(k)$ of exercise 6.8.1. Determine its autocorrelation function:

- by using the direct method (sect. 6.2.8);
- by computing the inverse Fourier transform of the power spectrum decompose into partial fractions).

Compare the two results.

6.8.3 Let us consider white noise filtered by a first-order recursive filter defined by

$$y(k) - 0.9\, y(k-1) = x(k)$$

The output signal of this filter is filtered again by another filter of same type, but with a coefficient of 0.8. What is the power spectrum of the output signal of the second filter?

6.8.4 Determine the autocorrelation function of the signal of exercise 6.8.3, at the output of the second filter:

- by using the direct method (sect. 6.2.7);
- by computing the inverse Fourier transform of its power spectrum.

6.8.5 What is the mean value of the variance of Gaussian white noise having a variance $\sigma_b^2 = 1$, from $K = 100$ samples? What is the confidence with which the real mean value can be found within the interval $[\hat{m}_b - 0.3, \hat{m}_b + 0.3]$?

6.8.6 The power spectrum of a random signal at a frequency f_0 is $\varphi_x(f_0)$. This value is estimated using the simple estimator by having f_0 coincide with a harmonic frequency f_n. Determine the confidence interval in which the real value $\varphi_x(f_0)$ can be found with 99% confidence.

6.8.7 Show that the simple estimator defined by (6.105) can be obtained from $X_K(f)$ given by (6.107).

6.8.8 By using (6.145) and the curves of Fig. 6.5, determine the window which leads to the smallest interval for a confidence of $1 - \alpha$, and for a given spectrum $\Phi_x(n)$.

6.8.9 Which spectral estimator must be used to bring out:

- two peaks that are of equal height and very close to each other?
- two peaks that are very different in height and not close to each other?

Chapter 7

Homomorphic Signal Processing

7.1 INTRODUCTION

7.1.1 Preliminary remarks

The great importance of linear systems in signal processing is obvious. They are relatively easy to analyze and to characterize, leading to powerful and elegant mathematical representations. Their implementation is also relatively simple, permitting a large number of processing operations. It has almost become traditional to use linear systems in problems where the linearity is obtained only by approximation, even if this approximation is poor.

In general, nonlinear systems are relatively difficult to analyze and to characterize mathematically. Some studies, even intensive ones, of this type of system exist, but are almost always reduced to very particular cases. Homomorphic processing, while limited in regard to the enormous number of nonlinear systems, has the advantage of being easy to analyze mathematically and having important practical applications of general interest. It was developed by Oppenheim [32, 33] in 1965. In the beginning, it was only a theoretical study based on linear algebra, but nowadays it is of great importance due to the increasing number of applications which have been implemented.

7.1.2 Organization of the chapter

In this chapter, we present the general trends of homomorphic system theory. Then, two particular classes of homomorphic systems, the multiplicative and the convolutional, are examined in detail. Finally, the practical applications of these systems are presented. The chapter closely follows the original work of Oppenheim. However, his approach of using the z-transform was deliberately omitted to present an approach using the discrete Fourier transform, because, in practice, it is the DFT which is used. The general theory is presented for one-dimensional digital signals. This is not a restriction, however, because no hypotheses are made concerning the nature and dimension of the signals.

7.2 GENERALIZED SUPERPOSITION

7.2.1 Preliminaries

Homomorphic processing stems from a generalization of linear system theory. It is based on the principle of generalized superposition. In this section, after a brief review of some fundamental properties of linear systems, the generalized superposition principle is described.

7.2.2 Review

Any cause and effect (or input-output) relationship can be considered as a system S, which acts on the input signal $x(k)$ to produce the output signal $y(k)$ (sect. 1.5).

The mathematical model associated with a system is a unique operator (or transformation), which maps the input signal $x(k)$ onto the output signal $y(k)$. This is noted by the following equation:

$$y(k) = S[x(k)] \tag{7.1}$$

Particular sets of systems can be defined by setting constraints on the operator S. The set of linear systems L is defined by the superposition principle (sect. 1.5.5). Let $y_1(k)$ and $y_2(k)$ be the output signals respectively corresponding to the excitations $x_1(k)$ and $x_2(k)$. A system L is linear if and only if for any arbitrary constant c, we have

$$L[x_1(k) + x_2(k)] = L[x_1(k)] + L[x_2(k)]$$
$$= y_1(k) + y_2(k) \tag{7.2}$$

and

$$L[c x_1(k)] = c L[x_1(k)] = c y_1(k) \tag{7.3}$$

Property (7.2) indicates that the system L transforms the sum of two excitations as if they were transformed separately and added afterward. Property (7.3) indicates the correspondence between the output and the input of the system L.

Linear systems, digital as well as analog, are used in signal processing for linear filtering (chap. 5). For example, if a signal is composed of two signals with different frequency bands, linear filtering can be used to separate these two signals.

7.2.3 Nonlinear filtering

Linear filtering techniques, in general, are simple, relatively old, commonly used, and therefore well known. However, they cannot be used if the signal is not combined with the others by addition. How then can we isolate, for example, a signal combined by multiplication or convolution? Although, in the beginning, the theoretical solution of this problem did not have practical fields of application, it is now a powerful and elegant tool, which is being used more and more. The principal application areas are audio signals and image processing.

7.2.4 Generalized superposition principle

To generalize (7.2) and (7.3), the rule for the combination of input signals will be denoted by □ and the rule for the combination of input signals with scalars by ■. For example, □ can be an addition, a convolution product, *etcetera*. Similarly, ○ will represent the combination of output signals and ● the combination of output signals with scalars. Equations (7.2) and (7.3) can then be generalized as follows:

$$H[x_1(k) \,\square\, x_2(k)] \;=\; H[x_1(k)] \circ H[x_2(k)] \qquad\qquad (7.4)$$

and

$$H[c \,\blacksquare\, x_1(k)] \;=\; c \bullet H[x_1(k)] \qquad\qquad (7.5)$$

where H represents the operator of the system. In the particular case in which □ and ○ are additions, and ■ and ● are multiplications, the class of linear systems is again obtained.

7.2.5 Homomorphic systems

To represent a system H satisfying (7.4) and (7.5), the input and output siganls are considered as vectors in a vector space [34], with the rules □ and ○ corresponding to the sum of vectors, and the rules ■ and ● corresponding to the product with scalars.

Thus, the transformation H is algebraically linear, in the sense of the algebraic structure of the vector space. It maps the vector space of the input signals onto the vector space of the ouput signals. To use the theory of vector spaces in the study of systems satisfying (7.4) and (7.5), the constraints of associativity and commutativity must also be verified by the input □ and ■, and output ○ and ● rules.

For example, the following relationships must be true:

$$x_1(k) \square x_2(k) = x_2(k) \square x_1(k) \tag{7.6}$$

and

$$x_1(k) \square [x_2(k) \square x_3(k)] = [x_1(k) \square x_2(k)] \square x_3(k) \tag{7.7}$$

Such systems, represented by a linear transformation between vector spaces, are called *homomorphic*. A homomorphic system H, with the input and output rules $\square$ and $\circ$, is outlined in Fig. 7.1.

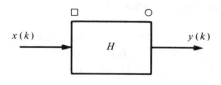

Fig. 7.1

7.2.6 Canonical form

A very important result proved by Oppenheim [32, 33] is the possibility of representing any homomorphic system by three systems in series, where the middle one is a conventional linear system. This is the so-called *canonical* form of homomorphic systems. It is shown in Fig. 7.2.

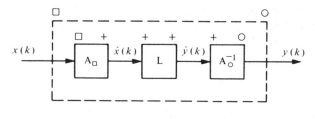

Fig. 7.2

For any input rule $\square$, the first nonlinear system $A_\square$ has a rule at its output which is $+$, that is the conventional linear combination. The system L is a conventional linear system. Finally, the system $A_\circ^{-1}$ passes from addition to $\circ$ at the output. The systems $A_\square$ and $A_\circ^{-1}$ are called *characteristic systems* of the particular class of homomorphic systems H represented by the rules $\square$ and $\circ$.

For the remainder of this chapter, the output signal of the characteristic system $A_\square$ and the input signal of the characteristic system $A_\circ^{-1}$ will be denoted by a circumflex ( ) (Fig. 7.2). For a given pair of rules □ and ○, only linear systems allow a distinction among different homomorphic systems H. This shows that if the systems $A_\square$ and $A_\circ^{-1}$ can be found, the desired processing can be performed using any of the methods for linear systems.

We will limit ourselves to cases where the rules at the output and input are indentical, and, in particular, to multiplication and the convolution product. In these two cases, the characteristic system and its inverse will be predetermined; only the linear system L will be chosen according to the particular problem under study.

7.3 MULTIPLICATIVE HOMOMORPHIC SYSTEMS

7.3.1 Combination of signals by product

In many applications of signal processing, the problem arises of extracting a signal combined with other signals by multiplication. Such problems require the modification of one of the signals without modifying the others, or the independent modification of each signal according to the specific needs of the problem. For example, a signal transmitted by using amplitude modulation is multiplied by a high-frequency carrier signal, which must be eliminated at the receiver. Another example is the case of an image which can be considered as the product of two signals: the *luminence* and the *reflection*. It is necessary to be able to act on one of them to enhance the contrast, or to modify the dynamic range. Other examples can be found in the study of transmission channels with attenuation and in studies of automatic control systems.

In this section, the ways of determining the characteristics of the multiplicative homomorphic system is presented as well as the different problems associated with the study of such systems and their use in signal processing. Consequently, the study will be limited to the set of homomorphic systems satisfying (7.4) and (7.5), with identical rules at the input and output, where □ denotes multiplication and ■ signifies exponentiation.

7.3.2 Characteristic system

The possible input signals are thus of the following form:

$$x(k) = [x_1(k)]^{a_1}[x_2(k)]^{a_2} \tag{7.8}$$

where a_1 and a_2 are two arbitrary constants.

The characteristic system $A_\bullet$ must satisfy the following equation:

$$A.[x_1(k)^{a_1}][x_2(k)^{a_2}] = a_1 A.[x_1(k)] + a_2 A.[x_2(k)] \tag{7.9}$$

It is well known that the logarithmic function satisfies such an equation if the two signals $x_1(k)$ and $x_2(k)$ are strictly positive for any k. In this case, the following equation holds:

$$\ln [x_1 (k)^{a_1} x_2 (k)^{a_2}] = a_1 \ln x_1 (k) + a_2 \ln x_2 (k) \tag{7.10}$$

This equation, which is only valid for positive signals, can be used, for example, in image processing, where the two-dimensional signals are positive because of the physical nature of light.

However, because most usual signals are bipolar (positive and negative), (7.10) is too restrictive. This restriction can be avoided if the complex logarithmic function, which is valid also for complex signals, is used. Let

$$x (k) = |x (k)| \; \exp \{j \arg [x (k)]\} \tag{7.11}$$

be the representation in polar coordinates of a complex signal $x(k)$. It can also be written as follows:

$$\begin{aligned} x (k) &= \exp [\ln |x (k)|] \exp [j \arg [x (k)]] \\ &= \exp [\ln |x (k)| + j \arg [x (k)]] \end{aligned} \tag{7.12}$$

The complex logarithm is thus given by (sect. 2.5.4):

$$\ln [x(k)] = \ln |x (k)| + j \arg [x (k)] \tag{7.13}$$

This is the inverse function of the complex exponential.

The canonical form of a multiplicative homomorphic system is illustrated in Fig. 7.3. In this figure, the signals $x(k)$, $\hat{x}(k)$, $y(k)$, and $\hat{y}(k)$ are complex.

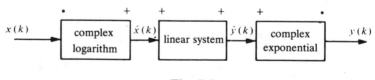

Fig. 7.3

7.3.3 Ambiguity of the phase

It must be noted that any multiple of 2π can be added to the imaginary part of the complex logarithm (7.13) without changing the result (7.12). This shows that the complex logarithm, as defined here, is not a one-to-one mapping. Nevertheless, each transformation involved in the global system must be one-to-one because, where no filtering is present, the output signal must be the same signal $x(k)$ as the input. To use the complex logarithm as the transformation of a characteristic system A, the ambiguity of the argument must first be removed, and then the general superposition principle must be satisfied. That is, if $x(k) = x_1(k)x_2(k)$, we must have

$$\ln [x(k)] = \ln [x_1(k) x_2(k)] = \ln [x_1(k)] + \ln [x_2(k)] \qquad (7.14)$$

This implies, using (7.13):

$$\ln |x(k)| = \ln |x_1(k)| + \ln |x_2(k)| \qquad (7.15)$$

and

$$\arg [x(k)] = \arg [x_1(k)] + \arg [x_2(k)] \qquad (7.16)$$

The ambiguity on the argument is removed by considering its principal value, that is, its value modulo 2π. This cannot be used here because, in general, (7.16) will not be satisfied.

7.3.4 Phase unwrapping

To remove the ambiguity and at the same time satisfy (7.16), the argument $\arg[x(k)]$ must be defined so that it is a continuous function of $x(k)$. However intuitive, this common-sense condition can be justified by using Riemann surfaces. The complete proof, which is not within the context of this book, can be found in [35]. Section 5.5.6 can also be examined for an approach using the z-transform.

Figure 7.4 shows the difference between a continuous argument and its variation obtained by considering only the principal value.

We can move from $\arg[x(k)]$ to its principal part $\mathrm{Arg}[x(k)]$ by computing $\arg[x(k)]$ modulo 2π. Starting from an initial condition, it is also possible to reconstruct the function $\arg[x(k)]$ from its principal part $\mathrm{Arg}[x(k)]$ if a sufficient number of values of $\mathrm{Arg}[x(k)]$ are available.

7.3.5 Comments

The restriction imposed on the argument thus excludes the application of multiplicative homomorphic systems to bipolar signal processing. However, it is always possible, in a preprocessing step, to transform a real bipolar signal into a strictly positive signal by adding a constant positive signal, which is greater than the largest negative value of the input signal. A real signal can also be made complex by adding an imaginary part with a small amplitude as compared

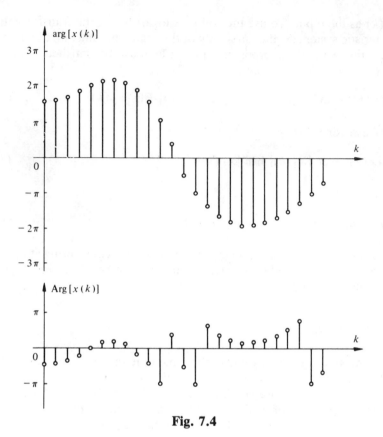

Fig. 7.4

to that of the real part. The characteristic system is thus a logarithmic amplifier, in principle, with a unity gain. Multiplicative homomorphic systems differ from each other by the linear system L (Fig. 7.3), which can, if necessary, process complex signals. The more general form of such a system composed of linear subsystems processing real signals is shown in Fig. 7.5.

7.3.6 Output signal

For an input signal of type (7.8), the output signal of the characteristic signal is of the form:

$$\hat{x}(k) = a_1 \ln[x_1(k)] + a_2 \ln[x_2(k)]$$
$$= a_1 \hat{x}_1(k) + a_2 \hat{x}_2(k)$$
$$= \text{Re}[\hat{x}(k)] + j \text{Im}[\hat{x}(k)] \tag{7.17}$$

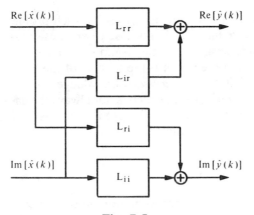

Fig. 7.5

7.3.7 Remarks

The constraints on the subsystems $L_{\alpha\beta}$ (Fig. 7.5) depend on the nature of the numbers a_1 and a_2. The inverse characteristic system is the complex exponential given by (7.12). The choice of the system L obviously depends on the particular problem under study. For example, in the separate processing of the components of a signal obtained by the product of two signals, one of low and the second of high frequency, the corresponding frequency spectra $\hat{X}_1(f)$ and $\hat{X}_2(f)$ must overlap very little. This is especially the case for images. A complete example of the application of multiplicative homomorphic systems is presented in section 7.6.1.

7.4 CONVOLUTIVE HOMOMORPHIC SYSTEMS

7.4.1 Combination of signals by the convolution product

In many applications, the problem arises of isolating a useful signal combined by convolution with other signals, or measuring some of its characteristics. Several examples can be listed: a natural signal observed only through a linear system, which may be the measurement device itself (distortion through measurement); a signal recorded in a reverberating environment, which is altered by unexpected echoes; the restoration of an image recorded out of focus, or with a camera that moved while recording. Other examples can be found in the enhancement of old audio records obtained with devices of poor quality; in the determination of the probability distribution of a signal distorted by an additive random signal; in voice processing, where the excitation of the impulse response of the voice channel must be isolated; and in the study of seismic signals.

One method that is widely used to solve such problems, often called *deconvolution,* is inverse filtering. In the case where a useful signal convolved with another signal must be isolated, the composite signal is filtered with a filter having frequency response which is the inverse of the Fourier transform of the disturbing signal. This requires *a priori* knowledge, as precise and detailed as possible, of the disturbing signal, which is not always the case. Another possible method is to use convolutional homomorphic systems. This section is devoted to their study and implementation.

7.4.2 Characteristic system

We will limit ourselves here to homomorphic systems satisfying (7.4) and (7.5), with identical rules at the input and output, and where the combination rule for the input signals is the convolution product. The combination rule of a signal $x(k)$ with an integer scalar l corresponds to l successive convolution products of $x(k)$ with itself. If l is not an integer, the same notion can nevertheless be generalized [32].

The input signals of such an homomorphic system are thus of the form:

$$x(k) = \sum_{l=-\infty}^{\infty} x_1(l)\, x_2(k-l) = x_1(k) * x_2(k) \tag{7.18}$$

This equation is the discrete version of the well known integral equation:

$$y(t) = \int_{-\infty}^{+\infty} y_1(\tau)\, y_2(t-\tau)\, d\tau \tag{7.19}$$

Mathematicians call this *Fredholm's integral equation of the first type.* Homomorphic processing provides a method for solving this equation for $y_1(t)$ or $y_2(t)$.

The characteristic system A_* of the canonical representation must satisfy the equation:

$$A_*[x_1(k) * x_2(k)] = A_*[x_1(k)] + A_*[x_2(k)] \tag{7.20}$$

to transform a convolution product into an addition. To determine the system A_*, we must remember that (sect. 1.6.9) if

$$x(k) = x_1(k) * x_2(k) \tag{7.21}$$

then,

$$X(f) = X_1(f)\, X_2(f) \tag{7.22}$$

where $X(f)$, $X_1(f)$, and $X_2(f)$ are the Fourier transforms of $x(k)$, $x_1(k)$, and $x_2(k)$, respectively. It can be seen that the Fourier transform can be considered here to be a homomorphic system with the convolution product as the input rule and multiplication as the output rule. Consequently, if the signals are replaced by their Fourier transforms in (7.20), the problem of determining the characteristic system is reduced to a known case. This is the case of multiplicative homomorphic systems studied in the previous section.

7.4.3 Canonical form

The canonical representation of convolutional homomorphic systems for signals represented by their Fourier transforms is given in Fig. 7.6.

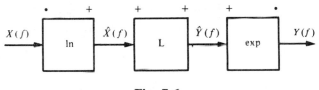

Fig. 7.6

Thus, the characteristic desired system A'_* is the series of two systems: one Fourier transform and one complex logarithm. The canonical form of convolutive homomorphic systems is shown in Fig. 7.7. Note that the canonical form show this figure and the linear system L_f act on signals which are functions of the frequency; therefore, the index f can represent the system L_f.

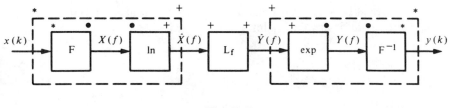

Fig. 7.7

7.4.4 Alternative

Another possible canonical form, equivalent to the previous case, is one in which the linear system acts on signals that are functions of time rather than frequency. This new form is once again obtained by using the Fourier transform.

The corresponding characteristic system A_* and its inverse A_*^{-1} which can be obtained from A_* are shown in Fig. 7.8.

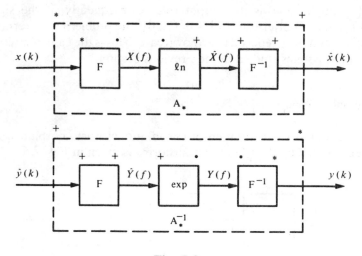

Fig. 7.8

From this figure, we can see that the direct and inverse Fourier transforms F and F^{-1} are used as conventional linear transformations as well as homomorphic transformations between the convolutional and multiplicative vector spaces.

7.4.5 Hypotheses

For practical reasons, it is reasonable to assume that the input signal $x(k)$ is real. If this signal is absolutely integrable, that is, if

$$\sum_{k=-\infty}^{+\infty} |x(k)| < \infty \tag{7.23}$$

then, the series:

$$\sum_{k=-\infty}^{+\infty} x(k) \, e^{-j2\pi fk} \tag{7.24}$$

converges uniformly toward a continuous function of f, which is by definition the Fourier transform $X(f)$ of the signal $x(k)$. The periodic function $X(f)$ of period one, generally complex, can be written as follows:

$$X(f) = \text{Re}\,[X(f)] + j\,\text{Im}\,[X(f)] \tag{7.25}$$

For real signals $x(k)$, $\text{Re}[X(f)]$ and $\text{Im}[X(f)]$ are continuous functions of f, which are even and odd, respectively. We shall assume in addition that $X(f)$ has no zeros.

7.4.6 Remarks

It is reasonable to require a characteristic system A_* (Fig. 7.8) which has a response $\hat{x}(k)$ to a real and absolutely integrable signal $x(k)$ is also an absolutely integrable real signal. This requires that $\hat{X}(f)$ be a continuous Fourier transform, the real part of which is an even function and the imaginary part is an odd one as a function of f.

It is clear that the same constraints must be imposed on the inverse characteristic system A_*^{-1}. Indeed, by definition, we must have

$$A_*^{-1}[A_*[x(k)]] = x(k) \qquad (7.26)$$

7.4.7 Ambiguity of the complex logarithm

The complex logarithm must be defined as in section 7.3.3, without ambiguity, and such that if

$$X(f) = X_1(f)\,X_2(f) \qquad (7.27)$$

then,

$$\hat{X}(f) = \ln[X(f)] = \ln[X_1(f)] + \ln[X_2(f)]$$
$$= \ln|X(f)| + j\arg[X(f)] \qquad (7.28)$$

is an even continuous function of f with an even real part $|X(f)|$ and an odd imaginary part $[X(f)]$.

The continuity of the real part $|X(f)|$ is assured by the continuity of $X(f)$ and the monotonic characteristic of the logarithm function. Because the real part of $X(f)$ is an even function and its imaginary part is odd, its modulus is an even function. Thus, $\text{Re}[\hat{X}(f)] = \ln|X(f)|$ is an even function of f.

For the same reasons as before (sect. 7.3.4), the argument $\arg[X(f)]$ must be defined so that it is a continuous function of $X(f)$. In the case of a real signal, it is also an odd function.

7.4.8 Theorem

The argument, defined as a continuous function, of the Fourier transform of a real signal is an odd function of frequency.

7.4.9 Proof

Let $\hat{X}'(f)$ be the derivative of the complex logarithm $\hat{X}(f) = \ln[X(f)]$ with respect to f. Then,

$$\hat{X}'(f) = \frac{d}{df} \ln[X(f)] = \frac{1}{X(f)} X'(f) \tag{7.29}$$

By using the definition of the derivative, we have

$$\hat{X}'(f) = \frac{d}{df} \operatorname{Re}[\hat{X}(f)] + j \frac{d}{df} \operatorname{Im}[\hat{X}(f)]$$

$$= \frac{d}{df} \ln|X(f)| + j \frac{d}{df} \arg[X(f)] \tag{7.30}$$

By identifying the imaginary parts of (7.29) and (7.30), the following result is obtained:

$$\frac{d}{df} \arg[X(f)] = \frac{\operatorname{Re}[X(f)]\dfrac{d}{df}\operatorname{Im}[X(f)] - \operatorname{Im}[X(f)]\dfrac{d}{df}\operatorname{Re}[X(f)]}{|X(f)|^2} \tag{7.31}$$

The numerator on the right-hand side of this equation can be replaced by

$$\operatorname{Re}^2[X(f)] \frac{d}{df}\left[\frac{\operatorname{Im}[X(f)]}{\operatorname{Re}[X(f)]}\right] \tag{7.32}$$

Finally, the following result is obtained:

$$\frac{d}{df} \arg[X(f)] = \frac{\operatorname{Re}^2[X(f)]}{|X(f)|^2} \frac{d}{df}\left[\frac{\operatorname{Im}[X(f)]}{\operatorname{Re}[X(f)]}\right] \tag{7.33}$$

This function is a continuous even function, due to hypothesis 7.4.5. With the initial condition $\arg[X(0)] = 0$, it becomes odd by integration with respect to f, but remains continuous.

7.4.10 Cepstrum

Theorem 7.4.8 thus ensures an output signal $\hat{x}(k)$, which is real and absolutely integrable, given by

$$\hat{x}(k) = \frac{-1}{j\,2\pi k} \int_{-1/2}^{1/2} \frac{1}{X(f)} \frac{dX(f)}{df} e^{j\,2\pi kf} \, df \quad \text{for } k \neq 0 \tag{7.34}$$

and

$$\hat{x}(0) = \int_{-1/2}^{1/2} \ln |X(f)| \, df \qquad (7.35)$$

Equation (7.35) is obtained by computing the inverse Fourier transform of (7.28) for $k = 0$, and taking into account the odd parity of the imaginary part of $\hat{X}(f)$. The signal $\hat{x}(k)$ is called *cepstrum* of the signal $x(k)$.

7.4.11 Direct relationship between a signal and its cepstrum

Let us consider (7.29), which can also be written as

$$X'(f) = \hat{X}'(f) \, X(f) \qquad (7.36)$$

The inverse Fourier transform for both sides of this equation leads to

$$k \, x(k) = \sum_{l=-\infty}^{+\infty} l \, \hat{x}(l) \, x(k-l) \qquad (7.37)$$

and, finally,

$$x(k) = \sum_{l=-\infty}^{+\infty} \left(\frac{l}{k} \right) \hat{x}(l) \, x(k-l) \qquad \text{for } k \neq 0 \qquad (7.38)$$

The particular value of $\hat{x}(k)$ for $k = 0$ is given by (7.35). If the input signals $x(k)$ are causal, then (7.38) becomes

$$x(k) = \sum_{l=-\infty}^{k} \left(\frac{l}{k} \right) \hat{x}(l) \, x(k-l) \qquad \text{for } k \neq 0$$

$$= \hat{x}(k) \, x(0) + \sum_{l=-\infty}^{K-1} \left(\frac{l}{k} \right) \hat{x}(l) \, x(k-l) \qquad (7.39)$$

After solving this equation for $\hat{x}(k)$, the recurrence relation representing the characteristic system A_* is obtained:

$$\hat{x}(k) = \frac{x(k)}{x(0)} - \sum_{l=-\infty}^{K-1} \left(\frac{l}{k} \right) \hat{x}(l) \frac{x(k-l)}{x(0)} \qquad (7.40)$$

7.4.12 Remarks

A system similar to that shown in Fig. 7.8 was introduced by Bogert *et al.* [36], in which the Fourier transform of the logarithm of the power spectrum of the input signal is proposed for detecting echos. They called this function *cepstrum*. Since then, it has been usual to call the signal $\hat{x}(k)$ the cepstrum of $x(k)$.

Once the characteristic system A_* and its inverse A_*^{-1} have been determined, the different convolutional systems differ from each other by the linear system of their canonical form. As shown before, the linear system can either act on the signal $\hat{X}(f)$, in which case it is denoted by L_f, or on the cepstrum $\hat{x}(k)$. The filtering of the cepstrum $\hat{x}(k)$ can be done by a linear invariant system.

It is also possible to define in a similar way a frequency-invariant linear system to filter the signal $\hat{X}(f)$. The output of such a filter is given by the following periodic convolution product:

$$\hat{Y}(f) = \int_{-1/2}^{1/2} \hat{X}(g) \; G_f(f - g) \; \mathrm{d}g \qquad (7.41)$$

where $G_f(f)$ is the frequency impulse response of the linear system L_f. It can be seen from this equation that, for a linear system L_f, the roles of the time and frequency domains are interchanged. The Fourier transform of (7.41) is simply

$$\hat{y}(k) = g_f(k) \; \hat{x}(k) \qquad (7.42)$$

where $g_f(k)$ is the inverse Fourier transform of $G_f(f)$. With (7.42), we can filter "low times" or "high times" in the same way as low or high frequencies with linear filters.

7.5 PROPERTIES OF THE CEPSTRUM

7.5.1 Exponential systems

For some particular signals, the cepstrum can be computed analytically. Consequently, it is possible to study some of its properties. The set of signals for which we have a simple analytical computation is that formed by a weighted sum of exponentials. They are of the form:

$$x(k) = \sum_n A_n \; a_n^k \; \epsilon(k) + \sum_n B_n \; b_n^{-k} \; \epsilon(-k-1) \qquad (7.43)$$

where $\epsilon(k)$ is the unit-step signal. The parameters A_n and B_n are the weights, and the coefficients a_n and b_n are numbers, possibly complex, of modulus less than one.

The Fourier transform of such a signal is given by

$$X(f) = \sum_n \frac{A_n}{1 - a_n \, e^{-j 2 \pi f}} + \sum_n \frac{B_n}{1 - b_n \, e^{j 2 \pi f}} \tag{7.44}$$

It can also be written in the following general form:

$$X(f) = \frac{A \prod_{n=1}^{z_i} (1 - \alpha_n \, e^{-j 2 \pi f}) \prod_{n=1}^{z_e} (1 - \beta_n \, e^{j 2 \pi f})}{\prod_{n=1}^{p_i} (1 - \gamma_n \, e^{-j 2 \pi f}) \prod_{n=1}^{p_e} (1 - \delta_n \, e^{j 2 \pi f})} \tag{7.45}$$

where A is a positive number and $|\alpha_n|$, $|\beta_n|$, $|\gamma_n|$, and $|\delta_n|$ are less than one.

7.5.2 Fourier transform of the cepstrum

The Fourier transform of the cepstrum is the complex logarithm of the transform $X(f)$. We have

$$\hat{X}(f) = \ln[X(f)]$$

$$= \ln A + \sum_{n=1}^{z_i} \ln(1 - \alpha_n \, e^{-j 2 \pi f}) + \sum_{n=1}^{z_e} \ln(1 - \beta_n \, e^{j 2 \pi f})$$

$$- \sum_{n=1}^{p_i} \ln(1 - \gamma_n \, e^{-j 2 \pi f}) - \sum_{n=1}^{p_e} \ln(1 - \delta_n \, e^{j 2 \pi f}) \tag{7.46}$$

The last four terms of this equation can be developed in power series. Given that we know

$$\ln(1 + x) = \sum_{k=1}^{\infty} (-1)^{k+1} \frac{x^k}{k} \qquad \text{with } |x| < 1 \tag{7.47}$$

thus, we obtain for the first two terms:

$$\ln(1 - \alpha_n \, e^{-j 2 \pi f}) = - \sum_{k=1}^{\infty} \frac{\alpha_n^k}{k} \, e^{-j 2 \pi k f} \qquad \text{with } |\alpha_n| < 1 \tag{7.48}$$

and

$$\ln(1 - \beta_n \, e^{j 2 \pi f}) = - \sum_{k=1}^{\infty} \frac{\beta_n^k}{k} \, e^{j 2 \pi k f} \qquad \text{with } |\beta_n| < 1 \tag{7.49}$$

7.5.3 Cepstrum

The right sides of (7.48) and (7.49) are respectively the Fourier transforms of the signals (sect. 1.3.2):

$$\hat{x}_{1n}(k) = -\frac{\alpha_n^k}{k} \, \epsilon(k-1) \tag{7.50}$$

and

$$\hat{x}_{2n}(k) = \frac{\beta_n^k}{k} \, \epsilon(-k+1) \tag{7.51}$$

Similar expressions can be derived for the two last terms $\ln[1 - \gamma_n \exp(-j2\pi f)]$ and $\ln[1 - \delta_n \exp(j2\pi f)]$. The cepstrum is thus given by

$$\hat{x}(k) = \begin{cases} \ln(A) & \text{for } k=0 \\[2mm] \displaystyle -\sum_{n=1}^{z_i} \frac{\alpha_n^k}{k} + \sum_{n=1}^{p_i} \frac{\gamma_n^k}{k} & \text{for } k>0 \\[4mm] \displaystyle \sum_{n=1}^{z_e} \frac{\beta_n^k}{k} - \sum_{n=1}^{p_e} \frac{\delta_n^k}{k} & \text{for } k<0 \end{cases} \tag{7.52}$$

The value $\hat{x}(0)$ in this expression is obtained by computing the inverse Fourier transform of $\hat{X}(f)$. By using the linearity of the Fourier transform for the term corresponding to $\ln(A)$, we obtain

$$\begin{aligned} \hat{x}_3(k) &= \int_{-1/2}^{1/2} \ln(A) \, e^{j2\pi kf} \, df \\ &= \begin{cases} \ln(A) & \text{for } k = 0 \\ 0 & \text{for } k \neq 0 \end{cases} \end{aligned} \tag{7.53}$$

Equation (7.52) allows us to derive three main properties of the cepstrum, which are valid for signals that are sums of exponential signals.

7.5.4 Property 1

If the input signals $x(k)$ are causal, then

$$x(k) = 0 \qquad \text{for } k < 0 \tag{7.54}$$

and

$$\hat{x}(k) = 0 \quad \text{for } k < 0 \tag{7.55}$$

For such input signals, the coefficients α_n and β_n are zero. Equation (7.52) clearly indicates that $\hat{x}(k) = 0$ for $k < 0$. In this case, (7.40) relating the signal and its cepstrum becomes

$$\hat{x}(k) = \frac{x(k)}{x(0)} - \sum_{l=0}^{K-1} \left(\frac{l}{k} \right) \hat{x}(l) \frac{x(k-l)}{x(0)} \tag{7.56}$$

From this equation, we may note that the cepstrum $\hat{x}(k)$ for $k < k_0$ depends only on the values of $x(k)$ for $k < k_0$. Consequently, the system A_* is causal in this case.

7.5.5 Property 2

If the signals $x(k)$ are noncausal, then

$$x(k) = 0 \quad \text{for } k < 0 \tag{7.57}$$

which means

$$\hat{x}(k) = 0 \quad \text{for } k < 0 \tag{7.58}$$

The proof is similar to the previous case. In this case, the coefficients γ_n and α_n are zero.

7.5.6 Property 3

If the coefficients γ_n and δ_n are set to zero in (7.45) and the products are developed, an expression of the following type is obtained:

$$X(f) = \sum_{k=1}^{N} \lambda_k \, e^{-j2\pi kf} \tag{7.59}$$

where $N = z_i + z_e$.

Equation (7.59) is the Fourier transform of a finite-length signal. Equation (7.52) shows that the cepstrum of a finite-length signal in this case is not of finite length.

7.6 APPLICATIONS OF MULTIPLICATIVE HOMOMORPHIC SYSTEMS

7.6.1 Variation of the dynamic range of audio signals

It is necessary to vary the dynamic range of audio signals when they are transferred from one carrier to another with incompatible dynamic ranges. This is frequently used to record audio signals of professional quality on magnetic tape. The dynamic range is generally reduced before recording and extended after reproduction.

Audio signals can be considered as amplitude modulated signals. For example, a speaker who is speaking more or less loudly modulates his voice; a musician playing piano or organ modulates the sound of his instrument.

An audio signal $x(k)$ can thus be considered as the product of a positive low-frequency envelope $e(k)$ with a bipolar high-frequency carrier signal $p(k)$. It is obvious that the carrier signal is not a sine wave in this case, but a composite signal occupying a wide band in the high-frequency range. Thus, we have the following equation:

$$x(k) = e(k)\, p(k) \tag{7.60}$$

At the output of the characteristic system, the signal $\hat{x}(k)$ is given by

$$\hat{x}(k) = \ln e(k) + \ln |p(k)| + j \arg [p(k)] \tag{7.61}$$

For audio signals, it is necessary to preserve the information about the sign of $p(k)$. The linear system chosen will thus act only on the real part of $x(k)$. Such a system is obtained by setting $L_{ir} = L_{ri} = 0$ and $L_{ii} = 1$ in the global system shown in Fig. 7.5.

7.6.2 Separation of the components

In general, the signals $e(k)$ and $p(k)$ are in overlapping frequency bands, but can be separated in a first approximation. This is confirmed experimentally by analyzing the spectrum of the human voice [37]. This spectrum decreases until a critical frequency f_0 is reached, and then stays approximatively constant for high frequencies. Neglecting the overlap, we can assign the almost constant high-frequency part of this spectrum to the signal $p(k)$ and the low-frequency part to the signal $e(k)$. The simplest filter, which acts differently on these two parts of the spectrum, is shown in Fig. 7.9.

The signal $y(k)$ at the ouput of the inverse characteristic system is thus of the form:

$$y(k) = e^a(k)\, p^b(k) \tag{7.62}$$

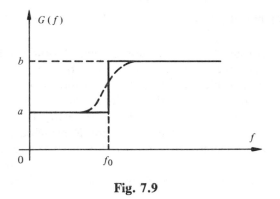

Fig. 7.9

To modify the dynamic range of audio signals, the value $b = 1$ is used in practice. If a is less than one, the multiplicative homomorphic system performs a compression of the dynamic range. If a is greater than one, it performs an extension of this range. For negative values of a, a very interesting phenomenon is obtained: the inversion of the audibility of the signals. Hardly perceptible signals become loud and very loud signals become soft. Other filters with a slower transition at f_0 can also be used.

7.6.3 Digital images

The second principal application of multiplicative homomorphic systems is in image enhancement. In general, a mathematical model is associated with every physical concept. This is also the case for images. A grey-level image can be characterized by a function $L(u, v)$, representing the brightness (sect. 8.4.4) at the pixel (u, v) of the image. The digital version of such an image is generally obtained by periodic two-dimensional sampling of the function $L(u, v)$ (sect. 8.5.3). We have

$$x(k, l) = L(k\Delta u, l\Delta v) \tag{7.63}$$

where Δu and Δv are the sampling periods along the u and v axes, respectively. They must be chosen in accordance with the two-dimensional sampling theorem, that is, to satisfy the following conditions (sect. 8.5.3):

$$\Delta u \leqslant \frac{1}{2F}, \quad \Delta v \leqslant \frac{1}{2G} \tag{7.64}$$

where F and G are the maximum frequencies for which the function $L(u, v)$ still has significant components.

7.6.4 Stockham's model

An image is formed by the reflection of light on objects. The structure of images can be divided into two parts. The first is the amount of light used to illuminate the objects, while the second is comprised of the reflection abilities of these objects when they reflect the original light beams. These elementary parts, which are two-dimensional signals, are respectively called *illumination* $x_i(k, l)$ and *reflection* $x_r(k, l)$. Their ranges are determined by physical rules. The illumination is a finite and strictly positive function. The reflection is positive and less than or equal to one. However, it is practically impossible to find a material which reflects less than 1% of the original light. Thus,

$$0 < x_e(k, l) < \infty$$
$$0.01 \leqslant x_r(k, l) < 1 \qquad (7.65)$$

These two components are related by the law of reflectivity to form the luminance $x(k, l)$ of the image. Because this physical law is multiplicative, the luminance is given by

$$x(k, l) = x_e(k, l) x_r(k, l) \qquad (7.66)$$

with

$$0 < x(k, l) < \infty \qquad (7.67)$$

This simple and elegant model was proposed by Stockham [38] in 1972, and has proved to be a powerful tool.

7.6.5 Homomorphic image processing

If such a signal $x(k, l)$ is introduced as the input of a multiplicative homomorphic system, the output of the characteristic system is given by

$$\hat{x}(k, l) = \ln [x(k, l)] = \ln [x_e(k, l)] + \ln [x_r(k, l)] \qquad (7.68)$$
$$= \hat{x}_e(k, l) + \hat{x}_r(k, l)$$

By virtue of (7.67), the real logarithm can be used here, thus avoiding the problem of the ambiguity of the complex logarithm. By using the definition of the linear system L, we have

$$\hat{y}(k, l) = \hat{y}_e(k, l) + \hat{y}_r(k, l) \qquad (7.69)$$

where $\hat{y}_i(k, l)$ and $y_r(k, l)$ are the illumination and the reflectance after filtering, respectively. The luminance $y(k, l)$ processed by the homomorphic system is thus given by

$$
\begin{aligned}
y(k,l) &= \exp[\hat{y}(k,l)] \\
&= y_e(k,l)\, y_r(k,l)
\end{aligned}
\tag{7.70}
$$

7.6.6 Remarks

Equation (7.70) shows that the output luminance and its components are all positive if the system L is real. This satisfies the requirements for real images (7.65). These requirements remain unfulfilled if the input luminance is processed directly by a conventional linear system. There is, in fact, a basic incompatibility between the multiplicative nature of images (7.66) and the additive structure of conventional linear systems (classical superposition principle). However, their compatibility is ensured with multiplicative homomorphic systems.

The choice of the linear system L, which acts on the output of the characteristic system A, depends on the properties of the signals $\hat{x}_e(k, l)$ and $\hat{x}_r(k, l)$. The reflection, which depends on the shape, the edges, and the texture of the objects, is primarily a high-frequency signal. The illumination, however, is a slowly varying signal, mostly present at low frequencies. Although the separation of these two components in the frequency domain is not obvious, a partial filtering can be performed nonetheless.

7.6.7 Modification of the contrast

The luminance varies within a relatively large range across the image. If these images must be stored on magnetic tape or transmitted over long distances on a transmission channel, it is generally necessary to reduce this dynamic range. However, this reduction can alter the quality of the image. An image having a dynamic range which has been reduced can be enhanced into one that is more pleasing to the observer by increasing its contrast. This operation, called *enhancement*, is an important problem in image processing (sect. 8.7). The contrast of an image is related to the abrupt changes of luminance along the edges of the objects. Thus, it is related to the reflection. To increase the contrast, the reflection must therefore be increased.

7.6.8 Choice of the linear system

Thus, the linear system L must attenuate low frequencies and amplify high frequencies. However, because of the two-dimensional nature of images, the linear system must also be two-dimensional. The desired effect can be obtained with a filter having a harmonic response $G(f, g)$ of circular symmetry. It can thus be expressed as a function of the radial frequency $f_r = (f^2 + g^2)^{1/2}$. The function $G(f, g)$ has the same form as that of Fig. 7.9. The common values of the parameters a and b are in this case 0.5 and 2, respectively.

It is clear that the same system can produce the opposite effects (increase the dynamic range and decrease the contrast) by interchanging the values of a and b. Other types of filter with less sharp transitions at f_0 are also possible. In this case, a and b respectively represent the value at the origin and the asymptotic value for $f = f_{max}$, where f_{max} is the maximum radial frequency for which the image still has significant components.

Two original images and their reproductions after being processed with $a = 0.5$ and $b = 2$ are shown in Fig. 7.10. To conclude, we note that the problem considered in this section is a two-dimensional generalization of the problem presented in section 7.6.2

7.7 APPLICATION OF CONVOLUTIVE HOMOMORPHIC SYSTEMS

7.7.1 Echo suppression

In various problems, for example, in those of measurement or communication, the useful signals propagate in a reverberating medium. In such cases, several versions of the desired signal, delayed with respect to one another, are added to form the received, or measured, signal. For example, this occurs in radar and sonar wave detection, in seismic signals, and in audio recording. In some cases, the structure of the echoes are observed in order to study the propagation medium. In other cases, where the vibration is considered as a disturbance, the echoes must be suppressed.

7.7.2 Model for the echoes

The signal $x(k)$ studied can be represented by

$$x(k) = x_1(k) + \sum_{i=1}^{M} \alpha_i \, x_1(k - k_i) \tag{7.71}$$

where $x_1(k)$ is the desired signal, α_i are the attenuation coefficients, and k_i are the different delays. In a real case, the k_i are all positive. The signal $x(k)$ can be considered as the convolution product of $x_1(k)$ with a signal of the form:

Original images

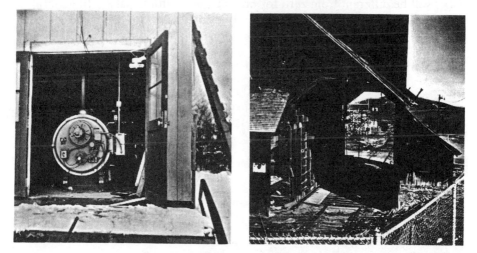

Processed images

7.10

$$x_2(k) = d(k) + \sum_{i=1}^{M} \alpha_i \, d(k - k_i) \tag{7.72}$$

where $d(k)$ is the unit sample. Thus,

$$x(k) = x_1(k) * x_2(k) \tag{7.73}$$

7.7.3 Homomorphic echo processing

To simplify the following computations, let us consider the case of a single echo, that is, the case $M = 1$. Thus,

$$x_2(k) = d(k) + \alpha_1 d(k - k_1) \tag{7.74}$$

The Fourier transform of this equation is given by

$$X_2(f) = 1 + \alpha_1 \exp(-j2\pi k_1 f) \tag{7.75}$$

The Fourier transform of the cepstrum of $x_2(k)$ is, thus,

$$\hat{X}_2(f) = \ln[1 + \alpha_1 \exp(-j2\pi k_1 f)] \tag{7.76}$$

This function is periodic with a period of $1/k_1$. Consequently, the cepstrum $\hat{x}_2(k)$ will be different from zero for only integer multiples of k_1. In general, the attenuation factor α_1 is less than one. By using (7.47), we can write

$$\hat{X}_2(f) = \sum_{k=1}^{\infty} (-1)^{k+1} \frac{\alpha_1^k}{k} \exp(-j2\pi k k_1 f) \quad \text{with } |\alpha_1| < 1 \tag{7.77}$$

By using the frequency scale change $f' = fk_1$ and the corresponding change in the time scale, the following result is obtained:

$$\hat{x}_2(k) = \sum_{n=1}^{\infty} (-1)^{n+1} \frac{\alpha_1^n}{n} d(k - nk_1) \tag{7.78}$$

7.7.4 Comments

Thus, if $\hat{X}_1(f)$ varies slowly compared to $\hat{X}_2(f)$, these two components can be separated with a frequency-invariant filter. Because the contribution of the echo to the cepstrum is a nonzero signal for only integer multiples of k_1, a comb filter may be used. This filter can be considered as a weighted function for the cepstrum with a unit weight everywhere, except in the regions that are multiples of k_1, in which case the weight is zero. This is illustrated in Fig. 7.11. In the cases for which the cepstrum $\hat{x}_1(k)$ is not totally attenuated, the peaks can be eliminated by interpolation.

Such a filter can be used only if the echo time k_1 is known. If this information is available beforehand, k_1 can be estimated from the cepstrum in the periodic repetition of period k_1 of the peaks issued from the echo.

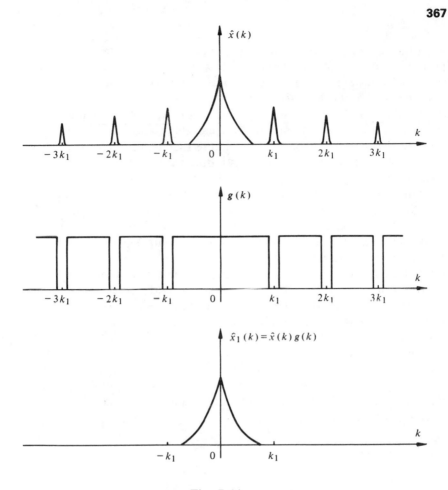

Fig. 7.11

7.7.5 Estimation of speech parameters

The electrical signal representing the audio wave emitted by the human voice channel has been thoroughly investigated. Convolutive homomorphic systems can be used to estimate various parameters of this relatively complex signal, or to separate its different components. Before presenting the use of homomorphic systems in the study of voice, it is useful to examine how this signal is formed.

7.7.6 Model for speech

Speech is formed by the excitation of the vocal track, which can be considered as a resonant tube of about 17 cm in length. If the vocal cords are bent, as in the case of vowels, the air going out of the mouth has the form of quasi-periodic impulses. Signals formed in this manner are called *voiced* signals. If the vocal cords are spread, either a quasirandom turbulence is produced in the voice channel by the decrease of its width (like the letter *s* in the word structure), or the channel is temporarily completely closed to increase the pressure and instantaneously opened to produce a decreasing transient sound (like the letter *p* in the word *"past"*). Signals formed in this manner are called *unvoiced* signals. The vocal track can be considered as a time-varying system, which imposes its transfer properties according to the form of the excitation at its input. If the excitation and the vocal track are assumed to be relatively independent, then speech production can be summarized by the model shown in Fig. 7.12 and proposed by Schafer. In this model, the source of the excitation is either an impulse generator producing periodic impulses with a period having an inverse called the *pitch frequency* of the sound, or a random number generator, which simulates quasirandom as well as decreasing transient turbulences. One of these sources is applied to the input of a time-varying digital filter, which simulates the vocal track. The coefficients of this filter change approximately every 10 ms to indicate a new configuration of the vocal track. A gain control between the source and the system adds further flexibility to the speech level. For the sounds produced with spread vocal cords, the digital filter includes a prefilter to take into account the shape of the time-limited impulses, which are not unit samples.

Let $g(k)$ and $h(k)$ be the impulse responses of the prefilter and the digital filter, respectively. If $p(k)$ is the excitation signal, the speech can be represented over a short interval

$$s(k) = p(k) * g(k) * h(k) \tag{7.79}$$

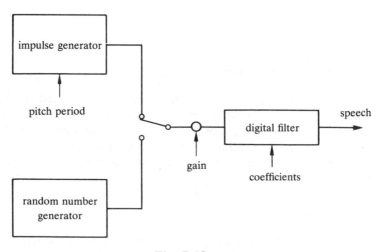

Fig. 7.12

To attenuate the effect of discontinuities at the interval boundaries, the signal $s(k)$ is multiplied by a window function $w(k)$. For any speech segment, we thus have

$$x(k) = s(k)\, w(k)$$
$$= [p(k) * g(k) * h(k)]\, w(k) \tag{7.80}$$

7.7.7 Homomorphic speech processing

If the function $w(k)$ is approximately constant with respect to the variations of the term $g(k) * h(k)$, then

$$x(k) \cong p_w(k) * g(k) * h(k) \tag{7.81}$$

with

$$p_w(k) = p(k)\, w(k) \tag{7.82}$$

If the period of the impulses $p(k)$ is k_0, it is possible to write

$$p_w(k) = \sum_{m=0}^{M-1} w(mk_0)\, d(k - mk_0) \tag{7.83}$$

where M is the number of impulses "seen" through the window $w(k)$. The Fourier transform of $p_w(k)$ is given by

$$P_w(f) = \sum_{m=1}^{M-1} w(mk_0)\, \exp(-j\, 2\pi f m k_0) \tag{7.84}$$

We can easily see that it is periodic with a period of $1/k_0$. Because the logarithm of a periodic function is also periodic, the Fourier transform of the cepstrum $\hat{p}_w(k)$ is a periodic function with a period $1/k_0$. Consequently, $\hat{p}_w(k)$ will be nonzero for only integer multiples of k_0. In practice, if the speech is sampled with a period of 10 kHz, k_0 will be within the interval [40, 200].

Moreover, the impulse response of the system formed by the prefilter and the digital filter has a length which is scarsely larger than 30 ms. The corresponding harmonic response is a slowly-varying function. The application of the complex logarithm and inverse Fourier transform leads to the cepstra $\hat{g}(k)$ and $h(k)$, which are not negligible for only a small number of samples in the vicinity of $k = 0$. Consequently, for relatively large values of k_0, the aliasing of $\hat{p}_w(k)$ with $\hat{g}(k)$ and $h(k)$ is practically negligible. Thus, to separate the components of $x(k)$, a low-pass filter can be used to obtain $g(k) * h(k)$ and a high-pass filter to obtain $p_w(k)$. Figure 7.13 shows the different stages of the homomorphic processing for a segment of speech.

7.7.8 Restoration of audio recordings

One of the most spectacular applications of convolutional homomorphic systems is without doubt the restoration of audio recordings. The two-dimensional generalization of this principle for image restoration has also given very satisfying results and will be examined in section 7.7.19. These two problems were originally studied by Stockham [40].

7.7.9 Model of audio recording

In general, the deconvolution of two signals requires *a priori* knowledge of one of them. However, with a slight restriction, the deconvolution of two signals can be handled without any prior knowledge. The only restriction is a consequence of the assumption that one of the two signals is much shorter than the other. This situation appears frequently in practice when a long signal is distorted by a time-invariant linear system. The audio recording of a singer satisfies this hypothesis. The audio wave $s(k)$ produced by the artist is recorded with a system having an impulse response $g(k)$ which is much shorter than $s(k)$. The recording $x(k)$ is given by

$$x(k) = s(k) * g(k) \tag{7.85}$$

This simple model is particularly applicable to old recordings, where the main distortion was produced by the acoustic recording device, which can be considered as a time-invariant linear system. On the other hand, the nonstationary behavior of the signal $s(k)$ and the lack of information about the system $g(k)$ do not permit the use of classical filtering techniques.

7.7.10 Segmentation of audio recording

The complex logarithm applied to the Fourier transform for both sides of (7.85) leads to

$$\ln[X(f)] = \ln[S(f)] + \ln[G(f)] \tag{7.86}$$

One possible method for estimating $G(f)$ is to use this equation for several recordings performed with the same recording device, and then to compute the mean value of both sides of (7.86) over the complete set of recordings. With a

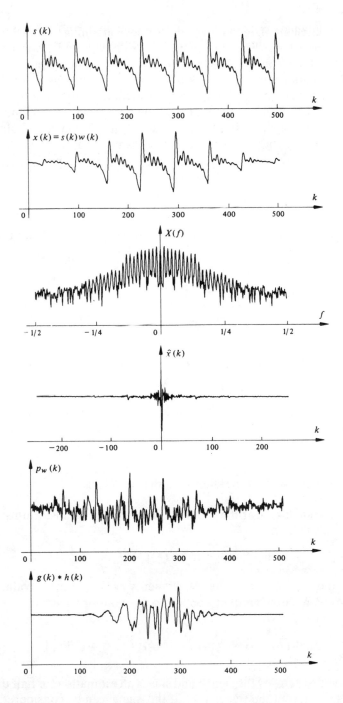

Fig. 7.13

sufficient number of recordings, each one containing a signal $s(k)$ relatively different from the others, the right-hand side will tend toward $\ln[G(f)]$. This procedure, known as *statistical filtering,* is unfortunately not directly applicable because it is difficult to find a number of old recordings made with the same device. However, this problem can be overcome by dividing the recording $x(k)$ into several adjacent segments $x_i(k)$ of medium length (for example, about one second). Each of these intervals contains a musical piece relatively different from the others, but recorded with the same device. If the length of these segments is large compared to the length of the response $g(k)$, the following approximate equation can be written:

$$x_i(k) \cong s_i(k) * g(k) \tag{7.87}$$

These signals $x_i(k)$ can also be considered as the product of the signal $x(k)$ with window functions $w_i(k)$:

$$x_i(k) = w_i(k)\, x(k)$$
$$= w_i(k)\, [s(k) * g(k)] \tag{7.88}$$

If the variations of $w_i(k)$ are slow as compared with those of $g(k)$, we can write

$$x_i(k) \cong [w_i(k)\, s(k)] * g(k)$$
$$\cong s_i(k) * g(k) \tag{7.89}$$

Depending on the shape of the window function chosen (rectangular or another), a superposition of the intervals must be considered to avoid boundary effects.

7.7.11 Homomorphic processing of audio recordings

The complex logarithm applied to the Fourier transform for both sides of (7.87) leads to

$$\ln[X_i(f)] \cong \ln[S_i(f)] + \ln[G(f)] \tag{7.90}$$

If the recording contains M segments $x_i(k)$, the mean value over M segments of (7.90) is given by

$$\frac{1}{M} \sum_{i=1}^{M} \ln[X_i(f)] \cong \frac{1}{M} \sum_{i=1}^{M} \ln[S_i(f)] + \ln[G(f)] \tag{7.91}$$

The first term of the right-hand side is an estimation for half the logarithmic power spectrum of the signal $x(k)$, if this signal can be considered as stationary (sect. 6.5.6). Although this is not really the case, we shall consider it as the second approximation in this study. This term can be estimated by using the same relation (7.91) applied to a new recording of the same musical work, but performed with modern equipment. The harmonic response $G(f)$ of modern recording devices is nearly constant and has no gain. Thus, the term $\ln[G(f)]$ in (7.91) applied to a modern record is zero. This permits us to obtain an estimation of the term:

$$\frac{1}{M} \sum_{i=1}^{M} \ln[S_i(f)] \cong \ln[S(f)] \tag{7.92}$$

By substituting (7.92) into (7.91) and solving for the term $\ln[G(f)]$, the following result is obtained:

$$\ln[G(f)] \cong \frac{1}{M} \sum_{i=1}^{M} \ln[X_i(f)] - \ln[S(f)] \tag{7.93}$$

Conversely, because the human ear is relatively insensitive to phase distortions, we can consider only the modulus in the complex logarithm while neglecting the phase. This leads to

$$\ln|G(f)| = \frac{1}{M} \sum_{i=1}^{M} \ln|X_i(f)| - \ln|S(f)| \tag{7.94}$$

Thus, the frequency response $G_r(f)$ of the restoration filter is given by

$$|G_r(f)| = \frac{1}{|G(f)|} \tag{7.95}$$

The phase-frequency response can be obtained by the Hilbert transform (sect. 5.5.3), or it may be neglected because of the insensitivity of the human ear to the phase.

7.7.12 Comments

To avoid the effect of surface noise in the frequency bands where the term $\ln|S(f)|$ is negligible, the response $G_r(f)$ is additionally reduced to be negligible in the same frequency bands. The inverse Fourier transform enables us to obtain the impulse response $g_r(k)$ of the digital restoration filter. This filter is convolved by using the fast convolution algorithm with the recording $x(k)$ to produce an estimation of the signal $s(k)$. Finally, the digital signal $s(k)$ is converted into an analog one to be heard. This principle was used by Stockham for the restoration of the recordings of Enrico Caruso made in 1907. The term $\ln|S(f)|$ was estimated by using a recent recording of the same musical work by Jussi Bjoerling. Despite the enormous number of approximations and simplifying hypotheses, the quality of the recording is considerably improved.

7.7.13 Image restoration

As in the case of one-dimensional signals, there are several methods for image restoration. Each has its own advantages and drawbacks, but they all have a common characteristic in that they require a more or less exact knowledge of the distortion contained in the image to be restored. The distortions of images frequently observed are caused either by camera motion, or by poor focusing or atmospheric disturbances. These distortions have the advantage of being capable of representation by the model of a stationary linear system.

7.7.14 Movement of the camera

Let us consider a photo with an exposure time of T units, during which uniform movement occurs, for example, in the horizontal direction, of the form $k = \alpha n$ where n represents the discrete time. Assuming that the shutter is instantaneously opened or closed, a point of the image originally at a given place will be displaced in the horizontal direction by a distance αT. The impulse response is consequently given by

$$g_c(k,l) = \sum_{m=-\infty}^{k} d\left(m - \frac{\alpha T}{2}, l\right) - d\left(m + \frac{\alpha T}{2}, l\right) \tag{7.96}$$

where $d(k, l)$ is the two-dimensional unit sample (sect. 8.2.3). A movement in another direction can be easily expressed by a rotation of the coordinate axes.

7.7.15 Poor focusing

Poor focusing can be approximately represented by a constant function, which is nonzero over a circular domain only. This approximation is derived from optical laws. Thus,

$$g_m(k,l) = \begin{cases} 1 & \text{for } k^2 + l^2 \leqslant a^2 \\ 0 & \text{for } k^2 + l^2 > a^2 \end{cases} \tag{7.97}$$

where the parameter a is a measurement of the amount by which the photo is out of focus.

7.7.16 Atmospheric turbulances

The last type of distortion which frequently occurs is caused by atmospheric disturbances (change of the refractive index, *etcetera*.). This distortion is represented, according to physical laws, by a two-dimensional Gaussian function:

$$g_a(k,l) = \exp\left[-b(k^2 + l^2)\right] \tag{7.98}$$

7.7.17 Model of distorted images

The distorted image is thus considered to be the output signal of a two-dimensional system characterized by its impulse response $g(k, l)$, excited by an input signal, which is the undistorted original image. Starting from the output signal, the response $g(k, l)$ must be estimated, and then the input signal must be recovered by inverse filtering.

Let $y(k, l)$ be the distorted image and $x(k, l)$ will be the original image to estimate. With the hypothesis that the distortion system is linear, we can write

$$y(k,l) = g(k,l) ** x(k,l) \tag{7.99}$$

For simplicity, the images will be assumed to be square with K samples along each axis, k and l. We shall also assume (this is the principal hypothesis of this application) that the response $g(k, l)$ is nonzero in only a square domain of $M \times M$ samples with

$$M \ll K \tag{7.100}$$

This condition is necessary to estimate $y(k, l)$ without any prior information about $g(k, l)$. The arbitrary restriction of $g(k, l)$ to a square domain is once again for simplicity. Condition (7.100) is practically a standard condition for image restoration.

7.7.18 Segmentation of the distorted image

To perform the statistical filtering, the square domain $K \times K$ of the images is divided into I square sections of dimension $N \times N$ with

$$M \ll N \ll K \tag{7.101}$$

For each section, an approximate equation of the following type is obtained:

$$y_i(k,l) \cong g(k,l) ** x_i(k,l) \quad \text{with } i = 1, ..., I \qquad (7.102)$$

The signals $y_i(k, l)$ can be considered as the product of the image $y(k, l)$ with two-dimensional window functions $w_i(k, l)$ of identical shape. Thus,

$$\begin{aligned} y_i(k,l) &= w_i(k,l)\, y(k,l) \\ &= w_i(k,l)\, [g(k,l) ** x(k,l)] \end{aligned} \qquad (7.103)$$

If the function $w_i(k, l)$ is practically constant over the defined domain of $g(k, l)$, then

$$\begin{aligned} y_i(k,l) &\cong [w_i(k,l)\, x(k,l)] ** g(k,l) \\ &\cong x_i(k,l) ** g(k,l) \end{aligned} \qquad (7.104)$$

If the function $w_i(k, l)$ is not rectangular, a superposition of the sections must be considered to avoid boundary effects. For example, for a Hanning window, a superposition over half of each section is necessary. In this case, (7.102) can be considered as a periodic convolution.

7.7.19 Homomorphic processing of the image sections

The complex logarithm of the two-dimensional Fourier transform for both sides of (7.102) leads to

$$\ln [Y_i(f,g)] \cong \ln [X_i(f,g)] + \ln [G(f,g)] \qquad (7.105)$$

By contrast with audio signals, in the case of images, the phase distortions cannot be neglected because the human eye is very sensitive to phase variations. By using (7.13), relation (7.105) can be written as follows:

$$\ln |Y_i(f,g)| \cong \ln |X_i(f,g)| + \ln |G(f,g)| \qquad (7.106)$$
$$\arg [Y_i(f,g)] \cong \arg [X_i(f,g)] + \arg [G(f,g)] \qquad (7.107)$$

7.7.20 Statistical filtering on the modulus

The mean value over I sections leads to

$$\frac{1}{I} \sum_{i=1}^{I} \ln |Y_i(f,g)| \cong \frac{1}{I} \sum_{i=1}^{I} \ln |X_i(f,g)| + \ln |G(f,g)| \qquad (7.108)$$

The first term on the right-hand side of this equation is an estimation of the logarithmic power spectrum of the original image (sect. 6.5.6). When the autocorrelation function of the images is studied, we may observe that this varies very little from one image to another. The power spectrum is related to the autocorrelation function by the Fourier transform.

Consequently, the term

$$\frac{1}{I} \sum_{i=1}^{I} \ln |X_i(f,g)| = \ln |X(f,g)| \qquad (7.109)$$

will not be very different from one image to another. Thus, a prototype spectral density can be computed on a clear image, which is not distorted and has statistical characteristics similar to those of the image to be restored. By combining (7.108) and (7.109), we obtain

$$\ln |G(f,g)| \simeq \frac{1}{I} \sum_{i=1}^{I} \ln |Y_i(f,g)| - \ln |X(f,g)| \qquad (7.110)$$

or

$$|G(f,g)| \cong \exp \left[\frac{1}{I} \sum_{i=1}^{I} \ln |Y_i(f,g)| - \ln |X(f,g)| \right] \qquad (7.111)$$

This relation permits us to obtain the modulus of the harmonic response of the distortion system as a function of the logarithmic spectrum of the image to be restored and the prototype spectrum $X(f, g)$ measured on a clear image.

7.7.21 Determination of the phase-frequency response

Unfortunately, for reasons explained in section 7.3.3, statistical filtering cannot be used for (7.107). The sum of the principal values of the phase is not equal to the principal value of the sum. Cannon [41] proposed the use of two-dimensional cepstrum to estimate the zeros of the function $G(f, g)$, and then to generate the appropriate phase 0 or π according to the sign of several lobes of $G(f, g)$. If, for example, the distortion is caused by camera movement, $g_c(k, l)$ is a rectangle of length αT. The corresponding harmonic response is of the form:

$$G_c(f,g) = \alpha T \frac{\sin \pi \alpha T f}{\pi \alpha T f} \qquad (7.112)$$

and has zeros at integer multiples of $1/\alpha T$. The logarithm of its modulus will have periodic peaks of period $1/\alpha T$ with an amplitude tending toward $-\infty$. The Fourier transform of such a function will consequently have an important peak

DISTORTED IMAGES   RESTORED IMAGES

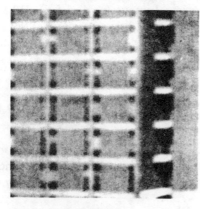 

by out of focus

by camera movement

by atmospheric disturbances

Fig. 7.14

at a distance αT from the origin, in a direction corresponding to the direction of the camera movement. The cepstrum of an image recorded out of focus has the same form, but with a circular symmetry in the peak distribution. The distribution and the position of these peaks allow us to discriminate an out-of-focus image from an image effected by camera movement. The cepstrum of an image distorted by atmospheric disturbances does not have these characteristic peaks, which is a characteristic in itself. Consequently, it is possible to implement an algorithm that looks for the distribution and positions of these peaks in the cepstrum, first to determine the type of distortion and then the phase of the corresponding harmonic response.

7.7.22 Examples

Figure 7.14 shows three images distorted by poor focusing, camera motion during exposure, and atmospheric turbulances, respectively. The results obtained by using the method previously described are shown in the same figure. The improvement obtained is obvious. The edges of the building are now sharper, the text of the Hilton advertisement is now readable. Even if the text of the third image is not readable, the improvement is obvious nonetheless. The perturbations observed in the results are not caused by the method itself, but by its implementation with a relatively small number of samples. It is useful to remember that these results have been obtained without any prior information about the nature of the distortion.

7.8 EXERCISES

7.8.1 The general form of the input signals of the multiplicative homomorphic systems is given by (7.8). Show that if the numbers a_1 and a_2 are complex, we must have $L_{rr} = L_{ii}$ and $L_{ri} = L_{ir}$ in the general form of the linear system (Fig. 7.5).

7.8.2 Prove property 7.5.6.

7.8.3 Let us consider the z-transform given by

$$X(z) = |A| \frac{\displaystyle\prod_{n=1}^{z_i} (1 - \alpha_n z^{-1})}{\displaystyle\prod_{n=1}^{p_i} (1 - \gamma_n z^{-1})} \quad \text{with } |z| > 1 \text{ and } |\alpha_n| < 1, |\gamma_n| < 1$$

Let $x(k)$ be the inverse z-transform of $X(z)$. Show that $x(k)$ and its cepstrum $\hat{x}(k)$ are of minimum phase.

7.8.4 Let us consider the z-transform given by

$$X(z) = |A| \frac{\displaystyle\prod_{n=1}^{z_e} (1 - \beta_n z)}{\displaystyle\prod_{n=1}^{p_e} (1 - \delta_n z)} \qquad \text{with } |z| < 1 \quad \text{and} \quad |\beta_n| < 1, |\delta_n| < 1$$

Let $x(k)$ be the inverse z-transform of $X(z)$. Show that $x(k)$ and its cepstrum $\hat{x}(k)$ are of maximum phase.

7.8.5 Let us consider the following z-transform:

$$X(z) = X_{\min}(z) X_{\max}(z)$$

with

$$X_{\min}(z) = \prod_{n=1}^{z_i} (1 - \alpha_n z^{-1})$$

and

$$X_{\max}(z) = A \prod_{n=1}^{z_e} (1 - \beta_n z)$$

and where $|\beta_n|$ and $|\alpha_n|$ are less than one. By using (7.40), show that only $N = z_i + z_e + 1$ values of the cepstrum $\hat{x}(k)$ are needed to completely determine the signal $x(k)$.

7.8.6 Let us consider a signal $x(k)$ and its cepstrum $\hat{x}(k)$. A new signal $y(k)$ is defined as follows:

$$y(k) = \begin{cases} x(k/K) \text{ with } k = nK, \ n = 0, \pm 1, \pm 2, \dots \\ 0 \qquad\qquad \text{otherwise} \end{cases}$$

Show that the cepstrum of $\hat{x}(k)$ is given by

$$\hat{y}(k) = \begin{cases} \hat{x}(k/K) \text{ with } k = nK, \ n = 0, \pm 1, \pm 2, \dots \\ 0 \qquad\qquad \text{otherwise} \end{cases}$$

7.8.7 Let us consider the following z-transform:

$$X(z) = A \frac{\prod\limits_{n=1}^{z_i} (1 - \alpha_n z^{-1}) \prod\limits_{n=1}^{z_e} (1 - \beta_n z)}{\prod\limits_{n=1}^{p_i} (1 - \gamma_n z^{-1}) \prod\limits_{n=1}^{p_e} (1 - \delta_n z)}$$

with

$$|\alpha_n|, |\beta_n|, |\gamma_n|, |\delta_n| < 1.$$

Show that the inverse z-transform $x(k)$ is the convolution product of a minimum phase and a maximum phase system. What can be said about the cepstrum $\hat{x}(k)$?

7.8.8 Show that if the cepstrum $\hat{x}(k)$ of exercise 7.8.7 is filtered by a linear system, such that

$$\hat{y}(k) = \begin{cases} 0 & \text{for } k < 0 \\ \hat{x}(0)/2 & \text{for } k = 0 \\ \hat{x}(k) & \text{for } k > 0 \end{cases}$$

then the corresponding signal $y(k)$ is of minimum phase.

7.8.9 How must the cepstrum $\hat{x}(k)$ of exercise 7.8.7 be filtered so that the signal corresponding to the filtered cepstrum is of maximum phase?

7.8.10 Let $x(k)$ be a mixed phase system having a z-transform that is the same as in exercise 7.8.7. Which linear system must be used to filter its cepstrum so that the signal corresponding to the filtered cepstrum is of minimum phase and has the same amplitude spectrum as the initial signal?

Chapter 8

Two-Dimensional Signal Processing

8.1 INTRODUCTION

8.1.1 Preliminary remarks

Two-dimensional digital signal processing is a very important field of digital signal processing. Its importance is essentially due to the fact that these signals represent images. Indeed, the term *image processing* is often used. Despite the tremendous amount of data involved in this kind of processing, the flexibility provided by digital methods and the progress in the related technology has continually encouraged the researchers in this field.

The number of methods and application areas has increased so much that nowadays image processing is becoming a discipline in itself. Several specialized books are devoted to the study of the methods and applications of this field. Some aspects of digital image processing have matured enough to be the subject of specialized books as well. However, other aspects are still under development. A considerable amount of research must yet be done.

8.1.2 Organization of the chapter

This chapter is only an introduction to the principal methods and applications of image processing. A large segment of the principal applications is summarized. For more details, the reader can consult the references [42–44].

In the first part, the mathematical formalism of two-dimensional digital signal processing is presented by generalizing the notions previously introduced. Because most of the processing invokes the human eye as final observer, the principal properties of the human visual system are also studied. Sampling and quantization problems are examined in detail and illustrated by examples.

In the second part, the methods of application are briefly presented. Historically, the first image processing was devoted to redundancy reduction. The number of bits used in the digital representation of the image must be reduced while preserving an acceptable quality. Another application field consists of restoring or enhancing a distorted image that is difficult to interpret. Finally, the current trends in image recognition and description are presented to introduce the reader to these research areas.

8.2 TWO-DIMENSIONAL DIGITAL SIGNALS

8.2.1 Definitions

A *two-dimensional digital signal* is a real or complex function of two independent integer variables. The range of variation for the independent variables is called *spread*.

8.2.2 Notation

In general, a two-dimensional digital signal is represented by

$$x(k,l) \tag{8.1}$$

If the variation ranges of the independent variables k and l are finite, then the signal $x(k, l)$ can be represented by a rectangular matrix:

$$x(k,l) \doteq X = \begin{pmatrix} x(k_1,l_1) & x(k_1,l_2) \ldots \ldots x(k_1,l_L) \\ x(k_2,l_1) & \vdots \\ \vdots & \vdots \\ \vdots & \vdots \\ x(k_K,l_1) & \ldots \ldots \ldots \ldots x(k_K,l_L) \end{pmatrix} \tag{8.2}$$

Such a signal is characterized by a finite spread.

8.2.3 Definition

The *two-dimensional unit impulse* is defined by

$$d(k,l) = \begin{cases} 1 & \text{for } k = l = 0 \\ 0 & \text{otherwise} \end{cases} \tag{8.3}$$

The *two-dimensional unit step* is defined by

$$\epsilon(k,l) = \begin{cases} 1 & \text{for } k \geqslant 0, \ l \geqslant 0 \\ 0 & \text{otherwise} \end{cases} \tag{8.4}$$

The *two-dimensional rectangular signal* is defined by

$$\text{rect}_{K,L}(k,l) = \begin{cases} 1 & \text{for } 0 \leqslant k \leqslant K - 1 \text{ and } 0 \leqslant l \leqslant L - 1 \\ 0 & \text{otherwise} \end{cases} \tag{8.5}$$

A two-dimensional signal $x(k, l)$ is called *separable,* if

$$x(k,l) = x_1(k) x_2(l) \tag{8.6}$$

8.2.4 Example

The two-dimensional rectangular signal is a separable signal. In fact, from definitions (1.7) and (8.5), it is possible to write

$$\text{rect}_{K,L}(k,l) = \text{rect}_K(k) \, \text{rect}_L(l) \tag{8.7}$$

8.2.5 Two-dimensional Fourier transform

The *two-dimensional Fourier transform* of a digital signal $x(k, l)$ is defined by

$$X(f,g) = \sum_{k=-\infty}^{+\infty} \sum_{l=-\infty}^{+\infty} x(k,l) \, e^{-j2\pi(fk+gl)} \tag{8.8}$$

This transform exists if and only if the right-hand side of (8.8) is finite, that is, if the series converges. The sufficient condition for the convergence of this series is (sect. 1.3.3):

$$\sum_{k=-\infty}^{+\infty} \sum_{l=-\infty}^{+\infty} |x(k,l)| < \infty \tag{8.9}$$

If this condition is satisfied, then the series (8.8) converges absolutely toward a continuous function of f and g.

8.2.6 Property

As in the one-dimensional case, the function $X(f, g)$ is periodic, of period one, in f and g. We have

$$X(f+1, g+1) = \sum_{k=-\infty}^{+\infty} \sum_{l=-\infty}^{+\infty} x(k,l) \, e^{\,j2\pi[(f+1)k+(g+1)l]}$$

$$= \sum_{k=-\infty}^{+\infty} \sum_{l=-\infty}^{+\infty} x(k,l) \, e^{\,-j2\pi(fk+gl)} \, e^{\,-j2\pi(k+l)}$$

$$= X(f,g) \tag{8.10}$$

8.2.7 Definition

Because the function $X(f, g)$ is periodic in both dimensions, any domain of unit surface in the (f, g) plane is sufficient to describe it entirely. Usually, the following domain is used:

$$-\tfrac{1}{2} \leqslant f \leqslant \tfrac{1}{2}$$
$$-\tfrac{1}{2} \leqslant g \leqslant \tfrac{1}{2} \tag{8.11}$$

which is called the *principal period*.

8.2.8 Inversion formula

Equation (8.8) can be interpreted as the Fourier series expansion of a two-dimensional periodic function. The samples of the signal $x(k, l)$ can be obtained from the transform $X(f, g)$ by using the following equation:

$$x(k,l) = \int_{-1/2}^{1/2} \int_{-1/2}^{1/2} X(f,g) \, e^{\,j2\pi(fk+gl)} \, df \, dg \tag{8.12}$$

This is the inverse Fourier transform for two-dimensional digital signals (sect. 8.9.1).

8.2.9 Property

In general, the transform $X(f, g)$ is a complex function:

$$X(f,g) = \text{Re}\,[X(f,g)] + j\,\text{Im}\,[X(f,g)] \tag{8.13}$$

In the case of a real signal $x(k, l)$, the real and imaginary parts are respectively given by

$$\text{Re}\,[X(f,g)] = \sum_{k=-\infty}^{+\infty} \sum_{l=-\infty}^{+\infty} x(k,l)\,\cos\,[\,2\pi(fk+gl)\,] \qquad (8.14)$$

and

$$\text{Im}\,[X(f,g)] = -\sum_{k=-\infty}^{+\infty} \sum_{l=-\infty}^{+\infty} x(k,l)\,\sin\,[\,2\pi(fk+gl)\,]$$

$$(8.15)$$

8.2.10 Definitions

The complex function $X(f, g)$ can also be expressed in terms of its modulus and argument:

$$X(f,g) = |X(f,g)|\exp\,\{\,j\arg\,[\,X(f,g)\,]\,\} \qquad (8.16)$$

The term $|X(f, g)|$ is called the *two-dimensional amplitude spectrum*. It represents the frequency repartition of the amplitude of the signal $x(k, l)$ in the plane (f, g). However, the notion of frequency used here is not associated with a temporal evolution. In general, the variables k and l represent distances.

For example, the variables may correspond to the two dimensions of an image plane. Therefore, the notion of *spatial frequency* is used in the two-dimensional case.

The term $\theta_x(f, g) = \arg[X(f, g)]$ is the *two-dimensional phase spectrum*. It expresses the frequency repartition of the phase of the signal $x(k, l)$ in the (f, g) plane.

The term $|X(f, g)|^2$ is the *energy spectrum*.

8.2.11 Example

Let us consider the computation of the Fourier transform of the two-dimensional unit sample $d(k, l)$. We have

$$X(f,g) = \sum_{k=-\infty}^{+\infty} \sum_{l=-\infty}^{+\infty} d(k,l)\,e^{-j2\pi(fk+gl)}$$

$$= \sum_{k=0}^{0} \sum_{l=0}^{0} 1 = 1$$

The amplitude and energy spectra are thus constant for any frequencies f and g. The phase spectrum is equal to zero.

8.2.12 Example

Let us consider the computation of the Fourier transform of the following signal $x(k, l)$ (see Fig. 8.1):

$$x(k,l) = \text{rect}_{K,L}\,(k + K/2, l + L/2)$$

$$= \begin{cases} 1 & \text{for} & -K/2 \leqslant k \leqslant K/2 - 1 \\ & & -L/2 \leqslant l \leqslant L/2 - 1 \\ 0 & \text{otherwise} \end{cases} \tag{8.17}$$

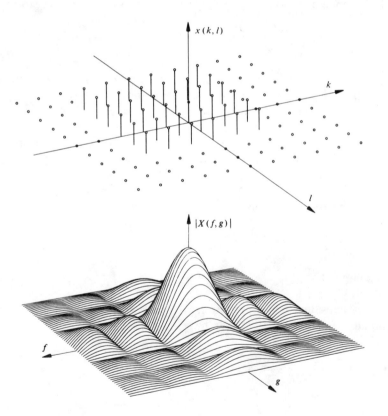

Fig. 8.1

From definition (8.8), we have

$$
\begin{aligned}
X(f,g) &= \sum_{k=-K/2}^{K/2-1} \sum_{l=-L/2}^{L/2-1} e^{-j2\pi fk}\, e^{-j2\pi gl} \\[2mm]
&= \sum_{k=-K/2}^{K/2-1} e^{-j2\pi fk} \sum_{l=-L/2}^{L/2-1} e^{-j2\pi gl} \\[2mm]
&= e^{j\pi fK}\, e^{j\pi gL} \sum_{k'=0}^{K-1} e^{-j2\pi fk'} \sum_{l'=0}^{L-1} e^{-j2\pi gl'} \\[2mm]
&= e^{j\pi(fK+gL)}\, \frac{1-e^{-j2\pi fK}}{1-e^{-j2\pi f}}\, \frac{1-e^{-j2\pi gL}}{1-e^{-j2\pi g}} \\[2mm]
&= e^{j\pi(f+g)}\, \frac{\sin(\pi fK)}{\sin(\pi f)}\, \frac{\sin(\pi gL)}{\sin(\pi g)} \qquad (8.18)
\end{aligned}
$$

The modulus of this function is shown in Fig. 8.1 for $K = L = 6$.

8.2.13 Two-dimensional discrete Fourier transform

Two difficulties are involved in the implementation of (8.8) and (8.12), which define the two-dimensional Fourier transform. First, f and g are continuous variables, which cannot be handled in a digital processing system. Second, it is impossible, in practice, to process an infinite number of samples $x(k, l)$. These difficulties can be overcome by replacing the continuous variables with discrete ones, and by limiting the spread of the signal $x(k, l)$ in the plane (k, l).

By setting

$$
f = m\,\Delta f \quad \text{and} \quad g = n\,\Delta g \qquad (8.19)
$$

in the principal domain (8.11) (sect. 3.2.3), with

$$
\Delta f = 1/M \quad \text{and} \quad \Delta g = 1/N \qquad (8.20)
$$

where M and N are integers, (8.12) takes the following form:

$$
x(k,l) \cong \frac{1}{MN} \sum_{n=-N/2}^{N/2-1} \sum_{m=-M/2}^{M/2-1} X(m,n)\, \exp\left[j\,2\pi\left(\frac{mk}{M} + \frac{nl}{N} \right) \right]
$$

$$
(8.21)
$$

We can see that the right-hand side of this equation is periodic of period M in k and of period N in l. Thus, the approximation in (8.21) becomes an equality if the spread of the signal $x(k, l)$ is limited to a domain of dimension $M \times N$. In this case, $x(k, l)$ can be extracted exactly from any period on the right-hand side of (8.21). This leads to

$$x(k,l) = \frac{1}{MN} \sum_{m=-M/2}^{M/2-1} \sum_{n=-N/2}^{N/2-1} X(m,n) \exp\left[j\, 2\pi \left(\frac{mk}{M} + \frac{nl}{N} \right) \right]$$

(8.22)

with

$$k_0 \leqslant k \leqslant k_0 + M - 1$$
$$l_0 \leqslant l \leqslant l_0 + N - 1$$

(8.23)

By taking (8.19) and (8.20) into account, the Fourier transform of a signal limited in spread to (8.23) can be written as follows:

$$X(m,n) = \sum_{k=k_0}^{k_0+M-1} \sum_{l=l_0}^{l_0+N-1} x(k,l) \exp\left[-j\, 2\pi \left(\frac{mk}{M} + \frac{nl}{N} \right) \right]$$

(8.24)

Equations (8.22) and (8.24) define the *two-dimensional discrete Fourier transform* (2-D DFT) for a signal $x(k, l)$ of limited spread.

8.2.14 Matrix form of the two-dimensional DFT

For many applications, it is useful to write (8.22) and (8.24) in a matrix form. Due to the separability of the kernel (sect. 4.2.12), the 2-D DFT and its inverse can be expressed by using the one-dimensional DFT. This transform is well known (chap. 3) and easily computed with the FFT. In applications where the detail of the formalism (8.22) or (8.24) is not required, the matrix notation leads to very simple and compact equations.

The spread of the signal $x(k, l)$ in the (k, l) plane and the spread of the Fourier transform in the (m, n) plane are limited to sets of $M \times N$ digital values or samples. Thus, the signal and its transform can be represented by matrices.

Let x and X be the matrices corresponding to the signal and its transform. The comparison of (4.31) and (8.24) shows that the kernel of the 2-D DFT is

$$\exp\left[-j\, 2\pi \left(\frac{mk}{M} + \frac{nl}{N} \right) \right]$$

(8.25)

This kernel is separable (see eq. (4.32)) because we can write

$$\exp\left[-j\,2\pi\left(\frac{mk}{M}+\frac{nl}{N}\right)\right] = \exp\left[-j\,2\pi\,\frac{mk}{M}\right]\exp\left[-j\,2\pi\,\frac{nl}{N}\right] \quad (8.26)$$

Each term of the product on the right-hand side is the kernel of a one-dimensional DFT. The corresponding transformation matrices are denoted (see eq. (4.116)) by F_M and F_N, respectively. The matrix form of (8.24) then becomes

$$X = F_N \; x \; F_M^H \quad (8.27)$$

The inverse transformation is given by

$$x = F_N^{-1} \; X \; (F_M^{-1})^H \quad (8.28)$$

8.2.15 Two-dimensional correlation: Definitions

The *two-dimensional crosscorrelation function* is a measurement of the similarity between two signals $x(k, l)$ and $y(k, l)$. Mathematically, it is expressed as

$$\overset{\circ}{\varphi}_{xy}(k,l) = \sum_{k'=-\infty}^{+\infty}\sum_{l'=-\infty}^{+\infty} x(k',l')\,y(k'+k,l'+l) \quad (8.29)$$

If the two signals $x(k, l)$ and $y(k, l)$ are identical, then the signal $\overset{\circ}{\varphi}_{xy}(k, l)$ is called *two-dimensional autocorrelation function*.

8.2.16 Direct computation method

The different steps in the evaluation of (8.29) can be listed as follows:
- The signal $y(k', l')$ is translated into the $k'l'$ plane to a point (k, l);
- The product $x(k'l')y(k' + k, l' + l)$ is performed sample by sample for all values of k' and l' in the intersection of the spreads of both signals;
- The values obtained are summed to form a sample of the signal $\overset{\circ}{\varphi}_{xy}(k, l)$.

These steps are repeated for other values of k and l.

8.2.17 Properties

The crosscorrelation function can also be computed, in an indirect way, by using the two-dimensional Fourier transform. Applying definition (8.8) to (8.29) leads to

$$\overset{\circ}{\Phi}_{xy}(f,g) = \sum_{k=-\infty}^{+\infty} \sum_{l=-\infty}^{+\infty} \overset{\circ}{\varphi}_{xy}(k,l)\, e^{-j2\pi(fk+gl)}$$

$$= \sum_{k=-\infty}^{+\infty} \sum_{l=-\infty}^{+\infty} \sum_{k'=-\infty}^{+\infty} \sum_{l'=-\infty}^{+\infty} x(k',l')$$

$$y(k'+k,l'+l)\, e^{-j2\pi(fk+gl)} \tag{8.30}$$

With the variable changes $u = k' + k$ and $v = l' + l$, we obtain

$$\overset{\circ}{\Phi}_{xy}(f,g) = \left[\sum_{k'=-\infty}^{+\infty} \sum_{l'=-\infty}^{+\infty} x(k',l')\, e^{j2\pi(fk'+gl')} \right]$$

$$\left[\sum_{u=-\infty}^{+\infty} \sum_{v=-\infty}^{+\infty} y(u,v)\, e^{-j2\pi(fu+gv)} \right] \tag{8.31}$$

The first double sum in this equation is the complex conjugate of the Fourier transform of $x(k, l)$. Consequently,

$$] \quad \overset{\circ}{\Phi}_{xy}(f,g) = X^*(f,g)\, Y(f,g) \tag{8.32}$$

The crosscorrelation function is obtained by computing the inverse transform of $\overset{\circ}{\Phi}_{xy}(f, g)$. Equation (8.32) is the two dimensional generalization of the property of section 1.4.3.

8.3 TWO-DIMENSIONAL DIGITAL SYSTEMS

8.3.1 Definitions

A *two-dimensional digital system* characterized by an operator S, acts on an input signal $x(k, l)$ to produce an output signal $y(k, l)$. This operation is formally represented by

$$y(k,l) = S[x(k,l)] \tag{8.33}$$

An important class of digital systems is that of linear systems. The operator characterizing these systems satisfies the well known equation (sect. 1.5.5):

$$S\left[a_1 x_1(k,l) + a_2 x_2(k,l)\right] = a_1 S\left[x_1(k,l)\right] + a_2 S\left[x_2(k,l)\right] \qquad (8.34)$$

8.3.2 Property

These systems are characterized completely by their response to a unit impulse. In fact, by expressing the input signal using unit impulses, we have

$$x(k,l) = \sum_{k'=-\infty}^{+\infty} \sum_{l'=-\infty}^{+\infty} x(k',l')\, d(k-k',l-l') \qquad (8.35)$$

If we substitute (8.35) into (8.33) and take account of (8.34), then this leads to

$$y(k,l) = \sum_{k'=-\infty}^{+\infty} \sum_{l'=-\infty}^{+\infty} x(k',l')\, S[d(k-k',l-l')] \qquad (8.36)$$

If $g(k, l, k', l')$ is the response of the system to the excitation $d(k - k', l - l')$, we have

$$y(k,l) = \sum_{k'=-\infty}^{+\infty} \sum_{l'=-\infty}^{+\infty} x(k',l')\, g(k,l,k',l') \qquad (8.37)$$

8.3.3 Definitions

One sub-class of linear systems plays an important role in image processing. This is the class of *linear, translation-invariant systems*. Let $y(k, l)$ be the response to $x(k, l)$. A linear system is called linear, translation-invariant if the response to the translated input $x(k - k_0, l - l_0)$ is $y(k - k_0, l - l_0)$, where k_0 and l_0 are two integers.

8.3.4 Property

For linear, translation-invariant systems, (8.37) takes the following particular form:

$$y(k,l) = \sum_{k'=-\infty}^{+\infty} \sum_{l'=-\infty}^{+\infty} x(k',l')\, g(k-k',l-l')$$

$$= x(k,l) * * g(k,l) \qquad (8.38)$$

As with (8.30) and (8.31), we can compute the Fourier transform of (8.38), leading to the result:

$$Y(f,g) = X(f,g)\,G(f,g) \tag{8.39}$$

8.3.5 Definitions

Equation (8.38) is called the *two-dimensional convolution product*. The Fourier transform $G(f, g)$ of the impulse response $g(k, l)$ is called the *two-dimensional frequency response* of the corresponding system.

8.3.6 Theorem

The necessary and sufficient condition to guarantee the stability of a two-dimensional system is

$$\sum_{k=-\infty}^{+\infty} \sum_{l=-\infty}^{+\infty} |g(k,l)| < \infty \tag{8.40}$$

8.3.7 Two-dimensional difference equation

A linear-invariant system by translation can also be characterized by a difference equation with constant coefficients of the form:

$$\sum_{k'=0}^{K_1} \sum_{l'=0}^{L_1} a_{k'l'}\, y(k-k', l-l') = \sum_{k''=0}^{K_2} \sum_{l''=0}^{L_2} b_{k''l''}\, x(k-k'', l-l'') \tag{8.41}$$

This equation is the two-dimensional generalization of (1.73).

8.3.8 Two-dimensional digital filtering

As in the one-dimensional case, the action of a two-dimensional digital linear, translation-invariant system can be interpreted as filtering. The signal $g(k, l)$ is the impulse response of the corresponding filter. This response, according to the desired filtering, can be of finite or infinite spread.

In the case of a filter having an impulse response which is limited to the spread ($0 \leq k \leq K - 1, 0 \leq l \leq L - 1$), the filtering can be performed using the following equation:

$$y(k,l) = \sum_{k'=0}^{K-1} \sum_{l'=0}^{L-1} g(k',l') \, x(k-k',l-l') \qquad (8.42)$$

This relation is a particular form of the general equation (8.38).

In the case of a filter having an impulse response of infinite spread, (8.38) cannot be used in practice. The filtering can then be performed using the following recursive equation, deduced from (8.41) by solving it for $y(k, l)$.

$$y(k,l) = \frac{1}{a_{00}} \left\{ \sum_{k''=0}^{K_2} \sum_{l''=0}^{L_2} b_{k''l''} \, x(k-k'',l-l'') - \right.$$
$$\left. \sum_{k'=1}^{K_1} \sum_{l'=1}^{L_1} a_{k'l'} \, y(k-k',l-l') \right\} \qquad (8.43)$$

In both cases, the digital version of (8.39) can be used to perform the filtering indirectly with the 2-D DFT. The choice among these three methods depends on the particular problem and the constraints upon its implementation (speed, memory size, number of operations to process, precision, *etcetera.*).

8.3.9 Remarks

The general problem of two-dimensional digital filtering consists of elaborating an impulse response $g(k, l)$, which, on one hand, has the desired frequency response and, on the other, is efficiently implemented. Due to this second constraint, realizable filters reach the desired frequency responses only approximately (Fig. 5.7).

Often, a two-dimensional digital filter is specified by its frequency response. If the indirect method employing the 2-D DFT can be used, the filtering can be performed by using the following equation:

$$Y(m,n) = X(m,n)G(m,n) \qquad (8.44)$$

The desired output signal is obtained by computing the inverse 2-D DFT of $Y(m, n)$. In the case where the desired frequency response leads to an impulse response of finite spread, this frequency response can be obtained by using the two-dimensional inverse Fourier transform. The filtering is thus performed by the convolution product (8.42).

If the spread of the impulse response $g(k, l)$ is infinite, then the set of coefficients a_{kl} and b_{kl} of (8.43) must be determined from the desired response $G(f, g)$. This can be done with the two-dimensional z-transform.

8.3.10 Two-dimensional z-transform: Definitions

The *two-dimensional z-transform* of a signal $x(k, l)$ is defined by

$$X(z_1, z_2) = \sum_{k=-\infty}^{+\infty} \sum_{l=-\infty}^{+\infty} x(k, l)\, z_1^{-k}\, z_2^{-l} \qquad (8.45)$$

where z_1 and z_2 are complex variables.

Series (8.45) must converge to ensure the existence of this transformation. That is, the following condition must be satisfied:

$$\sum_{k=-\infty}^{+\infty} \sum_{l=-\infty}^{+\infty} \left| x(k, l)\, z_1^{-k}\, z_2^{-l} \right| < \infty \qquad (8.46)$$

The set of the values z_1 and z_2, for which this condition is satisfied, defines the *convergence region* of $X(z_1, z_2)$.

8.3.11 Inverse z-transform

The inverse z-transform is obtained by generalizing the results of the one-dimensional case (sect. 2.3.1), which leads to

$$x(k, l) = \frac{1}{(2\pi j)^2} \oint_{C_1} \oint_{C_2} X(z_1, z_2)\, z_1^{k-1}\, z_2^{l-1}\, \mathrm{d}z_1\, \mathrm{d}z_2 \qquad (8.47)$$

where C_1 and C_2 are two contours surrounding the origin inside the convergence region. In contrast to the one-dimensional case, it is generally very difficult to determine the convergence region and the integration contours C_1 and C_2.

8.3.12 Relationship with the Fourier transform

The transform $X(z_1, z_2)$ is closely related to the Fourier transform $X(f, g)$. By setting

$$z_1 = r_1 \exp(j2\pi f) \quad \text{and} \quad z_2 = r_2 \exp(j2\pi g) \qquad (8.48)$$

and substituting them into (8.45), we obtain

$$X(r_1 e^{j2\pi f}, r_2 e^{j2\pi g}) =$$

$$\sum_{k=-\infty}^{+\infty} \sum_{l=-\infty}^{+\infty} x(k,l)\, r_1^{-k}\, r_2^{-l}\, e^{-j2\pi(fk+gl)} \tag{8.49}$$

Thus, the z-transform $X(z_1, z_2)$ of $x(k, l)$ is the Fourier transform of the signal $x(k, l)r_1^{-k}r_2^{-1}$. For $|z_1| = |z_2| = 1$, the z-transform becomes the Fourier transform.

By analogy with (8.39), it can be shown that the z-transform for both sides of (8.36) is given by

$$Y(z_1, z_2) = X(z_1, z_2)\, G(z_1, z_2) \tag{8.50}$$

8.3.13 Definition

The z-transform $G(z_1, z_2)$ of the impulse response $g(k, l)$ is called *transfer function* of the corresponding system.

8.3.14 Transfer function of a system defined by a difference equation

The transfer function is obtained by computing the z-transform for both sides of (8.41). We have

$$\left[\sum_{k'=0}^{K_1} \sum_{l'=0}^{L_1} a_{k'l'}\, z_1^{-k'} z_2^{-l'} \right] Y(z_1, z_2) =$$

$$\left[\sum_{k''=0}^{K_2} \sum_{l''=0}^{L_2} b_{k''l''}\, z_1^{-k''} z_2^{-l''} \right] X(z_1, z_2) \tag{8.51}$$

The transfer function is then given by

$$G(z_1, z_2) = \frac{Y(z_1, z_2)}{X(z_1, z_2)} = \frac{\displaystyle\sum_{k''=0}^{K_2} \sum_{l''=0}^{L_2} b_{k''l''}\, z_1^{-k''} z_2^{-l''}}{\displaystyle\sum_{k'=0}^{K_1} \sum_{l'=0}^{L_1} a_{k'l'}\, z_1^{-k'} z_2^{-l'}} \tag{8.52}$$

8.3.15 Remarks

If a system is specified by its frequency response, its impulse response can be computed with the inverse Fourier transform. Its transfer function is determined from its impulse response. $G(z_1, z_2)$ must then be expressed as a quotient of two polynomials in $z_1 z_2$ to obtain the coefficients $a_{k'l'}$ and $b_{k''l''}$ by identification with (8.52). In this case, the filtering can be implemented recursively according to (8.43).

8.3.16 Example

To illustrate the effect of a linear, translation-invariant system, let us consider an original image shown in Fig. 8.2. A low-pass filter is performed by using a system having an impulse response of the following type:

$$g(k,l) = \frac{1}{2\pi\sigma^2} \exp\left[-\frac{k^2 + l^2}{2\sigma^2} \right] \qquad (8.53)$$

In theory, the spread of this filter is infinite. For practical reasons, it was truncated to the domain ($|k| \le 9$ and $|l| \le 9$). Thus, it was possible to apply (8.42) to the signal $\hat{x}(k, l)$ the spread of which is $0 \le k \le 255, 0 \le 1 \le 255$. The result obtained with $\sigma = 4$ is shown in Fig. 8.2. In this image, which reminds us of an out-of-focus photo, the high spatial frequencies are strongly attenuated. There are no longer sharp transitions of the grey level in any direction.

8.3.17 Example

The second example is a high-pass filter (see Fig. 8.2).

(a) (b) (c)

Fig. 8.2

The filtering is performed by using a system having an impulse response of finite spread. The matrix form of this response is

$$g(k,l) = \begin{pmatrix} 1 & 1 & 1 \\ 1 & -8 & 1 \\ 1 & 1 & 1 \end{pmatrix} \tag{8.54}$$

The filtering is once again performed by using (8.42). The result is shown in Fig. 8.2. In this case, the slow variations of the grey level are visibly attenuated. The image obtained represents only the edges along which sharp-intensity transitions occur.

8.4 VISUAL PERCEPTION AND FUNDAMENTAL PROPERTIES OF THE HUMAN EYE

8.4.1 Short description of the human eye

The organ of vision, the eye, is a receiver of light signals. The eye focuses the light to form an image on the *retina,* which then analyzes this image and sends the message to the brain via the *optical nerve* and optical paths in the head. Very roughly, the eye can be considered as a camera (see Fig. 8.3). The image of an object is formed on the retina by the *lens.* The *pupil,* controlled by the *iris,* corresponds to the diaphragm.

The eye has a small volume (about 6.5 cm³) and an approximately rotational symmetry. Its weight is 7 grams for a diameter of about 24 mm. The retina is a neurosensorial layer which occupies a surface of approximately 12.5 cm². It contains two types of photoreceiver cells: *cones* and *rods.* The retina of

Synthetic description of the eye.

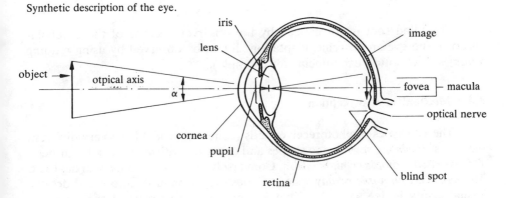

Fig. 8.3

a normal eye contains about 130 million rods and 6.5 million cones. The structure of the cones and rods is different enough to distinguish them under a microscope. This allows us to measure the distribution of the cones and rods on the retina. Figure 8.4 shows these distributions as a function of the perimetric angle α (see Fig. 8.3). The point of the retina corresponding to α = 0 is the *fovea*. In the vicinity of this point, only cones are present in significant numbers. Their density decreases very quickly as the distance from the fovea increases. The rods appear at about one degree from the fovea. Their density increases and reaches a maximum in the vicinity of 20 degrees. It then decreases for larger values of α. On the nasal side, at about 4 mm from the fovea, there is the *optical papilla*, which represents the head of the optical nerve. Because there are neither cones nor rods at this position, the papilla is insensitive to light. It is the *blind spot*. For this reason, the distribution of Fig. 8.4 has a dip at this point.

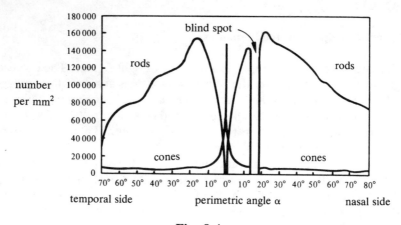

Fig. 8.4

The blind spot is not noticed by the observer because of the binocular vision of the eye and psychic adaptation. It can be observed by using reading schemes (Marriotte's experiment, for example).

8.4.2 Mechanism of perception

The two types of photoreceiver cells (cones and rods) have very different functions. Rods are sensitive to shapes and can work with very weak brightness. This is called *scotopic* or *night* vision. Conversely, cones work only with daylight. This is called *photopic* or *day* vision. Cones can distinguish colors and details. Visual acuity is due to cones. That is why a precise image of the details is obtained only if the eye "holds" them, which means, when the image is formed at the center of the fovea, where the cone density is largest. The dimensions of the retinal image of an object, seen with an angle of approximately one minute, are comparable to those of the cones. This roughly gives the resolution of the eye.

If the head is kept fixed, the eye axis can be displaced from 30 to 40 degrees about its mean position. This determines the *field of vision*. If the eye is fixed, the observer can see objects having an image formed outside the fovea. The visual field extends to about 100 degrees from the axis of the temporal side and to 70 degrees in other directions. This is the *peripheral vision*. It is relatively insensitive to details, especially after 20 to 25 degrees. By day, it is used only to orient the observation and detect movements. This can be seen by moving the hand at 45 degrees. The fingers cannot be counted, but their movement can be seen. The change from day to night vision (dark adaptation) cannot be performed instantaneously. The variation of the sensitivity of the cones and rods is performed progressively, with two different time delays. A first stage, of about 10 minutes length, involves the adaptation of the cones. The adaptation of the rods is performed only after 30 to 35 minutes. The readaptation to daylight also requires a significant amount of time. It is due mostly to light of short wavelength (blue).

8.4.3 Spectral sensitivity of the eye

Light is an electromagnetic wave. The energy produced by this wave is visible only within a very narrow frequency band. It is common to describe this band as a function of the wavelength λ. The visible spectrum is from about 400 nm (violet) to 700 nm (red). A light having a wavelength around the center of this interval produces a stronger sensation of brightness than those with wavelengths near the extremes. The eye's response to monochromatic light has been measured as a function of the wavelength. The International Light Committee has accepted (on the basis of experimental results) a typical response, corresponding to an average, normal observer. This response, measured in day vision, is shown in Fig. 8.5 for two values of the perimetrical angle α. The response for $\alpha = 2$ degrees primarily corresponds to foveal vision, while the response for $\alpha = 10$ degrees corresponds to extrafoveal vision. These curves show that the eye acts as a bandpass filter. The maximum sensitivity is obtained at a wavelength of 555 nm.

8.4.4 Definitions and photometric units

Several notions and units are involved in image formation and processing in relation to the response of the eye. It is useful to define them so as to avoid confusion which often arises ([45], p. 7).

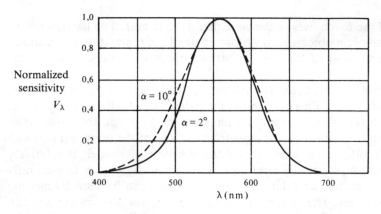

Fig. 8.5

The total power of the light (electromagnetic wave) received or emitted by a given surface is called the *luminous flux F*. It is measured in lumens (lm). The luminous flux received by a unit surface is called the *illumination E*. The unit of illumination is the lux (lx). For a given surface S, we have the equation:

$$F = \iint_{S} E \, dS \tag{8.55}$$

The relationship between lumens and lux is $1 \, lx = 1 \, lm/m^2$.

The luminous flux emitted per unit surface is the *emittance M*. It is also measured in lux.

The *luminous intensity I* of a point light source, in a given direction, is the luminous flux emitted per unit solid angle. It is measured in candela (cd), that is, in lumen per steradian. The notion of luminous intensity can be extended to nonpunctual sources. In this case, it is called the *luminance L*.

The luminance is the luminous intensity per unit surface perpendicular to the direction of the emission. It is measured in nit, which is candela per square meter.

The *luminous efficiency K* is the ratio between the luminous flux F and the emitted power flow Φ.

$$K = \frac{F}{\Phi} \qquad\qquad lm/w \tag{8.56}$$

This allows us to relate, in a very simple manner, radiometric to photometric quantities.

The quantities defined in this section represent values integrated over the entire visible spectrum. If these values are specified for a given wavelength, they have the index λ. They are then expressed per unit wavelength. Consequently, they can be considered as densities. We have equations of the form:

$$F = \int_0^\infty F'_\lambda \, d\lambda = \int_0^\infty K_\lambda \, \Phi_\lambda \, d\lambda \qquad (8.57)$$

The luminous efficiency density $K_{\lambda\lambda}$ is related to the relative sensitivity $V\lambda$ of the eye (see Fig. 8.5) by the following equation.

$$K_\lambda = 620 \, V_\lambda \qquad (8.58)$$

For a surface, let F_i, F_a, F_r, and F_t be the incoming, absorbed, reflected, and transmitted luminous fluxes, respectively. The three following factors can be defined:

- *absorption factor* or *absorbance* $\alpha = F_a/F_i$
- *reflection factor* or *reflectance* $\rho = F_r/F_i$
- *transmission factor* or *transmittance* $\tau = F_t/F_i$

Because $F_i = F_a + F_r + F_t$, we have

$$\alpha + \rho + \tau = 1 \qquad (8.59)$$

These factors are essentially involved when using optical filters and when analyzing the reproduction of images on photographic material. Logarithmic units are also used to define *optical densities*. They are measured either by transparence:

$$d_\tau = \log_{10}(1/\tau) = -\log_{10} \tau \qquad (8.60)$$

or by reflection:

$$d_\rho = \log_{10}(1/\rho) = -\log_{10} \rho \qquad (8.61)$$

8.4.5 Quantitative photography

An image (two-dimensional signal), by contrast with objects seen, must be presented on a physical support in order to be seen by the observer. This support can be, for example, a screen, a film, or photographic paper. It is well known from photography that the best results are obtained with film. The exposure to light, and subsequent development of a film, produces a layer of microscopic grains of silver, suspended in the emulsion. The concentration and density of these particles are controlled by the exposure, which is the product of the incident light E and the exposure time T. Thus,

$$\epsilon = E\,T \tag{8.62}$$

The higher is the density of these grains, the more opaque is the film. For example, if the transmittance τ of a uniformly illuminated film is 1%, (8.60) shows that the optical density is $d_\tau = -\log_{10}(1/100) = 2$. The optical density is related to the exposure ϵ by the ideal equation:

$$d_\tau = \gamma \log_{10} \epsilon \tag{8.63}$$

where γ is a proportionality constant.

This function, well known in photography, is called the *Hurter-Driffield* or *d-ϵ curve*. Current photographic materials are far from following this ideal law (8.63). In general, an unexposed and unprocessed emulsion, as well as its film, are not perfectly transparent. A constant level d_0, called the *veil*, must be added to (8.63).

$$d_\tau = d_0 + \gamma \log_{10} \epsilon \tag{8.64}$$

Moreover, the dynamic range of the emulsions remains necessarily finite. Therefore, the real curves diverge from the ideal curve (8.63) at the boundaries of this range. Figure 8.6 shows the characteristic curve of a real photographic material. The proportionality factor γ defines the *contrast*. This is the slope of the tangent at the inflection point. The factor γ is positive for a negative film and *vice versa*.

Above a certain exposure, the optical density cannot be increased. The curve thus reaches a saturation level. These notions are easily transferred to the case of photographic paper, where the transmittances must be replaced by reflectances.

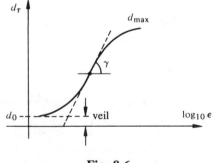

Fig. 8.6

8.4.6 Brightness perception

By observing an image, its reproduction quality can be judged by the transmission of the intensity of illumination emerging from its elementary surfaces. These are influenced by the immediate neighbors of the observed point, as well as by those farther away. The *apparent intensity* of a point of the image, the *physical intensity* (or *effective*) of which is given, depends on the local and general adaptation of the eye and the effects of simultaneous contrast.

Local adaptation to the immediate vicinity of the observed point is very quick. It allows us to see more clearly the details of the image in the neighboring regions.

General adaptation permits the eye to cover a very wide dynamic range of light. This range starts at about .0003 lux for the night sky and reaches 100,000 lux for full sunlight. General adaptation is much slower than local. The effect of simultaneous contrast is caused by the fact that the apparent intensity of a region does not only depend on its effective intensity. To understand this, it is sufficient to observe an opaque surface of a given color with two different backgrounds.

The reflectance of the surface does not change, but it appears darker with a lighter background. This effect is illustrated by Fig. 8.7. The luminances of the little squares are identical, but their apparent intensities differ due to the different effective luminances of their neighboring regions.

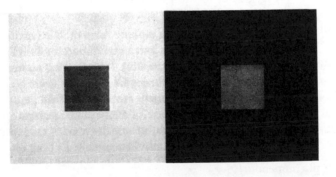

Fig. 8.7

These considerations show that the effective (objective measurable value) and apparent (subjective value) intensities should be distinguished. The apparent intensity is the subjective response of the eye to the light perceived. It is common to call *brightness B* the apparent luminance.

The progression of the luminances (effective) is a continuous function.

However, this is not the case for the progression of the corresponding brightness. The eye cannot distinguish two very close luminances. The smallest perceptible luminance difference is called a *just noticeable difference*. This can be determined from the Weber-Fechner experiment. In this experiment, a uniform visual luminance field L is divided into two parts along a line. The luminance of one part is increased by ΔL. The observer sees this variation only when ΔL reaches a precise nonzero value. Moreover, by repeating this experiment for several values of L, we can see that the ratio $\Delta L/L$ is constant:

$$\Delta L/L = C_W \tag{8.65}$$

Thus, we have Weber-Fechner's law. It is valid over a wide luminance interval, from one to about 1000 nits. The constant C_w in this interval is between 0.01 and 0.02, depending on the observer. A commonly made mistake is to integrate this simple law. In fact, the ratio $c_1 \Delta L/L$ is often interpreted as an elementary brightness ΔB. After integration, we obtain:

$$B = c_1 \log L + c_2 \tag{8.66}$$

This is a simple equation relating the apparent luminance (brightness B) with the effective luminance L, where c_1 and c_2 are two constants. Because of this equation, the eye is often said to be logarithmic. This is only approximately true. In fact, (8.65) is subject to the perturbations of the human visual system and the integration only increases their effects. On the basis of recent experimental results, a rule of the type:

$$B = c_3 L^n \tag{8.67}$$

seems to be more adequate. The exponent n in this equation primarily depends on the local field and the adaptation state of the eye. In general, n is between 0.3 for a dark local field without adaptation, and three for very light fields.

8.4.7 Mach phenomenon

The Mach phenomenon comes from the frequency (spatial) response of the eye. The perimetric angle α (see Fig. 8.3) is an important parameter in the study of acuity, due to the apparent revolution symmetry of the eye. This is why the spatial frequencies of an image are often expressed in cycles per degree or per radians, instead of cycles per meter. The frequency response (spatial) of the eye can be measured more or less precisely ([45]), chap. 12). Although the experimental results are sometimes very different from one another, due to delicate experimental conditions, it is unanimously accepted that the frequency response of the eye is a function of the form shown in Fig. 8.8. This response has a maximum sensitivity between five and 10 cycles per degree.

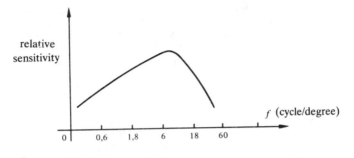

Fig. 8.8

We can see from this response that the eye has two types of functions. For spatial frequencies less than about five cycles per degree, it works like a differentiator. For spatial frequencies greater than about 10 cycles per degree, it works like an integrator.

The differentiator part gives rise to the *Mach phenomenon,* which is shown in Fig. 8.9. This figure shows grey bands between white and black.

The luminance is constant in each band, but the brightness of a band increases when approaching the darker neighboring band and decreases when approaching the lighter neighboring band. The luminances and brightnesses are also given in the same figure.

The Mach phenomenon can be explained quantitatively by Fourier anal-

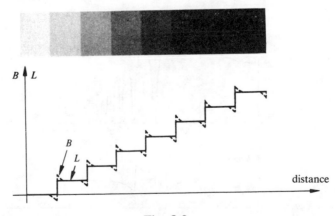

Fig. 8.9

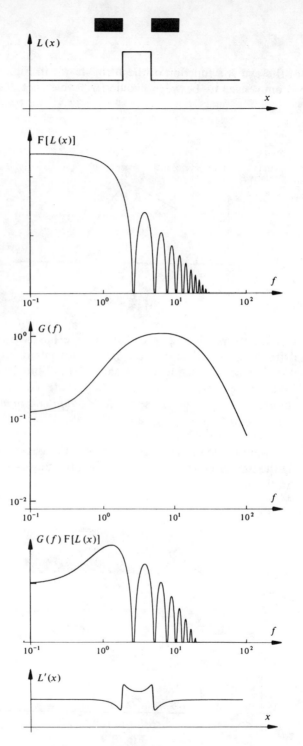

Fig. 8.10

ysis. Let us consider a luminance $L(x)$, which has two values along a given direction, and is uniform in the perpendicular direction (Fig. 8.10). The Fourier transform of the luminance is given in the same figure. This transform, assuming that the eye is characterized by a linear system, is multiplied by the frequency response $G(f)$ of the eye (fig. 8.10) to give the transform illustrated in Fig. 8.10. The inverse transform gives the distribution of the luminance $L'(x)$ (Fig. 8.10). This is, in fact, the apparent luminance—or brightness—perceived by observing the image of Fig. 8.10. The differentiation effect can be clearly seen in this figure.

The integrator function of the eye is often used in image reconstruction and display. A digital image is, in general, reconstructed by using regularly spaced points or symbols (Fig. 8.11). The structure of these points becomes visible when the image is sufficiently enlarge (Fig. 8.11). However, when the dimension of the elementary shape corresponds to a spatial frequency larger than about 10 cycles per degree, this structure becomes practically invisible.

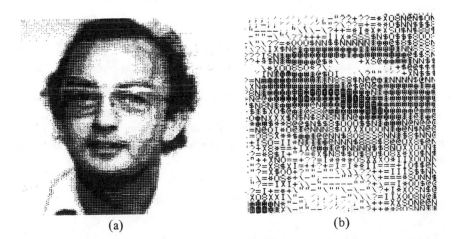

(a)   (b)

Fig. 8.11

8.5 SAMPLING, QUANTIZATION, AND RECONSTRUCTION OF TWO-DIMENSIONAL SIGNALS

8.5.1 Preliminary remarks

In general, images are continuous or analog functions. The luminance is a continuous function of the position in the image plane. This is universally accepted, depsite the granular (discrete) nature of the images at a microscopic level. Moreover, although the dynamic range of the luminance is physically limited, an infinite number of values can be taken in this band. For digital processing, the luminance must be sampled and quantized. Finally, an image processed digitally must be reconstructed in a form appropriate for presentation to the observer.

8.5.2 Ideal sampling

Let $x_a(u, v)$ be a two-dimensional analog function, and $X_a(f, g)$ be its Fourier transform, defined by

$$X_a(f,g) = \int_{-\infty}^{+\infty} \int_{-\infty}^{+\infty} x_a(u,v)\, e^{-j2\pi(fu+gv)}\, du\, dv \qquad (8.68)$$

In general, the sampling of the function $x_a(u, v)$ is performed by periodically taking samples along the u and v axes. The corresponding digital signal is expressed as

$$x_a(k\Delta u, l\Delta v) = x(k,l) = x_a(u,v)\Big|_{\substack{u=k\Delta u \\ v=l\Delta v}} \qquad (8.69)$$

where Δu and Δv are the sampling periods along the u and v axes, respectively.

To reconstruct the function $x_a(u, v)$ from the set of samples $x(k, l)$, Δu and Δv must satisfy some conditions. These conditions come from the *two-dimensional sampling theorem*. It is a simple generalization of the theorem studied in section 1.7.

8.5.3 Two-dimensional sampling theorem

An analog signal $x_a(u, v)$ limited in spatial frequencies to F and G (cycle per unit distance) can be reconstructed exactly from its samples $x_a(k\Delta u, l\Delta v)$ if and only if these were taken with periods Δu and Δv less than or equal to $1/2F$ and $1/2G$, respectively.

8.5.4 Proof

The sampled version $x_e(u, v)$ of an analog function $x_a(u, v)$ can be considered, in theory, as the product of $x_a(u, v)$ and a periodic sequence of Dirac impulses $e(u, v)$ (Fig. 8.12):

$$x_e(u,v) = x_a(u,v)\, e(u,v) \qquad (8.70)$$

with

$$e(u,v) = \sum_{k=-\infty}^{+\infty} \sum_{l=-\infty}^{+\infty} \delta(u - k\Delta u, v - l\Delta v) \qquad (8.71)$$

In the frequency domain, the simple product (8.70) of the spatial domain corresponds to the following convolution product:

$$X_e(f,g) = X_a(f,g) * * E(f,g)$$

$$= \int_{-\infty}^{+\infty} \int_{-\infty}^{+\infty} X_a(f',g') \; E(f-f',g-g') \, df' \, dg' \qquad (8.72)$$

It can be shown that

$$E(f,g) = \frac{1}{\Delta u \, \Delta v} \sum_{m=-\infty}^{+\infty} \sum_{n=-\infty}^{+\infty} \delta \left(f - \frac{m}{\Delta u}, g - \frac{n}{\Delta v} \right) \qquad (8.73)$$

By taking into account the property of convolution of the Dirac impulse ([44], p. 11):

$$Y(\alpha,\beta) * * \delta(\alpha - a, \beta - b) = \int_{-\infty}^{+\infty} \int_{-\infty}^{+\infty} Y(\alpha',\beta') \delta(\alpha - \alpha' - a, \beta - \beta' - b) d\alpha' \, d\beta$$

$$= Y(\alpha - a, \beta - b) \qquad (8.74)$$

then, (8.72) can be written as

$$X_e(f,g) = \frac{1}{\Delta u \, \Delta v} \sum_{m=-\infty}^{+\infty} \sum_{n=-\infty}^{+\infty} X_a \left(f - \frac{m}{\Delta u}, g - \frac{n}{\Delta v} \right) \qquad (8.75)$$

The function $X_e(f, g)$ is thus obtained by periodic repetition in both directions, of periods $1/\Delta u$ and $1/\Delta v$, of the Fourier transform $X_a(f, g)$ (Fig. 8.12). In fact, the function $X_e(f, g)$ is the two-dimensional Fourier transform $X(f, g)$ of the digital signal in the sense of definition (8.8).

$$x(k,l) = x_a(u,v) \Big|_{\substack{u = k\Delta u \\ v = l\Delta v}} \qquad (8.76)$$

Two-dimensional sampling, like the one-dimensional case, creates a sequence of secondary spectra in both directions, proportional to the spectrum $X_a(f, g)$. The signal $x_a(u, v)$ can be reconstructed if and only if these secondary spectra can be eliminated without modifying the spectrum $X_a(f, g)$. This operation, performed, in principle by using a spatial low-pass filter, is possible

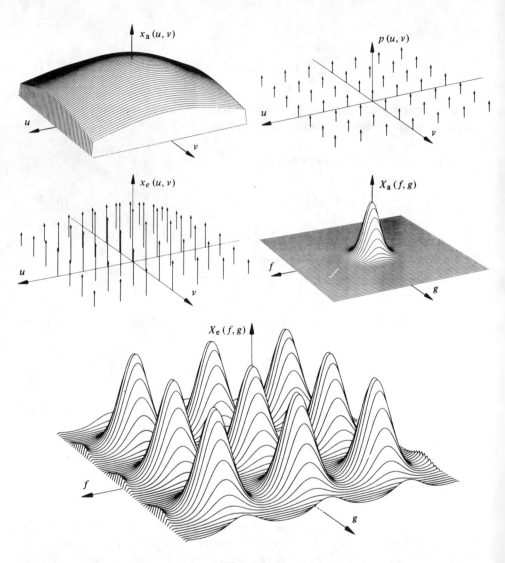

Fig. 8.12

only if $X_a(f, g)$ is zero for spatial frequencies greater than $F = 1/2\Delta u$ along the f axis, and greater than $G = 1/2\Delta v$ along the g axis, that is, if

$$X_a(f, g) = 0 \qquad \text{for } f > F = 1/2\,\Delta u$$
$$\text{and } g > G = 1/2\,\Delta v \qquad (8.77)$$

If this condition is not satisfied, aliasing occurs.

8.5.5 Aliasing errors

If the condition of section 8.5.3 is not satisfied, the secondary spectra overlap with the spectrum $X_a(f, g)$. Even if, using a rectangular low-pass filter, the main part of $X_a(f, g)$ is preserved, the signal at the output of this filter will be only a distorted version of $x_a(u, v)$. Figure 8.13 shows the case of a sampling which leads to aliasing of the secondary spectra (the condition of section 8.5.3 is thus violated).

These considerations are purely theoretical. They are even more difficult to implement than those of the one-dimensional case. A first difficulty is involved in measuring $X_a(f, g)$ from the given corresponding image $x_a(u, v)$. Without any appropriated tool or method, only more or less rough approximations of the spread of the spectrum $X_a(f, g)$ can be made. Then, in most cases, the reconstruction must be taken into account.

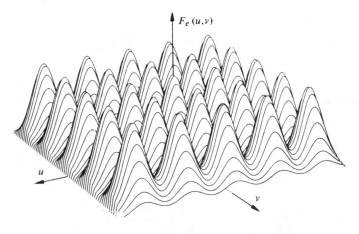

$F_e(u,v)$

u

v

Fig. 8.13

Processed images are generally destined to be observed. Because digital processing operations are often long and complex, it is generally preferred not to further complicate them by introducing reconstruction filters. The integrator effect of the eye and the physical carrier of the image are used for the reconstruction. In this case, a sampling period slightly smaller than the resolution of the eye (1' arc) must be chosen. For printed documents observed at a distance of about 30 cm, a sampling period of about 1/10 mm is required. By using the integrator effect of the eye, this period can be increased to 1/5 mm. Figure 8.14 shows a part of an original text in original size and its digital reproduction after two-dimensional sampling with a period of 0.2 mm in both directions (Fig. 8.14).

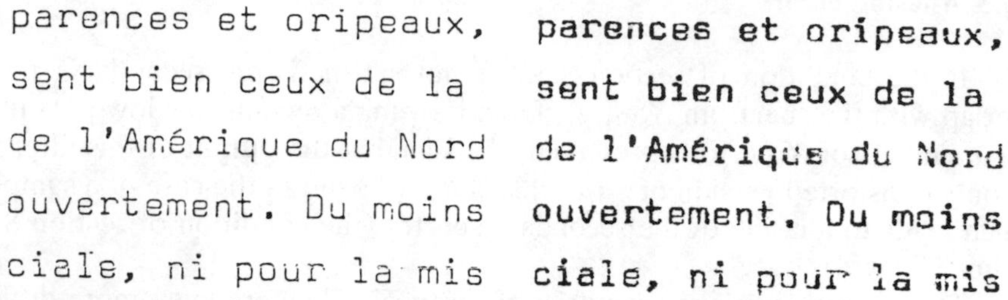

```
parences et oripeaux,     parences et oripeaux,
sent bien ceux de la      sent bien ceux de la
de l'Amérique du Nord     de l'Amérique du Nord
ouvertement. Du moins     ouvertement. Du moins
ciale, ni pour la mis     ciale, ni pour la mis
```
(a)   (b)

Fig. 8.14

The sampling effect is more clearly seen in the corresponding enlargements (Fig. 8.15).

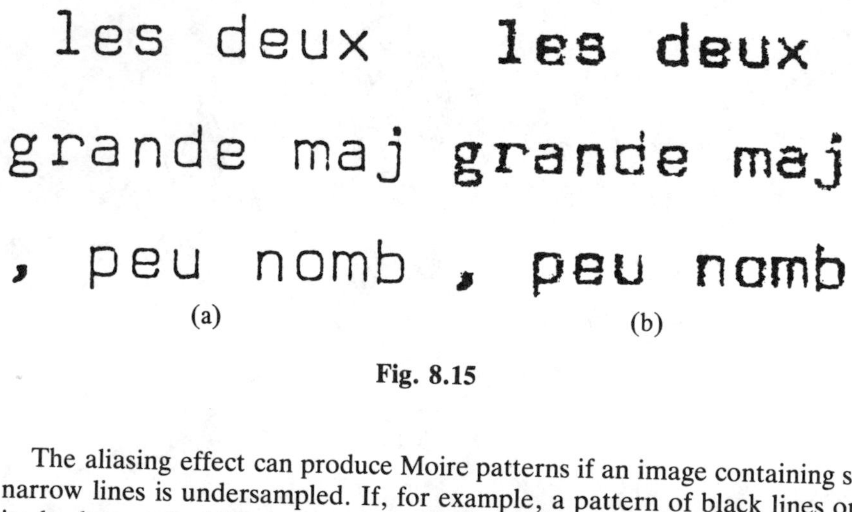

les deux   les deux

grande maj   grande maj

, peu nomb , peu nomb

(a)   (b)

Fig. 8.15

The aliasing effect can produce Moire patterns if an image containing sets of narrow lines is undersampled. If, for example, a pattern of black lines on a white background with spatial frequency f_r is undersampled, the reconstructed image after low-pass filtering shows a different line set in another direction and with another spatial repetition frequency f_r'. This is illustrated in Fig. 8.16.

8.5.6 Prefiltering

A sampling system analyzes a given image $x_a(u, v)$ using a light beam concentrated on one point of the image. The quantity of reflected or transmitted light at this point is a measurement of its luminance. Due to the imperfections of optical systems, this spot cannot be reduced to a point. Thus, the luminance

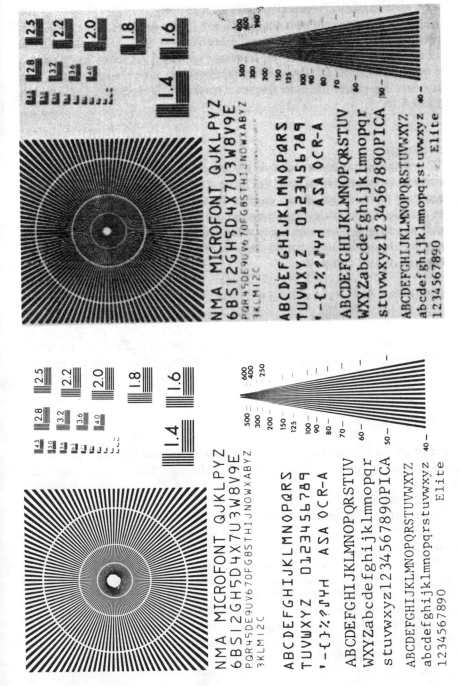

Fig. 8.16

measured represents the mean value over a small region around the targeted point. Let $y_a(u, v)$ be the function representing the shape of the aperture of the spot. The luminance measured at a point (u_0, v_0) is given by

$$p_a(u_0, v_0) = \iint_S x_a(\alpha, \beta)\, y_a(u_0 - \alpha, v_0 - \beta)\, d\alpha\, d\beta \qquad (8.78)$$

where S represents the spread, relatively small, of the function $y_a(u, v)$. In the frequency domain, the product:

$$P_a(f, g) = X_a(f, g)\, Y_a(f, g) \qquad (8.79)$$

corresponds to the convolution product (8.78) in the spatial domain. $P_a(f, g)$, $X_a(f, g)$, and $Y_a(f, g)$ are the two-dimensional Fourier transforms of the signals $p_a(u, v)$, $x_a(u, v)$ and $y_a(u, v)$, respectively. The real luminance is thus filtered by a filter having characteristics described by $Y_a(f, g)$.

For the beam shapes normally used, $Y_a(f, g)$ is generally the frequency response of a low-pass filter. If the spot is wide, the spread S of $y_a(u, v)$ is large. In this case, $Y_a(u, v)$ becomes a sharp low-pass filter. Some shapes of beam apertures (function $y_a(u, v)$) and their frequency responses are shown in Fig. 8.17.

If the form of $Y_a(f, g)$ is known and if the filtering (8.79) is bothersome, the aperture defect can be corrected by inverse filtering. The frequency response of this correction filter is simply the inverse of $Y_a(f, g)$.

8.5.7 Sampling grids

For practical reasons (design of image acquisition and display equipment), two-dimensional sampling is almost always performed with a square or quasisquare grid.

In this grid, the coordinates of the sampling points are of the form $(k\Delta u, l\Delta v)$. For a square grid, we clearly have $\Delta u = \Delta v$. Any point of this grid has eight neighbors.

Half of these points, in the diagonal directions, are at a distance $\Delta u \sqrt{2}$ of the center, while the others are at Δu. In some applications requiring spread or surface measurements using the sequence of some points, the existence of two different neighbor distances can give rise to complications. Appropriate weighting methods must be used to take this fact into account.

Another possibility for the sampling is to use a regular hexagonal grid. In this case, each point has six neighbors at the same distance. This type of grid has been used, with appropriate operators, in pattern recognition for biological image processing [46].

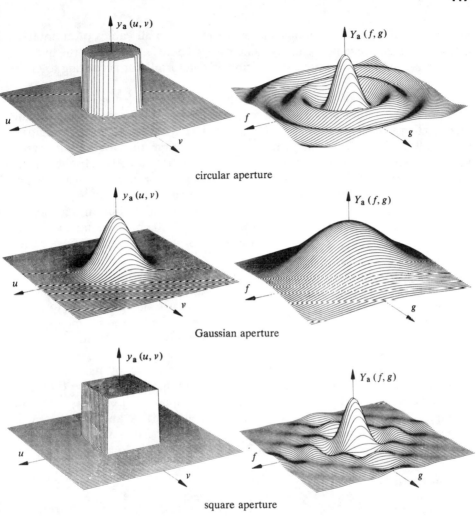

circular aperture

Gaussian aperture

square aperture

Fig. 8.17

A similar grid can be obtained from a square grid by shifting the odd (or even) lines by $\Delta u/2$.

Sampling grids can be very large. For example, to reproduce a television picture, with no apparent distortion, a matrix of about 300×300 points is required. To reach the quality of photographs, on a post card, a matrix of about 600×900 points is necessary.

8.5.8 Luminance quantization

For digital processing, the image samples must be quantized. In principle, this is performed by splitting the dynamic range of the luminance of an image into a finite number of intervals and by attributing to all values of an interval the same luminance value. The problem involved here is to find out the number of intervals and how they must be distributed (uniformly, or according to given rules).

In the regions of an image where the luminance varies very slowly from one point to another (regions where the spatial frequencies are small), the quantization will create a constant luminance level. Between two such regions, there will be a step from one level to another, whereas the original luminance varies continuously. In general, these steps create artificial edges in the image, which are very disturbing. They can, in some cases, produce artificial shapes and objects, which do not exist in the original image.

The choice of the number of intervals and their distribution depends on two key factors. The first is the eye of the observer (subjective factor), while the second is the physical support on which the quantized image must be reproduced (objective factor).

To satisfy the requirements of the eye, a quantization interval amplitude must be chosen slightly smaller than the just noticeable difference (sect. 8.4.6). This avoids the creation of false contours. It is sometimes preferable to chose a uniform distribution of the quantization intervals over the entire dynamic range (intervals of equal width). This choice is motivated by the need for simplicity in practical implementations.

The physical support determines the dynamic range by its photometric properties. In general, it is characterized either by its transmittance (film), or by its reflectance (photographic paper or opaque supports). Let us consider, for example, the case of a film having the transmittances τ_1 and τ_2, when it is respectively unexposed and exposed to saturation. The dynamic range of the luminance of this film, in optical density units, is given by

$$D = d_{\tau_2} - d_{\tau_1} = -(\log_{10} \tau_2 - \log_{10} \tau_1) \tag{8.80}$$

The just noticeable difference C_w (8.65) is given in percent. To express it in optical density units, the term $1 + C_w$ must be considered. This allows us to avoid negative densities by shifting the origin of the logarithmic function to one. The just noticeable difference, expressed in optical density units, is thus $\log_{10}(1 + C_w)$. The number of quantization levels N_q is then given by

$$N_q = \frac{D}{\log_{10}(1 + C_W)} = \frac{\log_{10}\tau_1 - \log_{10}\tau_2}{\log_{10}(1 + C_W)} \tag{8.81}$$

8.5.9 Comments

Equation (8.81) is valid if the Weber-Fechner law is also valid, that is, when the luminances are within the interval [1,1000] nits. Moreover, depending on the value of the constant C_w for different observers, very different values for N_q may be obtained. To satisfy the majority of observers, the smallest value of C_w must be used. Finally, we must not forget that (8.81) was established with the hypothesis of a uniform distribution of the quantization intervals. Figure 8.18 shows an original image and its quantized versions with 256, 128, 64, 32, 16, 8, 4, and 2 grey levels. Even if these results are distorted by the printing, the artificial edges are visible (mostly in the background) even on the image quantized with 64 levels.

8.5.10 Example

To illustrate quantization and (8.81), let us consider the example of a film of dynamic range of $D = 2$ optical density units. This kind of film is frequently

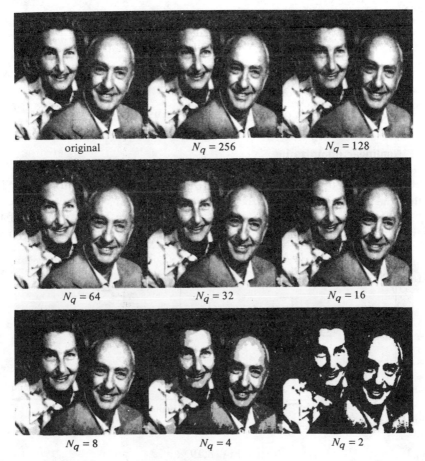

Fig. 8.18

used in amateur photography. With a mean value of $C_w = 0.015$, we have

$$N_q = 2/\log_{10} (1.015) = 309.3$$

However, N_q represents a number of intervals and therefore must be an integer. It can be rounded to 309 or 310.

Usually, the luminance of an image sample is represented by a binary word. The important parameter in practice is the number of bits in this word. To obtain it, the lowest power of 2 closest, greater than, or equal to N_q must be found. The spread is then the logarithm (base 2) of this number. In this case, the power is $2^9 = 512$. The number of bits searched is thus $B_2 = \log_2 512 = 9$.

If we consider the case of observers who require $C_w = 0.01$, then

$$N_q = 2/\log_{10} (1.01) = 462.8$$

The same number of bits is sufficient in this extreme case.

If the other extreme case $C_w = 0.02$ is considered, then

$$N_q = 2/\log_{10} (1.02) = 232.5$$

In this case, binary words of $B_q = 8$ bits can be used.

8.5.11 Nonlinear quantization

When the distribution of the quantization intervals is uniform, the law relating the original and quantized luminances is linear, constant by interval (Fig. 8.19). In this case, the same importance is given to all regions of the dynamic range. Because the quantized image is generally for an observer, a law

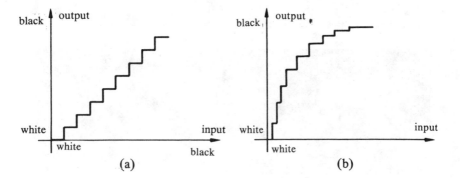

Fig. 8.19

more adapted to the properties of the eye can be used. This is motivated by economic reasons, the goal being to decrease the number of bits per sample.

We have seen that, on one hand, the eye is very sensitive in the dark (night vision with rods), and that, on the other hand, the brightness is a nonlinear function of the luminance (eq. (8.66) and (8.67)). It is thus possible to use a quantization rule with more dark levels than light levels (Fig. 8.19). The most commonly used nonlinear quantization rule is a logarithmic one. Figure 8.20 shows the original image of Fig. 8.18 quantized on 16 levels with linear and logarithmic rules.

(a)   (b)

Fig. 8.20

In some applications a so-called optimal quantization rule is used, which is based on the statistics of the luminance and the mean square error, rather than the requirements of the eye as criterion. Copied from the quantization of one-dimensional signals, in very different contexts and criteria, this method does not give satisfactory results. The eye does not follow the mean square error criterion.

8.5.12 Reconstruction of digital images

The reconstruction of one-dimensional analog signals, processed digitally, is performed using interpolative filters (sect. 1.7). Thus, the secondary spectra caused by the sampling are eliminated. This remains valid, in principle, for two-dimensional signals, that is, for images. However, the implementation of two-dimensional interpolative filters, analog as well as digital, is much more complicated and costly than in the one-dimensional case. Even if, analogically (optically), a spatial low-pass filter can be made by defocusing the image a little, the frequency response of such a filter (Fig. 8.17, circular case) is not convenient, due to the residual oscillations.

The most commonly used method consists of choosing a sampling period small enough so that the interpolation is completely performed by the integrating effect of the eye. The drawback of this method is that it can lead, according to the frequency (spatial) contents of the image, to an oversampling and thus to a large number of digital values.

8.6 REDUNDANCY REDUCTION

8.6.1 Definition

Let us consider a digital image $x(k, l)$ obtained by sampling and quantizing the luminance $L(u, v)$ of an original image.

Let us assume, moreover, a square sampling grid of $N \times N$ points and a B_q bits quantization of each sample. If the N^2 samples obtained are statistically independent, $N^2 B_q$ bits are required to represent the digital image.

For example, considering the common values $N = 256$ and $B_q = 8$, more than 500,000 bits must be used to represent a small-scale photograph (about 5 cm $\times$ 5 cm). For a photograph of 20 $\times$ 20 cm, the number of bits becomes greater than 8 million. The representation of a digital image in the form described above is called the *canonical form* of a digital image. (It is sometimes called a *PCM coded image* (pulse code modulation)). Due to the large number of bits used, the transmission and storage of a digital image using its canonical form is very costly. It is thus very useful to try to reduce the number of bits required to represent a digital image by exploiting the dependancy between adjacent samples. This operation is called *redundancy reduction* or *source coding of the image.*

8.6.2 Classification of the reduction method redundancy

From their properties, the methods for the redundancy reduction can be classified in two different ways.

In the first classification, two classes are distinguished:

- *exact methods:* These are methods permitting the exact retrieval of the samples of the original digital image.
- *psychovisual methods:* These are methods which produce a distortion of the original images by using the properties of the eye. It is clear that, to be valid, a psychovisual method must introduce distortions which are invisible, or at least tolerable to the eye.

In the second classification, the implementation fields are distinguished:

- *spatial methods:* These are methods which act directly on the samples of. an image in the spatial domain.
- *transformed methods:* These are methods which act on a transformation (generally linear) of the original image.

These two classifications are not exclusive. Moreover, a single method can be exact or psychovisual, according to its implementation.

8.6.3 Differential methods

Differential methods are spatial methods. The idea is to derive from the sequence of the signal samples $x(k, l)$ another sequence, by differentiation, which has a smaller dynamic range. Due to inherent correlation, adjacent samples in an image have very similar values. For example, by computing differences such as $x(k, l) - x(k, l - 1)$ along a row, a sequence of values, which are much smaller than the original values $x(k, l)$, is obtained. The sequence of the differences can be quantized with fewer levels, which reduces the number of bits per sample.

The differences can be computed in several ways. They can be obtained from the difference between successive rows or columns, or from the difference between rows followed by the difference between columns (or *vice versa*).

These methods have the advantages of being simple, easy, and inexpensive to implement. Their performance depends on the type of image. In general, a reduction by a factor of two is easy to obtain. As with all differential techniques, they have the drawback of being very sensitive to transmission errors. These errors propagate on the rows or columns, and have a very unpleasant effect on the eye. Differential methods can be exact or psychovisual, depending on the quantization rule for the differences.

8.6.4 Predictive methods

In predictive methods, the value $\hat{x}(k, l)$ of a sample $x(k, l)$ is predicted from the values of the neighboring samples. They can be spatial as well as transformed. In the transformed case, the samples of the transform are considered. The difference $e(k, l) = x(k, l) - \hat{x}(k, l)$ is called *prediction error*.

Because of the correlation between samples, it is possible, in general, to predict a value $\hat{x}(k, l)$ very close to $x(k, l)$. The dynamic range of the error signal $e(k, l)$ is much smaller than that of $x(k, l)$. The error signal can thus be quantized with much fewer levels. This significantly reduces the number of bits per sample.

The number of samples used for the prediction defines the *order* of the prediction. For example, a predictor of order 12 uses 12 neighboring samples to predict $\hat{x}(k, l)$. These samples can be chosen in different ways. By considering a scanning from left to right, row by row, they can be on the same row or grouped on several rows. They can also be on one or several columns.

The best estimation $\hat{x}(k, l)$, using the mean square error, is obtained with a nonlinear function of the neighboring samples. However, linear estimators are most often used because they are simple and well adapted to analytical computations. In this case, the mean square error can be used as a criterion because the estimation $\hat{x}(k, l)$ is not displayed, but is simply a tool. Depending on the number of quantization levels used for the signal $e(k, l)$, predictive methods can produce distortions. The two different distortions are saturation and oscillation effects. Saturation is produced in regions of the image where the spatial frequencies are high. In these regions, the error $e(k, l)$ is relatively large. If it is greater than the highest quantization level, the sharp luminance transitions look saturated. Oscillations are produced in regions of the image where the spatial frequencies are very low. In these regions, the sign of the error $e(k, l)$ changes often, but the amplitude of $e(k, l)$ remains very low. This produces an oscillation effect, which is like a salt-and-pepper noise.

The reduction in redundancy which can be obtained with these methods depends on the images, but also on the order of the prediction and the quantization of the error signal. For predictors of order three or less, a reduction by a factor between two and three is very common.

8.6.5 Synthetic-highs method

This psychovisual method uses the Mach phenomenon in a very elegant way and is one of the most efficient redundancy reduction methods. Its drawback is the tremendous amount of processing required. In this method, the digital image $x(k, l)$ is divided into two parts: a (spatial) low-frequency image representing the general luminance of the regions and an high frequency one representing the edges and sharp transitions of the luminance.

The low-frequency image is obtained simply by low-pass filtering of the original image. According to the two-dimensional sampling theorem, the low-frequency image can be represented with much fewer bits than the original image. For example, if the digital image is a $N \times N$ square matrix, the low-frequency image, depending on the filtering, can be represented by a $N/10 \times N/10$ matrix. In addition, each sample of the low-frequency image can be quantized with fewer levels than the samples $x(k, l)$.

The high-frequency image is obtained by high-pass filtering. The most commonly used filters are those employing either the discrete Laplacian or the discrete gradient operator.

If the gradient is used, two high-frequency images are obtained, respectively representing the horizontal (with respect to l) and the vertical (with respect to k) gradients. Polar coordinates can be used by computing the modulus and the argument. The edges are then detected on the high-frequency image (in modulus) using a threshold and a tracking of the points where the modulus is large. In a picturesque way, this can be characterized as following the mountain tops or ridge riding. This leads to an image in which the edges are thin. This second image with the resampled low-frequency image represent the compact form of the original image for transmission or storage.

For reconstruction, the low-frequency image is filtered with an interpolating filter. The edge image is filtered by a filter having characteristics that depend solely on those of the low-pass filter used to obtain the low-frequency image.

The artificial production of an high-frequency image from the thin edge leads to synthetic luminances; hence, the name of the method. Figure 8.21 shows the block diagram of this method. The sum of the low and high frequency images gives the reconstructed image. With this method, reduction up to a factor of 10 can be obtained with tolerable distortions. Figure 8.22 shows the different images involved in this method.

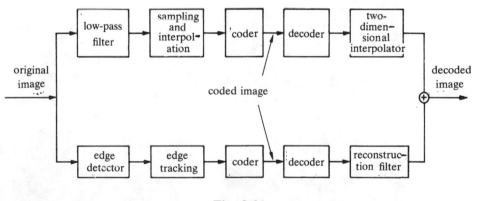

Fig. 8.21

8.6.6 Bit-planes methods

The samples of a digital image $x(k, l)$ are represented in the canonical form by binary words, each of B_q bits. If, in this representation, the jth bit of each sample is isolated, a two-level structure called *bit-plane* is obtained. The image is entirely described by the set of B_q bit planes. Figure 8.23 shows a digital image quantized with 256 levels and its eight bit planes.

In each bit plane we can apply redundancy reduction methods specially

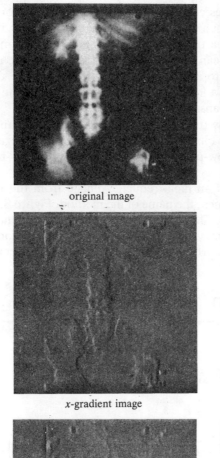

original image

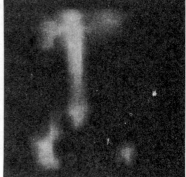

low-frequency image

x-gradient image

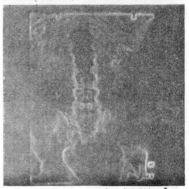

modulus of the gradient

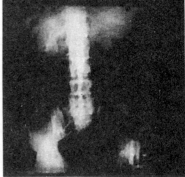

reconstructed x-gradient

reconstructed image

Fig. 8.22

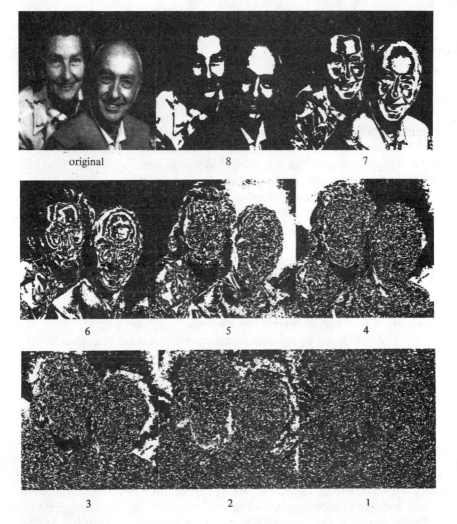

original   8   7

6   5   4

3   2   1

Fig. 8.23

developed for binary images. These methods are too specialized to be described in this text. To give an idea, one of the simplest is briefly presented here. The bit planes have the same dimensions $N \times N$ as the digital image. Each bit plane can be considered as formed by the juxtaposition of $M \times K$ blocks (with M and $K \leq N$). Each block thus has MK bits. To obtain a reduction, the most frequent block configurations must be represented with the smallest number of bits, while guaranteeing unique decoding. The statistical analysis of the block configurations in the bit planes of several images show that two particular configurations are the most frequent. They are the blocks formed only with zero bits (represented by black in Fig. 8.23), and blocks formed only with one bits (represented by white in Fig. 8.23). The most frequent configuration (blocks formed by zero bits) is represented simply by 0. For these blocks, $M \times K - 1$ bits are saved. The complementary bit 1 followed by a second prefix bit permits us to identify the second most likely configuration represented by 11 and all the other configurations. These are represented by the MK bits of the block after the prefix 10. $M \times K - 2$ bits are saved for the blocks formed by 1, but two additional bits are used for the less likely blocks. In general, a reduction up to a factor of three can be obtained with this method.

As previously described, the block coding method is exact. The original digital image can be entirely and exactly reconstructed. To increase the reduction, this method can be transformed into a psychovisual one. To this end, the number of blocks formed only with 0 and only with 1 must be increased. This is achieved by considering, for example, all $M \times K$ blocks having only D one bits as blocks formed only with 0. A similar operation is performed for blocks having only D zero bits.

The parameter D is called the *distortion factor*. The case $D = 0$ corresponds to the reduction without distortion. For the values of D up to three, with $M = K = 4$, the distortions are practically invisible. They remain tolerable for $D = 4$, but become obvious for $D = 5$ or 6. In these last cases, they can be strongly attenuated by low-pass filtering. The reduction obtained is then about five.

8.6.7 Run-length coding

When a row of the matrix representing a digital image is considered, several successive samples on this row can have the same value. The set of such samples is called a *run*. It is totally characterized by three parameters: the starting address, its spread, and its value. The binary representation of these parameters leads to an average reduction of about three times. This method is applied very efficiently to binary images, for which the value of the run need not be specified. It is determined by an initial condition and the color alternance from one run to the next. For binary images, the reduction factor can easily reach 10.

8.6.8 Roberts' method

A possibility for reducing the number of bits per sample in the canonical form of a digital image is to use a less precise quantization. This has the drawback, as seen in section 8.5.8, of producing artificial edges. They come from the quantization noise, which is correlated with the samples. Roberts [47] proposed adding a pseudorandom noise (sect. 1.2.8) to the samples before the quantization. This noise is uniform and has a dynamic range equal to the elementary quantization interval. It is, moreover, independent of the samples. An identical noise is subtracted in the reconstruction. The quantization of the modified samples no longer produces artificial edges. With this method, it is possible to obtain images of acceptable quality with only three bits per sample.

8.6.9 Transform methods

Transformation methods for redundancy reduction act on the digital image not in the spatial domain but in the transformed domain on a transform (linear, or otherwise) of the original image. In general, linear transforms are preferred over nonlinear ones because they are well suited for analytical studies. The set of nonlinear transforms presented in the previous chapter offers no significant advantage for the reduction of redundancy.

Starting from a set of correlated samples of an image $x(k, l)$, the idea is to obtain a set of other samples correlated as little as possible. The linear transformation, which leads to a set of uncorrelated coefficients is the Karhunen-Loeve transform (sect. 4.7.11). It is the limit of what a linear transform can give. It is rarely used in practice because no fast algorithm is available to compute it efficiently. This ideal case is thus approximated by other linear transforms, which have fast computation algorithms, as described in chapter 4.

The general formalism of a linear transform for two dimensional signals is the following (sect. 4.2.11):

$$X(m,n) = \sum_{k=0}^{N-1} \sum_{l=0}^{N-1} a(m,n,k,l)\; x(k,l) \tag{8.82}$$
$$\text{with } m, n = 0, ..., N-1$$

In this expression, $a(m, n, k, l)$ is the general element of an array $A[m, n, k, l]$, called the *kernel*. Almost all two-dimensional transforms have symmetric and separable kernels (sect. 4.2.12).

$$A[m,n,k,l] = A'[m,k]\, A'[n,l] \tag{8.83}$$

Among the most frequently used transforms are the Fourier, Hadamard, and Haar transforms.

After computing the transform (8.82), the coefficients must be quantized hoping that they require less bits per coefficient than the canonical form of the original image $x(k, l)$. The reduction which can be obtained depends on the particular image used. It is not very large if other techniques are not used.

A first possiblity consists of neglecting the least important coefficients, while guaranteeing an acceptable reconstructed image by inverse transform. If the coefficients whose magnitude is less than a given threshold are eliminated, then their addresses, or those of the remaining coefficients, must also be indicated. Run length coding can be used for this purpose. In some cases, depending on the type of image and transformation, the coefficients which can be neglected are in well defined regions of the transform domain. It is then possible to avoid indicating the addresses of the remaining or eliminated coefficients.

Another idea, very interesting economically, consists of transforming adjacent blocks of dimension $N \times M$ (with $M \ll N$) instead of the entire $N \times N$ image. In comparison with the previous case, for an equal number of remaining coefficients, the quality of the reconstructed image is in general better. For equal quality, more coefficients can be eliminated. Another advantage is the savings in memory size and computation times.

Other ideas, more or less complex, can be used in quantizing the remaining coefficients, such as using optimized nonlinear rules, adaptive, or mixed methods.

Despite the simplicity offered by linear transformations, there is still no theory which justifies these ideas and predicts the performances. The proper choice of the parameters, methods, and ideas require common sense and imagination. The best results obtained with transform methods are reductions of a factor about six or seven.

8.7 IMAGE RESTORATION AND ENHANCEMENT

8.7.1 Definitions

The quality of an image can be distorted when its support is changed (reproduction, transmission, storage, redundancy reduction, *et cetera*.) An image can also be of low quality due to poor focusing during its acquisition. The set of methods developed to compensate the known or estimated distortions to restore the initial quality is called *image restoration*.

The set of methods which modify the appearance of an image so that an observer or a machine can extract desired information more easily is called *image enhancement*.

Unfortunately, the notions of restoration and enhancement are often confused in the literature.

8.7.2 Comments

To judge the modifications, or "improvements," brought to an image, a quality criterion must be used. However, the quality of an image depends mostly on its goal. For example, the quality of a television image cannot be judged in the same way as the quality of a radiograph. Despite current efforts, there is still no criterion which is flexible, simple to implement mathematically, and satisfactory for the subjective judgement of the eye.

Many of the methods for image restoration have been developed in terms of an objective quality criterion, such as the mean square, absolute mean, or maximum absolute errors. Unfortunately, these criteria do not follow the subjective judgement of the eye. In each specific case, an appropriate criterion must be chosen to meet the fixed goal.

In the previous chapter, homomorphic methods for image restoration and enhancement were presented. They will not be reconsidered in this section.

8.7.3 General problem of image restoration

The detailed analysis of commonly encountered distorted images shows that the most frequent distortions can be represented by the model of a linear translation invariant system. Thus, the distorted image $y(k, l)$ is represented by:

$$y(k,l) = \sum_{k'=-\infty}^{+\infty} \sum_{l'=-\infty}^{+\infty} g(k',l') \, x(k-k',l-l') + b(k,l) \qquad (8.84)$$

where $x(k, l)$ is the ideal image, $g(k, l)$ is the impulse response of the distortion, and $b(k, l)$ is an additive noise characteristic of the distortion system. The block diagram of the model (8.84) is shown in Fig. 8.24.

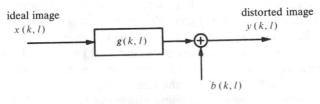

Fig. 8.24

More complex models than (8.84) can be established and studied by introducing, for example, an additional multiplicative noise. In photography, such noise characterizes the film grain and can be considered as an additive noise using the multiplicative homomorphic model.

The problem of image restoration is the following: given the distorted image $y(k, l)$ and possibly some prior information, how can the original image $x(k, l)$ be obtained, or approximated, as well as possible? Remember that the homomorphic restoration method does not require prior information about $g(k,$

l) (sect. 7.7.13). For other methods, $g(k, l)$ can be determined, either from the physical phenomenon which caused the distortion, or from the distorted image itself. In this latter case, it must be known beforehand that some regions of the distorted image correspond to a point, a straight line, or to some other simple shape in the ideal image.

8.7.4 Restoration by inverse filtering

Let us first consider the convolutional model (8.84) without additive noise, that is, with $b(k, l) = 0$. In this case, the two-dimensional Fourier transform leads to

$$Y(f,g) = G(f,g) \, X(f,g) \tag{8.85}$$

In the discrete case, we have

$$Y(m,n) = G(m,n) \, X(m,n) \tag{8.86}$$

If $y(k, l)$ and $g(k, l)$ are known, then, $Y(m, n)$ and $G(m, n)$ can be computed by using the FFT. Thus,

$$X(m,n) = Y(m,n)/G(m,n) \tag{8.87}$$

This is the classical problem of deconvolution by inverse filtering. The distorted image is filtered by a filter whose frequency response is $1/G(m, n)$. The inverse transform of (8.87) gives the desired image. However, some problems arise when (8.87) is used. In the frequency domain, $G(m, n)$ can be equal to zero. Also, for the same values of m and n, $Y(m, n)$ and $G(m, n)$ can be simultaneously zero, leading to an indeterminate result. This shows that despite the simplicity of (8.85) and (8.87), the ideal image $x(k, l)$ can be obtained only in an approximative manner, by using gimmicks to avoid discontinuities and indeterminancies. In the presence of noise, the Fourier transform leads to

$$X(m,n) = [Y(m,n) - B(m,n)]/G(m,n) \tag{8.88}$$

For small values of $G(m, n)$, the ratio $B(m, n)/G(m, n)$ can be large, and thus significantly influence the transform $X(m, n)$.

In practice, (8.88) is evaluated only for frequencies where the signal-to-noise ratio is high. Thus, the frequency response of the inverse filter is not $1/G(m, n)$, but another function of m and n. In general, the response $G(m, n)$

is low-frequency. To avoid the influence of noise in the higher frequencies, a restoration filter with the following response is often used:

$$
G'(m,n) = \begin{cases} 1/G(m,n) & \text{for } m^2 + n^2 \leqslant o^2 \\ 1 & \text{for } m^2 + n^2 > o^2 \end{cases}
\tag{8.89}
$$

where o is a radial frequency beyond which the noise is dominant.

8.7.5 Restoration by least squares

The inverse filtering presented above creates some practical problems, overcome by more or less arbitrary gimmicks. Another way of solving the restoration problem consists of seeking an estimation $\hat{x}(k, l)$ of the ideal image $x(k, l)$ to minimize a difference measurement. A measure which is mathematically very good, but unsatisfactory for the subjective judgement of the eye is the mean square error. The filter that minimizes the mean square error:

$$
\epsilon^2 = E\left[(x(k,l) - \hat{x}(k,l))^2\right]
\tag{8.90}
$$

in the statistical sense, is called the Wiener filter [48]. The restoration method that uses the Wiener filter is called the least-squares restoration.

The frequency response of the Wiener filter will not be established here, due to its very specialized nature. A detailed description can be found in reference [48]. This response has the following form:

$$
G_w(m,n) = \frac{G^*(m,n)\,\Phi_x(m,n)}{|G(m,n)|^2\,\Phi_x(m,n) + \Phi_B(m,n)}
\tag{8.91}
$$

where $(\Phi_x(m, n)$ and $(\Phi_B(m, n)$ are the spectral power densities of the ideal image and of the noise, respectively. Equation (8.91) is established with the hypothesis of statistical independence between the noise and the ideal image. Without noise $(\Phi_B(m, n) = 0)$ the Wiener filter is identical to the inverse filter.

The prior knowledge required by this method is the impulse response of the distortion $g(k, l)$ and the spectral power densities (or, equivalently, the autocorrelation functions) of the noise and the ideal image.

8.7.6 Restoration by least squares under constraints

It is possible to eliminate the requirement of *a priori* knowledge of the spectral densities or correlation functions of the previous method by developing a method of constrained optimization. It is possible to set as a constraint that the residual signal (see eq. (8.84)):

$$z(k,l) = y(k,l) - \sum \sum g(k',l') \, x(k-k',l-l') \tag{8.92}$$

have the same second moment as the noise $b(k, l)$:

$$\sum_k \sum_l z^2(k,l) = \sum_k \sum_l b^2(k,l) = e \tag{8.93}$$

Because of inherent correlation, the function $x(k, l)$ does not vary significantly from one point to another. It is thus possible to minimize a measurement of slow variation related to the second derivative of the form:

$$\sum_k \sum_l [x(k-1,l) + x(k,l-1) + x(k+1,l) + x(k,l+1) - 4\,x(k,l)]^2 \tag{8.94}$$

The restoration problem is then to find a signal $x(k, l)$, which minimize the quantity (8.94), while satisfying constraint (8.93). The resolution of this problem is relatively long and complex. It will be omitted in this text. The frequency response of the restoration filter is of the form [49]:

$$G_H(m,n) = \frac{G^*(m,n)}{|G(m,n)|^2 + \lambda\,|C(m,n)|^2} \tag{8.95}$$

where $C(m, n)$ is the Fourier transform of the extended Laplacian operator derived from (8.94) and λ is a parameter obtained by iteration to satisfy (8.93).

8.7.7 Enhancement by modification of the grey-level scale

In reproducing a monochromatic image, the grey-level scale must usually be preserved. Figure 8.25 shows an ideal input-output relationship. Due to

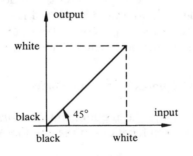

Fig. 8.25

technical imperfections in the optical device or the sensitive surface, the ideal rule may not be respected in some areas of the image plane. In such a case, the luminances are weighted differently, according to the position of the corresponding sample. An image of the following type is then obtained:

$$y(k,l) = p(k,l) \; x(k,l) \tag{8.96}$$

where $x(k, l)$ is the ideal image and $p(k, l)$ is the weight function. This can be determined by calibrating the system with a uniform light.

The ideal image is thus simply

$$x(k,l) = y(k,l)/p(k,l) \tag{8.97}$$

Another modification of the grey scale consists of changing the input-output equation in the same way for each sample in order to improve the contrast. Increasing of the contrast can often bring out details of the image, thus accelerating the observation and the interpretation. The general problem in this case is the following. We have an image $x(k, l)$ which has a dynamic range within the interval $[x_1, x_2]$. An image $y(k, l)$ is desired with a dynamic range within the interval $[y_1, y_2]$ by modifying the grey level distribution. Thus, it is a one-to-one mapping from $[x_1, x_2]$ onto $[y_1, y_2]$ (Fig. 8.26). In principle, any monotonic function $y = f(x)$ can be used to emphasize or attenuate some parts of the interval $[x_1, x_2]$.

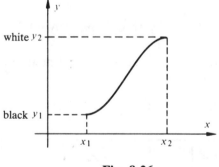

Fig. 8.26

The increase of the contrast is obtained in general with s-shaped curves $y = f(x)$. Such a curve compresses wide regions toward black or white.

8.7.8 Enhancement by histogram modification

Let us assume that $x(k, l)$ is an image in its canonical form, that is, a set of $N \times N$ digital values, each quantized with $Q = 2^{Bq}$ samples. Let p_i be the number of samples having the grey level value $i(i = 1, ..., Q)$. A very useful function in image processing is the representation of the p_i as a function of i. This function is called the *histogram* of the image. To decrease the quantization error, to compare two images obtained with different light conditions, or to measure certain properties of an image, the corrresponding histogram is frequently modified. In general, the histogram is equalized to give all grey levels an equal weight. Consequently, this increases the contrast. Indeed, by equalizing the histogram, the samples of the very dense levels are forced to occupy other, less dense grey levels. Thus, the grey-level scale is widened in these regions. This enlargement is balanced by a compression in the less dense regions.

The equalization of a histogram is equivalent to establishing an equation of the form $y = f(x)$, where y represents the modified image $y(k, l)$. Therefore, the discrete version of the equivalence condition in probability must be considered. This condition is given by (sect. 1.2.11 and vol. 15 chap. VI.):

$$p(y)\, dy = p(x)\, dx \qquad (8.98)$$

Let $\Delta^i x$, $\Delta^i y$ be the intervals corresponding to the level i and p_{ix}, p_{iy} the number of samples in these intervals. The discrete version of (8.98) is then

$$p_{iy}\, \Delta^i y = p_{ix}\, \Delta^i x \qquad (8.99)$$

Because p_{iy} must be constant for any i, the distribution of the levels along y is given simply by

$$\Delta^i y = p_{ix}\, \Delta^i x / C \qquad \text{with } C = p_{iy} = N^2/Q \qquad (8.100)$$

An example of histogram equalization is illustrated in Fig. 8.27.

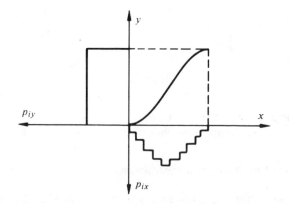

Fig. 8.27

8.7.9 Unsharp masking

The enhancement by unsharp masking consists simply of adding to the original image the difference between this image and a low-pass filtered version of it. At the origin, it is a photographic process in which a mask obtained by the superposition of the film and its negative, a little out of focus, is superimposed onto the original film. The effect of this operation is to enhance the high frequencies, that is, to increase the contrast.

To simplify the representation, this method is illustrated in the one-dimensional case by considering an image of the type $x(k, l) = \text{rect}_N(k)$ (Fig. 8.28). The slightly low-pass version $x'(k, l)$ of this signal is shown in the same

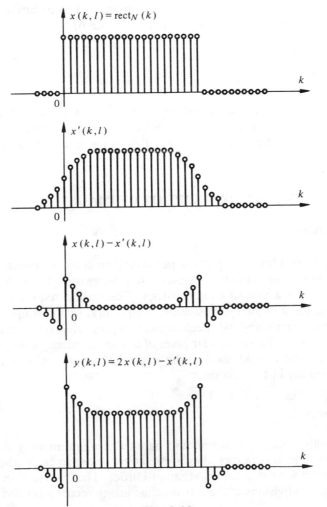

Fig. 8.28

figure. By adding the difference of these two signals to $x(k, l)$, the result $y(k, l)$ is obtained.

8.7.10 Enhancement by filtering

In general, the contrast of an image can be increased by linear high-pass filtering. The unsharp masking method can be considered as a particular case. Among the most frequently used two-dimensional filters is the digital Laplacian filter.

The enhanced image is obtained by the difference between the original image and its Laplacian. The advantage of the Laplacian is it is simple to implement. It suffices to perform the two-dimensional convolution product between the image and the following matrix (sect. 8.8.6):

$$\nabla^2 = \begin{pmatrix} 0 & 1 & 0 \\ 1 & -4 & 1 \\ 0 & 1 & 0 \end{pmatrix} \tag{8.101}$$

In the case of images altered by high-frequency noise, a low-pass filter may be used instead of a high-pass filter to enhance the image. Even if the contrast is decreased by this operation, the image is enhanced by the noise attenuation.

8.8 RECOGNITION AND DESCRIPTION OF IMAGES

8.8.1 Definitions

Image recognition is a process performed to classify images or to give a structural description of them. Classification requires the measurement of properties and the extraction of characteristics. Description requires, in addition, the segmentation and the establishment of structural relationships. The extraction of characteristics and the measurement of properties on an image created for classification are two particular cases of a domain called *pattern recognition*. This domain, which could itself be the subject of several volumes, will not be studied separately in this section.

8.8.2 Remarks

Originally, image processing and pattern recognition were developed as two different domains. In fact, they are relatively close to each other and there is a non negligible intersection along their border. This intersection is observed in two major subdivisions of these two fields: image recognition and description for one, and two-dimensional pattern recognition for the other. A given number of preprocessing, description, and classification methods are common to both disciplines.

In the two last sections, the processing methods act on one image to produce another (restored, enhanced, or compressed). The methods examined in this section act on images to produce a description instead of the images.

8.8.3 Examples

Image recognition and description are involved, for example, in the following practical applications:

- by processing an image representing mechanical parts, their description (shape, position, *et cetera.*) is given to a robot, which can pick them up according to a given program:
- in high-energy physics, the bubble chamber images are processed to detect and localize particular collisions between elementary particles;
- the aerial images of a region are analyzed to identify and locate forest, fields, rivers, *et cetera;*
- microscope photographs of blood cells are processed to classify chromosomes according to a given order;
- manuscripts or edited texts are analyzed to recognize the characters.

8.8.4 Matching

To detect the presence of a specific pattern in an image, a standardized version of the pattern is matched to the image. Thus, the regions of the image where the samples have a given grey-level structure must be found. Unfortunately, in practice, this structure is not identical to the structure of the desired pattern due to noise and grey scale modifications. We must thus maximize a measurement of the similarity. A measurement which can be used is the two-dimensional crosscorrelation. If $x(k, l)$ and $y(k, l)$ respectively represent the image and the desired pattern, the pair (k, l), for which the function:

$$\overset{\circ}{\varphi}_{xy}(k,l) = \sum_{k'} \sum_{l'} x(k',l') \, y(k'+k,l'+l) \tag{8.102}$$

is maximal, is determined. This function can be computed efficiently by using the indirect method and the FFT. However, we must remember that the measure (8.102) is invariant by translation, but depends on the relative orientation of $x(k, l)$ with respect to $y(k, l)$. Consequently, it will enable us to detect only the patterns in $x(k, l)$, which have the same orientation as $y(k, l)$. To be independent of the orientation, the computation of the function $\overset{\circ}{\varphi}_{xy}(k, l)$ must be performed for all possible orientations of $y(k, l)$. This, then, becomes a very costly operation with respect to the number of elementary operations to perform. This difficulty can be partially avoided by performing the crosscorrelation in two steps. In a first step, the function (8.102) is computed with an initial resolution by taking a small number of samples for example on $y(k, l)$. The initial resolution is used in the second step only at the points where the previous crosscorrelation is above a given threshold.

In some simple cases where the images have two levels and the patterns can be characterized by contours, some characteristics that are invariant by translation, rotation, or even by zooming can be measured. For this, we can use a set of coefficients deduced from the Fourier series expansion of the complex function which represents the contour in the image plane. One of the axes of this plane is taken as the imaginary axis.

8.8.5 Segmentation

Images are composed of regions having different local properties. These properties can be the grey-level distribution, the energy content in a frequency band, the value of a coefficient $X(m, n)$, et cetera. By grouping image points which have a given common property, uniform regions are obtained. This operation is called *segmentation*. The appropriate method for segmentation depends on the property used for the determination of the regions. For example, if all points having grey levels within a given interval must be detected, two thresholds are used to eliminate the other points.

The use of thresholds is very common for the segmentation of simple images. The detection of dark objects on light backgrounds is a typical example. In these simple cases, the choice of the threshold is often suggested by examining the grey level histogram which has peaks corresponding to the light and dark regions of the image. A sensible choice of the threshold is the grey level corresponding to the local minimum between two peaks of the histogram. It is possible that a single threshold is not satisfactory for all regions of the image. In this case, adaptive thresholds based on local histogram can be considered. It is also possible to perform special preprocessing, such as filtering or grey-level modification, to improve the results of thresholding. Another segmentation method can be based on the discontinuities of the grey levels. These indicate in general the border between the regions of an image. In image processing, a discontinuity or sharp variation of the grey level is called an edge. The segmentation is then performed by edge extraction. In fact, the human visual system works in the same way (see the low-frequency part of Fig. 8.8). In principle, it is a high-pass filter performed in general by a linear filter or a nonlinear operator. The most commonly used operators are the Laplacian and the gradient. Other operators obtained by generalizing the gradient can also be used.

These two operations (thresholding and edge detection) can be considered as parallel methods. They are applied to all points of the image without any influence of an intermediate result on the next computation. If a parallel processing system is available, these operations can be performed simultaneously on all the points of the image. Another method which can be used for edge detection is *edge tracking*. It is essentially a sequential method. The computations are performed only on certain points of the image. In this method, the edges are tracked according to the indications of a local high-pass operator. At each point, the neighboring point where the grey-level variations are the largest is sought. In picturesque terms, the "mountains tops" are followed (ridge riding).

8.8.6 Properties of regions

A property of a region is a function which maps this region of the image into numbers. The regions obtained by segmentation must be described by their properties. Several types of properties can be distinguished. Geometrical properties involve functions such as spread, surface, curvature, convexity, angle, *et cetera*. Physical properties involve functions such as the grey level at a point, the average grey level over a zone, the total energy or partial energy contained in a frequency band, *et cetera*. A region can also be described by its textural properties. These are represented by statistical measurements such as mean value, variance, and joint probability of the grey levels. The texture can be considered as a randomly disturbed periodic shape.

To determine the property of a region, first the property and the region must be defined. Then, this property must be expressed in an analytical form compatible with a digital processing. For example, if an edge is to be detected using the Laplacian:

$$\nabla^2 = \frac{\partial^2 f(u,v)}{\partial u^2} + \frac{\partial^2 f(u,v)}{\partial v^2} \tag{8.103}$$

then the derivatives must be replaced by differences. In this case, we have

$$\frac{\partial f(u,v)}{\partial u} \longrightarrow x(k+1,l) - x(k,l)$$

$$\frac{\partial^2 f(u,v)}{\partial u^2} \longrightarrow x(k+1,l) - 2x(k,l) + x(k-1,l) \tag{8.104}$$

The discrete Laplacian is thus given by

$$\nabla^2 = x(k+1,l) - 2x(k,l) + x(k-1,l) + x(k,l+1) - 2x(k,l) + x(k,l-1) \tag{8.105}$$

In matrix form, the corresponding operator is, therefore,

$$\nabla^2 = \begin{pmatrix} 0 & 1 & 0 \\ 1 & -4 & 1 \\ 0 & 1 & 0 \end{pmatrix} \tag{8.106}$$

It can then be implemented using the well known relation (8.38).

8.8.7 Structural relationships

The description of an image can be completed by establishing structural relationships between its different regions and their properties. These relationships, in general, not obvious and difficult to establish, can be performed at several levels. For this reason, it is preferable to organize them in a hierarchy. One possible decomposition is the following. Assume that an image represents objects on a background. The objects are formed by regions bounded by edges. The edges are formed by the samples. At each of these levels, structural relationships can be established, thus building a description "language." For example, if we consider the image of a cube, illuminated on one side, lying on a table, we can define the following relationships:

- the cube is on the table;
- the left side of the cube is darker than the front one;
- the top of the cube is lighter than its left side, *et cetera*.

8.8.8 Application example

To illustrate some notions introduced in this section, let us consider a concrete application: the development of visual perception for an industrial robot. The simplest case is the determination of the position and orientation of a known, flat object [50]. An appropriate sensor (for example a camera followed by a digitizer) can provide the image of this object on the working plane in a canonical form. Flat objects can be represented by their contours. By applying an edge detector to the original image, a two level image representing the edges of the corresponding object is obtained (Fig 8.29). The operator used in this case is the gradient (in modulus). The localization of this object can be characterized by two properties: the coordinates in the working plane of a central point and the rotation angle about this point. The obvious central point is here the center of gravity of the contour. Its coordinates are obtained by computing the mean values of the coordinates of the contour points. Then, the contour is scanned radially by the center of gravity with constant angular steps. A digital signal $x(k)$ is formed by measuring the distance between the center of gravity

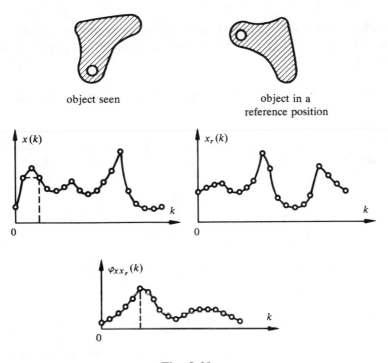

Fig. 8.29

and the contour for each step k. For N angular steps, N samples of the signal are thus obtained. Finally, the rotation angle is obtained by comparing this signal along a circle with the corresponding signal obtained from the same object in a reference position. The angle for which these two signals are superimposed is the desired rotation angle.

This circular comparison is equivalent to computing the one-dimensional crosscorrelation function of the previous signals, considered as periodic of period 2π (Fig. 8.29).

This method, which works successfully in practice, is an elegant and simple solution of the general problem studied in section 8.8.4. If several identical pieces are in the visual field, a segmentation must be performed before applying this method. If the pieces are not identical, geometrical properties such as contour spread, surface, or curvature, can be used to identify them.

8.9 EXERCISES

8.9.1 By using the orthogonality of complex exponentials, prove the inversion formula (8.12).

8.9.2 Show that two-dimensional crosscorrelation is not commutative. Express $\mathring{\phi}_{xy}(k, l)$ in terms of $\mathring{\phi}_{yx}(k, l)$. What does this equation represent in the (k, l) plane?

8.9.3 Prove (8.39). Is this result sufficient to affirm the commutativity of two-dimensional convolution?

8.9.4 By generalizing theorem 1.6.20 in two dimensions, show that (8.40) is a necessary and sufficient condition for a two-dimensional invariant linear system to be stable.

8.9.5 Compute the Fourier transform of the following signal:

$$x(k, l) = a^{|k|} b^{|l|} \quad \text{with} \quad |a| \text{ et } |b| < 1$$

8.9.6 Show that if a two-dimensional signal has circular symmetry, then its Fourier transform does as well:

8.9.7 What is the z-transform of the two-dimensional unit impulse?

8.9.8 Establish the expression of the two-dimensional shifting theorem by looking for the relationship between the z-transform of a signal $x(k, l)$ and its translated version $x(k - k_0, l - l_0)$. What does this relationship become in the case of the Fourier transform?

8.9.9 Let $x(k, l)$ be a signal representing an image. What is the effect of the linear system characterized by

$$y(k, l) = \frac{1}{MN} \sum_{m=0}^{M-1} \sum_{n=0}^{N-1} x(k - m, l - n)$$

on the image, where $y(k, l)$ is the output signal?

8.9.10 A film has a dynamic range of $D = 3.6$ optical density units. Of how many bits must the luminance of the samples of an image quantized on this device be composed so that the quantification error remains invisible? Use the two extreme values of the constant C_w.

8.9.11 An image is quantized with 12 bits. What is, in optical density units, the dynamic range of the support to be used so that the quantization error is invisible?

8.9.12 A monochromatic image is linearly quantized from zero (white) to 255 (black) with 256 levels. This image has to be requantized using a logarithmic rule of the type

$$y = \frac{255 \log_a (x + 1)}{\log_a (256)}$$

where x and y are the luminances before and after the second quantization, respectively. The luminance y can be represented with only eight bits. How many different grey levels are obtained in the logarithmic representation? Choose between 256, < 256, > 256.

CHAPTER 1

1.8.1 The signal $z(k)$ has a Gaussian probability density with a mean value of zero and a variance equal to 1.

1.8.2

$$x(k+1) = A + x(k) \qquad \text{with} \qquad x(0) = B = \text{cste}$$
$$y(k) = x(k) \bmod C \qquad \text{where } C \text{ is a constant}$$

$y(k)$ is a saw-tooth signal.

triangular signal $z(k) = \left| y(k) - \dfrac{C}{2} \right|$

1.8.5 The length is $M + N - 1$

$$y(k) = \begin{cases} N\left(1 - \dfrac{|k-N+1|}{N}\right) & \text{for } 0 < k < 2N-2 \\ 0 & \text{otherwise} \end{cases}$$

1.8.6

$$X(f) = \frac{1}{1 - a \exp(-j2\pi f)} \qquad |a| < 1$$

1.8.10 The Fourier transform is the same, but the main interval is K times longer. After filtering the initial signal is re-obtained with $K - 1$ samples interpolated between each initial sample.

1.8.11

$$G(f) = \frac{1}{M} \frac{\sin \pi f M}{\sin \pi f} \exp[j\pi f(1 - M)]$$

1.8.12

$$G(f) = \frac{1}{1 - a \exp(-j2\pi f)}$$

1.8.13

$$G(f) = \frac{1 - b \cos 2\pi f + jb \sin 2\pi f}{1 - a \cos 2\pi f + ja \sin 2\pi f}$$

$$|G(f)| = \text{cste} \quad \text{if} \quad b = 1/a$$

1.8.14 $y(k) = x(k) \exp(j2\pi f_0 k)$

1.8.15 See equation (5.49) and figure 5.12.

CHAPTER 2

2.8.1

- $X(z) = \dfrac{1}{1 - (1/2) z^{-1}}$ with $|z| > 1/2$
- $X(z) = 1$
- $X(z) = 1/z$ $\qquad\qquad |z| \neq 0$
- $X(z) = 1/z^{k_0}$ $\qquad\quad \begin{cases} z \neq 0, & k_0 > 0 \\ z \neq \infty, & k_0 < 0 \end{cases}$
- $X(z) = \dfrac{1}{1 - 2z}$ $\qquad\quad |z| < 1/2$
- $X(z) = \dfrac{2z}{1 - 2z}$ $\qquad\quad |z| < 1/2$

2.8.2

$$X(z) = \frac{1 - z^{-k}}{1 - z^{-1}}$$

2.8.3 $x(k) = (2^{k+1} - 1)\, \epsilon(k)$

2.8.4 2^K

2.8.5 For stability we must have $|a| < 1$, $|G(f)| = 1$.

2.8.7

$$G(z) = \frac{1/2}{(z^{-1}-1)(z^{-1}-1/2)}$$

The system is causal if $|z| > 2$, but it is unstable. $g(k) = (2^{k+1} - 1)\,\epsilon(k)$.

2.8.8

$$Y(z) = z^2\,\frac{d^2 X(z)}{dz^2} + z\,\frac{dX(z)}{dz}$$

2.8.11

$$G(z) = (3/8)\left[\frac{1}{1-3z^{-1}} - \frac{1}{1-(1/3)z^{-1}}\right]$$

The system is causal if $|z| > 3$, but it is not stable.

$$g(k) = (3/8)\,[3^k - (1/3)^k]\,\epsilon(k).$$

CHAPTER 3

3.9.1 There is ambiguity for $k_0 > N$. In this case we must use $k_0' = k_0 \bmod N$.

3.9.3 If N is even, there are 2 real coefficients. If N is odd, there is 1 real coefficient. These results are not valid for purely imaginary coefficients.

3.9.4

$$X_{2p}(n) = \begin{cases} 2X_{1p}\left(\dfrac{n}{2}\right) & \text{for } n \text{ even} \\ 0 & \text{for } n \text{ odd} \end{cases}$$

3.9.5 $10^4/1024$ Hz $\cong 10$ Hz

3.9.7

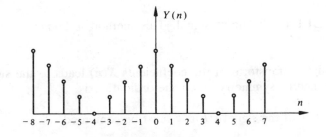

Fig. 3.24

3.9.8 12 samples from $k = 8$.

3.9.9

$$y(k) = \left[\sum_{i=-\infty}^{+\infty} (1/2)^{k+iN} \, \epsilon(k+iN) \right] \text{rect}_N(k)$$

3.9.10 Hanning with $N = 400$.

3.9.11 $y(k) = x(-k)$

3.9.13

$$X(n) = \frac{1}{2} \left[\frac{\sin \pi N \left(\dfrac{n}{N} - f_0 \right)}{\sin \pi \left(\dfrac{n}{N} - f_0 \right)} + \frac{\sin \pi N \left(\dfrac{n}{N} + f_0 \right)}{\sin \pi \left(\dfrac{n}{N} + f_0 \right)} \right]$$

Since the frequency f_0 is not an integer multiple of $1/N$, the result is subject to the distribution phenomenon.

CHAPTER 4

4.8.1

$$Y_n(m) = \sum_{k=0}^{N-1} x(k) \, W_N^{n(m-k)}$$

$$X(n) = Y_n(m) \big|_{m=N}$$

4.8.2

$$X(n) = \sum_{k=0}^{N-1} W_N^{nk} \, x(N-k)$$

For the flow chart, see section 4.8.3.

4.8.3 The DFT of the input signal. This method is known as the Goertzel algorithm.

4.8.4 The direct transform of the coefficients $X(n)$ leads to the signal $x(-k)$. $x(k)$ is obtained by symmetry about the ordinate axis.

4.8.5 The DFT of the signal $x(k)\rho^{-k}$ must be computed.

4.8.6 The FFT is more efficient.

4.8.7 Yes, the butterfly of the first stage must be preserved during all computations. Thus we have only ± 1 as a multiplicative coefficient.

4.8.8

$$X(2r) = \sum_{k=0}^{N/2-1} [x(k) + x(k+N/2)] \, W_N^{-2rk}$$

$$X(2r+1) = \sum_{k=0}^{N/2-1} [x(k) - x(k+N/2)] \, W_N^{-k} \, W_N^{-2rk}$$

$$r = 0, ..., N/2 - 1$$

4.8.9

$$X(z_R) = B^{-R^2/2} \sum_{k=0}^{N-1} g(k) \, B^{(r-k)^2/2}$$

$$= g(k) * B^{k^2/2}$$

4.8.10 See exercise 1.8.10.

CHAPTER 5

5.6.2

$$|G(f)| = \sqrt{2 - 2\cos(2\pi fL)}$$

$$\theta_g(f) = \arg[G(f)] = \mathrm{arctg}\left(\frac{\sin(2\pi fL)}{1 - \cos(2\pi fL)}\right)$$

5.6.3

$$G(z) = \frac{1 + z^{-1} + z^{-2}}{3 \, z^{-m}}$$

$$g(k) = \frac{1}{3} \, \mathrm{rect}_3(k+m)$$

The system is causal if $m \leq 0$.

The system is always stable. From (1.103) we have:

$$T = \sum_{K=-\infty}^{+\infty} |g(k)| = \sum_{K=m}^{m+2} \left|\frac{1}{3}\right| = 1 < \infty$$

Si $m = 0$ $\qquad G(z) = \dfrac{1 + z^{-1} + z^{-2}}{3}$

$$G(f) = \frac{1 + e^{-j2\pi f} + e^{-j4\pi f}}{3}$$

Si $m = 1$ $\qquad G(z) = \dfrac{z + 1 + z^{-1}}{3}$

$$G(f) = e^{j2\pi f} + 1 + e^{-j2\pi f}$$

The system is a digital averaging filter (low-pass).

5.6.4 z is replaced by z/ρ with $\rho = 1 - 1/2^8$. Equation (5.63) becomes:

$$w(k) = (1/16)\,[x(k) - 0{,}939\,x(k-16)]$$

The difference equations are given by:

$$y_1(k) = -1{,}662\,[w(k) - 0{,}996\,w(k-1)] + 0{,}762\,y_1(k-1) - 0{,}992\,y_1(k-2)$$
$$y_2(k) = 1{,}414\,[w(k) - 0{,}996\,w(k-1)] - 0{,}992\,y_2(k-2)$$
$$y_3(k) = -1{,}112\,[w(k) - 0{,}996\,w(k-1)] - 0{,}762\,y_3(k-1) - 0{,}992\,y_3(k-2)$$

5.6.5

$$y(k) = A\,x(k) + B\,y(k-1) + C\,y(k-2)$$

with

$$A = \frac{1{,}103\,\Delta t^2}{1 + 1{,}098\,\Delta t + 1{,}103\,\Delta t^2}$$

$$B = \frac{2 + 1{,}098\,\Delta t}{1 + 1{,}098\,\Delta t + 1{,}103\,\Delta t^2}$$

$$C = \frac{-1}{1 + 1{,}098\,\Delta t + 1{,}103\,\Delta t^2}$$

for

$$\Delta t = 1, A = 0.345, B = 0.968 \text{ and } C = -0.312$$

for

$$G(f) = \frac{0.345}{1 - 0.968 \exp(-j2\pi f) + 0.312 \exp(-j4\pi f)}$$

5.6.6 For $\Delta t = 1$, we have:

$$y(k) = 0.155\,[x(k) + x(k-2)] + 0.310\,x(k-1) + 0.773\,y(k-1) \\ - 0.394\,y(k-2)$$

and

$$G_1(f) = \frac{1.103 + 2.206 \exp(-j2\pi f) + 1.103 \exp(-j4\pi f)}{7.109 - 5.498 \exp(-j2\pi f) + 2.801 \exp(-j4\pi f)}$$

For $\Delta t = 1/8$, we have:

$$y(k) = 0.004\,[x(k) + x(k-2)] + 0.07\,x(k-1) + 1.864\,y(k-1) \\ - 879\,y(k-2)$$

and

$$G_2(f) = \frac{1.103 + 2.206 \exp(-j2\pi f) + 1.103 \exp(-j4\pi f)}{307.210 - 572.7 \exp(-j2\pi f) + 269.9 \exp(-j4\pi f)}$$

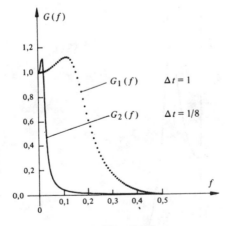

Fig. 5.29

5.6.8

$$G'(z) = \frac{0{,}0272 - 0{,}0462\,z^{-1} + 0{,}0462\,z^{-2} - 0{,}0272\,z^{-3}}{0{,}443 + 0{,}337\,z^{-1} + 0{,}312\,z^{-2} + 0{,}0923\,z^{-3}}$$

It is a high-pass filter.

5.6.9 $g_2(k)$ must be used because the corresponding filter computes weighted differences.

5.6.10 The type of filter is changed by the cyclic shift of its impulse response. Only the phase frequency response varies from one case to the other.

5.6.11 The filter is unstable because its poles are on a circle of radius 2. It is causal because the convergence region of its transfer function is outside a circle. To stabilize this filter, it must be put in series with 4 first order all-pass filters. Thus:

$$G'(z) = \frac{1}{16}\;\frac{1}{(z^2 - 0{,}354\,z + 0{,}25)\,(z^2 + 0{,}354\,z + 0{,}25)}$$

CHAPTER 6

6.8.1

$$\Phi_y(f) = \frac{\sigma_b^{\,2}}{2{,}25 - 3\cos 2\pi f + \cos 4\pi f}$$

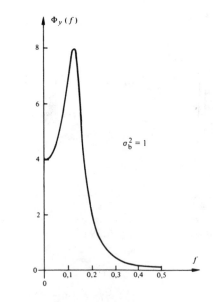

Fig. 6.30

6.8.2

$$\varphi_y(f) = \frac{\sigma_b{}^2}{5(\sqrt{2})^{|k|}} [12 \cos(|k|\pi/4) + 4 \sin(|k|\pi/4)]$$

6.8.3

$$\Phi_y(f) = \frac{\sigma_b{}^2}{4.41 - 5.85 \cos 2\pi f + 1.44 \cos 4\pi f}$$

6.8.4 $\varphi_y(k) = 169.18\sigma^2(0,8)^{|k|} - 79.38\sigma^2(0,9)^{|k|}$

6.8.5 $\text{Var}[\hat{m}_x] = 0.01$, Confidence $= 0.9973$.

6.8.6 $[0.43S_x(f_0), 10S_x(f_0)]$

6.8.8 Among the available windows, the one which has the lowest energy must be chosen.

6.8.9 To see two close peaks, the smoothed estimator must be chosen. For the best dynamic range resolution, the modified estimator must be chosen.

CHAPTER 7

7.8.7 The cepstrum is the sum of minimum phase and maximum phase signals.

7.8.9 The cepstrum must be completely attenuated on the positive half of the k axis and half attenuated at the origin.

7.8.10

$$G(z) = \prod_{n=1}^{p_e} \frac{(1 - \delta_n z)}{(1 - \delta_n^* z^{-1})} \prod_{n=1}^{z_e} \frac{(1 - \beta_n^* z^{-1})}{(1 - \beta_n z)}$$

CHAPTER 8

8.9.5

$$X(f,g) = \frac{(1 - a^2)}{(1 - 2a \cos(2\pi f) + a^2)} \frac{(1 - b^2)}{(1 - 2b \cos(2\pi g) + b^2)}$$

8.9.7 $X(z_1, z_2) = 1$

8.9.8

$$Y(z_1, z_2) = X(z_1, z_2) z_1^{-k_0} z_2^{-l_0}$$
$$Y(f, g) = X(f, g) e^{-j2\pi f k_0} e^{-j2\pi g l_0}$$

8.9.9 The effect of this system is to create a defocusing by low pass filtering with a filter whose magnitude frequency response is given by:

$$|G(f, g)| = \left| \frac{\sin(\pi f M)}{\sin \pi f} \frac{\sin(\pi g N)}{\sin \pi g} \right|$$

8.9.10 9 bits for $C_w = 0.02$, 10 bits for $C_w = 0.01$.

8.9.11 17.7 optical density units for $C_w = 0.01$, 35.2 optical density units for $C_w = 0.02$.

8.9.12 < 256.

Bibliography

[1] B. GOLD and C. RADER, *Digital Processing of Signals*, McGraw-Hill, New York, 1969.

[2] K. STEIGLITZ, *An Introduction to Discrete Systems*, Wiley, New York, 1974.

[3] A.V. OPPENHEIM and R.W. SCHAFER, *Digital Signal Processing*, Prentice-Hall, Englewood Cliffs, NJ, 1975.

[4] L.R. RABINER and B. GOLD, *Theory and Application of Digital Signal Processing*, Prentice-Hall, Englewood Cliffs, NJ, 1975.

[5] M. ABRAMOVITZ and I.A. STEGUN, *Handbook of Mathematical Functions*, National Bureau of Standard (USA), no. 55, 1968.

[6] H. CARTAN, *Théorie élémentaire des fonctions analytique d'une ou plusieurs variables*, Hermann, Paris, 1961.

[7] E.I. JURY, *Theory and application of the z-transform method*, R.E. Krieger Publishing Co., New York, 1973.

[8] S.D. STEARNS, *Digital Signal Analysis*, Hayden, Rochelle Park, NJ, 1975.

[9] J.R. RAGAZZINI and G.F. FRANKLIN, *Les systèmes asservis échantillonnés*, Dunod, Paris, 1962.

[10] T.G. STOCKHAM JR., High Speed Convolution and Correlation, *AFIPS Proceedings — 1966 Spring Joint Computer Conference*, vol. 28, 1966, pp. 229–233.

[11] R.B. BLACKMAN and J.W. TUKEY, *The Measurement of Power Spectra*, Dover, New York, 1958.

[12] J.F. KAISER, Digital Filters, chap. 7, *System Analysis by Digital Computer*, F.F. Kuo and J.F. Kaiser (eds), Wiley, New York, 1966.

[13] A. PAPOULIS, *The Fourier Integral and its applications*, McGraw-Hall, New York, 1962.

[14] J.W. COOLEY and J.W. TUKEY, An Algorithm for the Machine Calculation of Complex Fourier Series, *Math. of Computation*, vol. 19, April 1965, pp. 297–301.

[15] J.J. SYLVESTER, Thoughts on Inverse Orthogonal Matrices, Simultaneous Sign Successions and Terrelated Pavements in Two or More Colors, with Applications to Newton's Rule, Ornamental Tile Work, and the Theory of Numbers, *Philosophical Magazine*, 34, Series 4, 1867, pp. 461–475.

[16] C. RUNGE and H. KÖNIG, Vorlesungen über numerisches Rechnen, *Die Grundlehren der mathematischen Wissenschaften*, 11, Springer, Berlin, 1924.

[17] K. Stumpff, *Tafeln und Aufgaben zur harmonischen Analyse und der Periodogrammrechnung*, Springer, Berlin, 1939.

[18] J.J. Godd, The Interaction Algorithm and Practical Fourier Series, *Journal of the Royal Statistical Society*, vol. 20, series B, 1958, pp. 361–371.

[19] H.C. Andrews and K.L. Caspari, Degrees of Freedom and Modular Structure in Matrix Multiplication, *IEEE Trans. on Computer*, vol. C-20, no. 2, Feb. 1971, pp. 133–141.

[20] C.K. Yuen, Walsh Functions and Gray Code. *IEEE Trans. on Electromagnetic Compatibility*, vol. EMC-13, August 1971, pp. 68–73.

[21] H. Reitboeck and T.P. Brody, A transformation with Invariance under Cyclic Permutation for Applications in Pattern Recognition, *Inform. & Cont.*, vol. 15, 1969, pp. 130–154.

[22] N.T. Fine, The Generalized Walsh Functions, *Trans. of the Amer. Math. Soc.*, vol. 69, 1950, pp. 66–77.

[23] H.C. Andrews, *Computer Techniques in Image Processing*, Academic Press, New York, 1970.

[24] C. Rader, Discrete Fourier Transform when the Number of Data Samples in Prime, *Proc. of the IEEE*, vol. 56, June 1968, pp. 1107–1108.

[25] W.R. Crowther and C. Rader, Efficient Coding of Vocoder Channel Signal using Linear Transformation, *Proc. of the IEEE*, vol. 54, Nov. 1966, pp. 1594–1595.

[26] G.S. Moschytz, Linear Integrated Networks, *Fundamentals and Designs*, (2nd ed.), Van Norstrand Reinhold Co., New.York, 1974.

[27] L. Weinberg, Networks Design by Use of Modern Synthesis Techniques and Tables, *Proc. National Electronic Conference*, Chicago, vol. 12, 1956, pp. 794.

[28] W.H. Beyer, *Handbook of Tables for Probability and Statistics*, The Chemical Rubber Co, Akron, OH, 1966.

[29] M.S. Bartlett, On the Theoretical Specification and Sampling properties of autocorrelated time series, *Journal of the Royal Statistical Society*, vol. B-8, no. 27, 1946.

[30] G.M. Jenkins and D.G. Watts, *Spectral Analysis and its Applications*, Holden-Day, San Francisco, 1969.

[31] P.D. Welch, The Use of Fast Fourier Transform for the Estimation of Power Spectra, *IEEE Trans. Audio Electroacoustics*, vol. AU-15, June 1967, pp. 70–73.

[32] A.V. Oppenheim, Superposition in a Class of Nonlinear Systems, *Techn. Rept. 432*, Research Laboratory of Electronics, MIT, Cambridge, MA, March 1965.

[33] A.V. Oppenheim, Generalized Superposition, *Information and Control*, vol. 11, nos. 5–6, Nov.–Dec. 1967, pp. 528–536.

[34] L. Franks, *Signal Theory*, Prentice-Hall, Englewood Cliffs, NJ, 1969, chap. 2.

[35] E. KREYSZIG, *Advanced Engineering Mathematics,* Wiley, New York, 1968.

[36] B.P. BOGERT, M.J.R. HEALY, and J.W. TUKEY, The Frequency Analysis of Time Series for Echoes: Cepstrum, Pseudo-Autocovariance, Cross-Cepstrum and Saphe Cracking, *Proc. Symp. Time Series Analysis,* M. Rosenblatt (ed.), Wiley, New York, 1963, pp. 209–243.

[37] A.V. OPPENHEIM, R.W. SCHAFER, and T.G. STOCKHAM, Nonlinear Filtering of Multiplied and Convolved Signals, *Proc. IEEE,* vol. 56, no. 8, Aug. 1968, pp. 1264–1291.

[38] T.G. STOCKHAM, JR., Image Processing in the Context of a Visual Model, *Proc. IEEE,* vol. 60, no. 7, July 1972, pp. 828–842.

[39] B. BAXTER, *Image Processing in the Human Visual System,* Ph.D. Thesis, University of Utah, 1975.

[40] T.G. STOCKHAM, T.M. CANNON, and R.B. INGEBRETSEN, Blind Deconvolution through Digital Signal Processing, *Proc. IEEE,* vol. 63, no. 4, April 1975, pp. 678–692.

[41] T.M. CANNON, *Digital Image Deblurring by Nonlinear Homomorphic Filtering,* Ph.D. Thesis, University of Utah, 1974.

[42] H.C. ANDREWS, *Computer Techniques in Image Processing,* Academic Press, New York, 1970.

[43] T.S. HUANG (ed), *Picture Processing and Digital Filtering,* Springer-Verlag, Berlin, 1975.

[44] A. ROSENFELF and A.C. KAK, *Digital Picture Processing,* Academic Press, New York, 1976.

[45] T.N. CONNSWEET, *Visual Perception,* Academic Press, New York, 1970.

[46] M. INGRAM and K. PRESTON JR., Automatic Analysis of Blood Cells, *Scientific American,* pp. 72–82.

[47] L.G. ROBERTS, Picture Coding using Pseudo-random Noise, *IRE Trans. Information Theory,* vol. IT-8, 1962, pp. 145–154.

[48] H.L. VAN TREES, *Detection, Estimation and Modulation Theory,* Part 1, Wiley, New York, 1968.

[49] B.R. HUNT, The Application of Constrained Least Squares Estimation to Image Restoration by Digital Computer, *IEEE Trans. Computers,* vol. C-22, 1973, pp. 805–812.

[50] F. DE COULON and P. KAMMENOS, Polar Coding of Planar Objects in Industrial Robot Vision, *Neue Tecknik,* 1977.

Select Bibliography

The Traité d'Électricité, listed below by volume number, is published by the Presses Polytechniques Romandes (Lausanne, Switzerland) in collaboration with the École Polytechnique Fédérale de Lausanne. The title of each volume is given with the year of publication in parenthesis. English translations by Artech House are denoted by an asterisk with the year of publication in parenthesis.

Vol.	Author	Title
I	Frédéric de Coulon & Marcel Jufer	Introduction à l'électrotechnique (1981).
II	Philippe Robert	Materiaux de l'électrotechnique (1979).
III	Fred Gardiol	Electromagnétisme (1979).
IV	René Boite & Jacques Neirynck	Theorie des reseaux de Kirchhoff (1983).
V	Daniel Mange	Analyse et synthèse des systèmes logiques (1979). *Analysis and Synthesis of Logic Systems (1986).
VI	Frédéric de Coulon	Theorie et traitement des signaux (1984). *Signal Theory and Processing (1986).
VII	Jean-Daniel Chatelain	Dispositifs à semiconducteur (1979).
VIII	Jean-Daniel Chatelain & Roger Dessoulavy	Electronique (1982).
IX	Marcel Jufer	Transducteurs électromécaniques (1979).
X	Jean Chatelain	Machines électriques (1983).
XI	Jacques Zahnd	Machines séquentielles (1980).
XII	Michel Aguet & Jean-Jacques Morf	Energie électrique (1981).
XIII	Fred Gardiol	Hyperfréquences (1981). *Introduction to Microwaves (1984).
XIV	Jean-Daniel Nicoud	Calculatrices (1983).
XV	Hansruedi Bühler	Electronique de puissance (1981).
XVI	Hansruedi Bühler	Electronique de réglage et de commande (1979).
XVII	Philippe Robert	Mesures (1985).
XVIII	Pierre-Gérard Fontolliet	Systèmes de télécommunications (1983). *Telecommunication systems (1986).
XIX	Martin Hasler & Jacques Neirynck	Filtres électriques (1981). *Electric Filters (1986).
XX	Murat Kunt	Traitement numérique des signaux (1980). *Digital Signal Processing (1986).
XXI	Mario Rossi	Electroacoustique (1984).
XXII	Michel Aguet & Mircea Ianovici	Haute tension (1982).

Index